Satellite remote sensing in climatology

STUDIES IN CLIMATOLOGY SERIES
Edited by Professor S. Gregory, University of Sheffield

A. M. Carleton, *Satellite Remote Sensing in Climatology*

J. Hobbs, P. Aceituno, H. Bridgman, J. Lindesay, R. Allan, *Climates of the Southern Hemisphere*

B. Yarnal, *Synoptic Climatology in Environmental Analysis*

Other titles are in preparation

Satellite remote sensing in climatology

ANDREW M. CARLETON

Belhaven Press
London

CRC Press
Boca Raton   Ann Arbor   Boston

© A. M. Carleton, 1991

First published in Great Britain in 1991 by
Belhaven Press (a division of Pinter Publishers),
25 Floral Street, London WC2E 9DS

British Library Cataloguing in Publication Data
A CIP catalogue record for this book is available from
the British Library
ISBN 1 85293 039 X

First published in the United States of
America in 1991 by the CRC Press, Inc.,
2000 Corporate Blvd., N.W.,
Boca Raton, Florida, 33431

Library of Congress Cataloging in Publication Data
Carleton. Andrew M. (Andrew Mark)
Satellite remote sensing in climatology/Andrew M. Carleton.
p. c.m
Includes bibliographic references and index.
ISBN 0-8493-7720-X
1. Climatology—Remote sensing.   2. Clouds—Remote sensing.
3. Astronautics in meteorology.   I. Title.
QC981.C347 1991
551.6—dc20   91-14303
                                                 CIP

Typeset by Joshua Associates Ltd, Oxford
Printed and bound by Biddles Ltd, Guildford and Kings Lynn

To my Mother,
on the occasion of her
Grand Tour of North America:
Summer 1990

Contents

List of figures

List of tables

List of abbreviations and acronyms

AgRISTARS	Agricultural Resource Inventory Survey Through Aerospace Remote Sensing (NOAA)
AMSU	Advanced Microwave Sounding Unit (planned; Eos)
AMTEX	Air Mass Transformation Experiment
ANMRC	Australian Numerical Meteorology Research Centre (Melbourne)
ATS	Applications Technology Satellite (US)
AVHRR	Advanced Very High Resolution Radiometer (NOAA)
CAB	composite average brightness
CAC	Climate Analysis Center (Washington, DC)
CCL	convective condensation level
CCN	cloud condensation nuclei
CFP	cloud-free path
Cirrus IFO	Cirrus cloud intensive field observation
CISK	conditional instability of the second kind
CMB	composite minimum brightness
CMT	composite maximum temperature
CTT	cloud top temperature
CZCS	Coastal Zone Color Scanner (Nimbus-7)
DMS	dimethylsulphide
DMSP	Defense Meteorological Satellite Program (US)
EMS	electromagnetic spectrum
ENSO	El Niño/Southern Oscillation
EOF	empirical orthogonal function
Eos	Earth Observing System
EPI	ESOC precipitation index (Europe)
ERB	Earth radiation budget
ERBE	Earth Radiation Budget Experiment (US)
ERS-1	Earth Remote Sensing satellite (Europe)
ESMR	Electrically Scanning Microwave Radiometer (Nimbus-5)
ESOC	European Space Operations Centre
ESSA	Environmental Science Services Administration (US)
FACE	Florida Area Cumulus Experiment
FGGE	First GARP Global Experiment
FIFE	First ISLSCP Field Experiment
FIRE	First ISCCP Regional Experiment
FOV	field of view
FRONTIERS	Forecasting Rain Optimized using New Techniques of Interactively Enhanced Radar and Satellite (UK Meteorological Office)
GARP	Global Atmospheric Research Program
GATE	GARP Atlantic Tropical Experiment

List of abbreviations and acronyms

GCM	general circulation model
GMS	Geostationary Meteorological Satellite (Japan)
GOES	Geostationary Operational Environmental Satellite (US)
GOFS	Global Ocean Flux Study
GPI	GOES precipitation index
GVI	global vegetation index
HBTM	hybrid bispectral threshold method
HCMM	Heat Capacity Mapping Mission (NASA)
HIR	High Resolution Infrared (DMSP)
HIRS	High-resolution Infrared Radiation Sounder
HMMR	High-resolution Multi-frequency Microwave Radiometer (proposed; Eos)
HRC	highly reflective cloud
HVIS	High Resolution Visible (DMSP)
IFO	intensive field observation
INSAT	Indian geostationary satellite
IR	infrared
IRIS	Infrared Interferometer Spectrometer (Nimbus-4)
ISCCP	International Satellite Cloud Climatology Project
ISLSCP	International Satellite Land Surface Climatology Project
ISV	Intraseasonal variation
ITCZ	Intertropic Convergence Zone
ITOS	Improved TIROS Operational Satellite (NOAA)
LLJ	low-level jet
MCC	meso-scale cellular convection
MCO	mostly (cloud) covered
MCS	meso-scale convective system
MCSST	multi-channel sea surface temperature (product)
METEOSAT	European geostationary meteorological satellite
MIZ	marginal ice zone
MIZEX	Marginal Ice Zone Experiment
MONEX	Monsoon Experiment (South Asia, Australia)
MOP	mostly open (nephanalysis)
MPA	meso-scale precipitation area
MRIR	Medium Resolution Infrared Radiometer
MSS	Multispectral Scanner System (Landsat)
MSU	Microwave Sounding Unit
NASA	National Aeronautics and Space Administration (US)
NDVI	normalized difference vegetation index
NEMS	Nimbus-E (-5) Microwave Spectrometer
NESDIS	National Environmental Satellite Data Information Service (US)
NESS	National Environmental Satellite Service (US)
NFOV	narrow field of view (of a sensor)
NMC	National Meteorological Center (Washington, DC)
NOAA	National Oceanic and Atmospheric Administration (US)
OLR	outgoing longwave radiation
OPF	ocean polar front

PFJ	polar front jet
PVA	positive vorticity advection
R&D	research and development
RT	real time
SAGE	Stratospheric Aerosol and Gas Experiment
SASS	Seasat-A Satellite Scatterometer
SBUV	Solar Backscatter Ultraviolet radiometer
SCAMS	Scanning Microwave Spectrometer (Nimbus-6)
SCR	Selective Chopper Radiometer (Nimbus-5)
SD	snow depth
SIR	Shuttle Imaging Radar
SLP	sea level pressure
SMMR	Scanning Multichannel Microwave Radiometer (Nimbus-7, Seasat-A)
SMS	Synchronous Meteorological Satellite (US)
SPCB	South Pacific cloud band
SR	Scanning Radiometer (NOAA)
SSM/I	Special Sensor Microwave/Imager (DMSP)
SST	sea surface temperature
STJ	subtropical jet
SZA	solar zenith angle
T_a	air temperature
T_B	brightness temperature
T_{BB}	equivalent blackbody temperature
TECB	tropical–extratropical cloud band
THIR	Temperature-Humidity Infrared Radiometer (Nimbus-4)
T_H, T_V	horizontally and vertically polarized brightness temperatures
TIROS	Television and Infrared Observation Satellite
TIR	thermal infrared
TIROS-N	New generation of TIROS (NOAA) polar orbiters
TOA	top of the atmosphere (flux)
TOGA	Tropical Ocean-Global Atmosphere (experiment)
TOMS	Total Ozone Mapping Spectrometer (Nimbus-7)
TOVS	TIROS-N Operational Vertical Sounder
TRMM(S)	Tropical Rainfall Measuring Mission (planned) (satellite)
T_s	Surface temperature
3-D/RT NEPH	3-D/real time nephanalysis (US Air Force)
UTH	upper trophospheric humidity (product)
UV	ultraviolet (radiation)
VAS	VISSR Atmospheric Sounder (GOES)
VHRR	Very High Resolution Radiometer (NOAA)
VISSR	Visible Infrared Spin-Scan Radiometer (GOES)
V_T	thermal wind vector
WFOV	wide field of view (of a sensor)
WOCE	World Ocean Circulation Experiment

Editor's preface

From being a somewhat disregarded minority interest within the field of meteorology, or equally a minority interest in associated fields such as geography, climatology has become—within the last 10 to 15 years—a central focus of innumerable disciplines concerned with the present and future environment of this planet. No longer seen as being concerned solely with average weather, or with general descriptions of recent conditions, it spans time scales from the historical and geological past to probable events in the future: requires massive computing facilities for its statistical and mathematical modelling approaches; draws on the information from and technology of space satellites; and focuses on problems of prime importance to the future well-being of mankind.

Reflecting this growth in scale, quantity, scientific depth and applied relevance of research in climatology has been an equivalent growth in publications. The number of international journals specifically concerned with research papers in this field has increased considerably; conferences on climatic issues have proliferated, organized by a range of official or semi-official bodies, with volumes of proceedings or overall reports tending to be published for each of them; and various series of books concerned with specific aspects of the overall climatological field have appeared.

It is therefore reasonable to ask why another such series should be created, or at least where it fits into the overall picture of contemporary climatology. This series is primarily concerned with global climates at the present, with how they have fluctuated or changed over the recent past, and with the impact of such conditions on human society. This necessarily includes considerations of data sources, methodologies and theories as well as reviews and discussions of the evidence for the climatic conditions themselves. Such volumes will thus help to delimit and define the framework in the context of which the results of modelling studies, estimates of future climatic conditions, and assessments of environmental change, need to be evaluated. So this series does not comprise introductory textbooks, but rather monographs or reviews of the current state of knowledge and understanding, essential for the researcher and perhaps also for the final-year specialist student.

This first volume in the series is concerned with the vital data source that has helped to transform much of climatology over the past two decades—satellite remote sensing. The author, Andrew Carleton of Indiana University, is not presenting another book on space satellites, on remote sensing techniques as a whole, or on climatology in a general sense. Instead he concentrates on several critical substantive areas within climatology that have been transformed thanks to the data now available through space satellite images—cloud climatologies and systems; water vapour and the estimation of precipitation; and the Earth's radiation budget. With the numerous visual illustrations, and the very comprehensive bibliography, the textual presentation in this volume fully satisfies the general objectives presented at the end of the previous paragraph, and in this way sets the model for future contributions.

S. Gregory
15 October 1990

Preface

This is a monograph on satellite-based climatology and the remote sensing techniques that are used to extract climatologically useful information. It is not intended to be a textbook in remote sensing or a text in climatology. Instead, it focuses on the overlap between these two very large bodies of knowledge. It is assumed that the reader is at least generally familiar with the basic concepts and processes giving rise to weather and climate. At the same time, he or she will not find herein intricate derivations of radiation transfer theory; or an orbiter-by-orbiter account of the US meteorological satellite program; or a highly detailed discussion of the inception, development and decay of El Niño Southern Oscillation events. While these subjects are mentioned and even discussed in context, they do not constitute the key themes that I have decided to emphasize in this climate research monograph. The themes, which are identified in Chapter 1 and discussed in detail in subsequent chapters, involve primarily the contributions of satellite data to the development of climate theory. These topics include the role of clouds in the Earth radiation budget and in climate; the role of the Earth's surface characteristics in climate and its variations, such as desertification and drought; and the signature of climate teleconnections in clouds and cloud systems, precipitation and water vapor patterns. A theme that cuts across several of these foci involves the role of satellites in monitoring the climate system and providing 'baseline' climatologies within which the impact of humans might be more easily discerned. The latter is increasingly part of the Global Change initiative. Examples of the ways in which satellite data have played, and continue to play a major role in the monitoring of the climate system for anthropogenic impacts are in issues such as stratospheric ozone depletion; global warming; and deforestation. As we move into the 1990s, satellites will make ever-increasing contributions to these climate and environmental issues, and probably to others not yet apparent. I hope that the topics discussed in this monograph show to the reader, whether he or she be a climatologist, geographer, atmospheric scientist, hydrologist, or other specialist with an interest in satellite remote sensing and its applications, that we share some common bonds, an uncommon history and a truly exciting future.

Please note that the notation used in non-standard equations presented in this book is as it appeared in the original article. In order to prevent confusion arising from changes in notation between equations from different sources, the notation is defined at the outset of each new equation or equation series.

Andrew M. Carleton
Bloomington, Indiana
June 1990

Acknowledgements

Many individuals have helped, quite unwittingly, to bring this project to fruition by their profound influence on my professional upbringing. Their insights into climatology have proven invaluable, and are doubtless only imperfectly translated by me in this monograph. They also, in their different ways, have taught me anew how to read, write and do sums! The individuals concerned are: Bruce Mason (Adelaide); Peter Lamb; Roger Barry; Neil Streten; Bill Budd; Uwe Radok; Ann Henderson-Sellers; Harry van Loon; David Greenland; Erik Rasmussen; Dave Bromwich; John Walsh; Bill Black, and the late Werner Schwerdtfeger. I am most appreciative to all of the above for their guidance.

There are others who have helped contribute to this monograph simply by providing unfailing friendship and support over the years. I would like to acknowledge particularly the following: Barry and Carol McInnes; Rob and Linda Crane; Diane Whalley; Brent Yarnal; Bruce Rhoads; David Easterling; Doug Horn; Jerry Stasny. Last in this group, but definitely not least, I include my an-aerobics instructor, Jackie Ragan, for teaching me how to breathe—it's come in really handy.

I also appreciate the insights into various aspects of satellite climatology that were provided by graduate students in the course of their thesis research at Indiana University: Jim DeGrand; Mark Fitch; Rich Brinegar; Maxine Perry; Duane Carpenter; Cynthia Berlin; and David Arnold. You are the satellite climatologists of the 1990s—be proud!

This monograph would not have been possible without the amazing cartographic skills of J. Michael Hollingsworth, who painstakingly redrew all the diagrams for this work and still managed to smile each time I walked through the door with a new batch of diagrams to be done; to the tireless efforts of those at Belhaven Press, especially Jane Evans and Iain Stevenson; and to Professor Stan Gregory, the series coordinator. To the last two individuals in particular, I am grateful for their keeping the faith, even as deadlines on the manuscript submission date came and went. To all these persons, and others that I may have inadvertently omitted, I express my deepest gratitude. Of course, they are in no way responsible for any errors or misinterpretations that may remain in the text: those are entirely my own creation.

Absolutely finally, I thank myself for expertly typing the manuscript; not to mention the 45 pages of references (single-spaced), and 'me Mum' for having the incredible foresight to send me to summer typing school when I was 14 years old; doubtless in preparation for this undertaking.

This monograph was written with partial support from the following: National Science Foundation grants DPP-7920853, DPP-8816912, SES-8603470; National Aeronautics and Space Administration grants NAG-5-142 and JPL 957872; National Oceanic and Atmospheric Administration grant COMM-NA89RAH09086 and Univ. of California, Davis # 9012 14–IND.

Acknowledgements

The author and publishers are grateful to the following for permission to reproduce copyright material:

Academic Press and the authors for Fig. 6.9, from Figs. 15 and 16 in the chapter by G. Ohring and A. Gruber 'Satellite radiation observations and climate theory', *Advances in Geophysics, 25*, pp. 277 and 288 (1983). American Association for the Advancement of Science (AAAS) for Fig. 1.16, from Fig. 1 of the article 'Nimbus-7 Coastal Zone Color Scanner: system description and initial imagery' by W. A. Hovis, D. K. Clark, F. Anderson, R. W. Austin, W. H. Wilson, E. T. Baker, D. Ball, H. R. Gordon, J. L. Mueller, S. Z. El-Sayed, B. Sturm, R. C. Wrigley, and C. S. Yentsch, in *Science, 210*, p. 60 (1980). American Geophysical Union (AGU) for the following: Fig. 1.4, from Fig. 2 in the article by K-N Liou and S-C Ou, 'The role of cloud microphysical processes in climate: an assessment from a one-dimensional perspective', in *Journal of Geophysical Research, 94*, p. 8603 (1989); Fig. 1.15, from Fig. 1 in the article by P. A. Davis, E. R. Major and H. Jacobowitz, 'An assessment of NIMBUS-7 ERB shortwave scanner data by correlative analysis with narrowband CZCS data', in *Journal of Geophysical Research, 89*, p. 5079 (1984); Fig. 2.1, from Fig. 6 in the article by A. M. Carleton, 'Synoptic sea ice–atmosphere interactions in the Chukchi and Beaufort seas from Nimbus-5 ESMR data', in *Journal of Geophysical Research, 89*, p. 7252 (1984); Fig. 2.20, from Fig. 1 in the article by V. R. Taylor and L. L. Stowe, 'Reflectance characteristics of uniform Earth and cloud surfaces derived from NIMBUS-7 ERB',in *Journal of Geophysical Research, 89*, p. 4988 (1984); Fig. 2.22, from Figs. 1 and 2 in the article by W. J. Emery and P. Schuessel, 'Global differences between skin and bulk sea surface temperatures', in *EOS (Transactions of AGU), 70*, p. 210; Fig. 3.21, from Fig. 2a in the article by A. Henderson-Sellers, G. Seze, F. Drake and M. Desbois, 'Surface-observed and satellite-retrieved cloudiness compared for the 1983 ISCCP Special Study Area in Europe', in *Journal of Geophysical Research, 92*, p. 4023 (1987); Fig. 3.27, from Figs. 1a,b in the article by G. E. Woodbury and M. P. McCormick, 'Zonal and geographical distributions of cirrus clouds determined from SAGE data', in *Journal of Geophysical Research, 91*, p. 2777 (1986); Fig. 4.10, from Fig. 8, and Table 11 from Table 2 in the article by I. Velasco and J. M. Fritsch, 'Mesoscale convective complexes in the Americas', in *Journal of Geophysical Research, 92*, pp. 9601 and 9596 (1987); Fig. 5.7a,b, from Figs. 2 and 3 in the article by H. D. Chang, P. H. Hwang, T. T. Wilheit, A. T. C. Chang, D. H. Staelin, and P. W. Rosenkranz, 'Monthly distributions of precipitable water from the NIMBUS 7 SMMR data', in *Journal of Geophysical Research, 89*, p. 5330 (1984); Fig. 6.1, from Fig. 10 in the article by V. Ramanathan, 'The role of earth radiation budget studies in climate and general circulation research', in *Journal of Geophysical Research, 92*, p. 4082 (1987); Figs. 6.3, 6.5, 6.6, 6.7 and 6.8, from Figs. 3a, 6, 7, 8, 9, 11, in the article by G. L. Stephens, G. G. Campbell, and T. H. Vonder Haar, 'Earth radiation budgets', in *Journal of Geophysical Research, 86*, pp. 9743, 9747, 9748, 9750, and 9752 (1981); and Table 14, from Table 1 in the article by S. Q. Kidder and T. H. Vonder Haar, 'Seasonal oceanic precipitation frequencies from Nimbus 5 microwave data', in *Journal of Geophysical Research, 82*, p. 2085 (1977)—all the above copyright by the American Geophysical Union. The American Meteorological Society (AMS) for the following: Fig. 1.1, from Fig. 1 in the article by K. B. Katsaros, I. Bhatti, L. A. McMurdie and G. W. Petty, 'Identification of atmospheric fronts over the ocean with microwave measurements of water vapor and rain', in *Weather and Forecasting, 4*, p. 450 (1989); Fig. 1.6, from Fig. 6 in the article by P. M. Kuhn, 'Airborne observations of contrail effects on the thermal radiation budget', in *Journal of the Atmospheric Sciences, 27*, p. 942 (1970); Fig. 1.10, from Fig. 8 in the article by W. B. Rossow, 'Measuring cloud properties from space: a review', in *Journal of Climate, 2*, p. 211 (1989); Figs. 1.11 and 2.14, from Figs. 2 and 7 in the article by D. W. Reynolds, T. H. Vonder Haar, and L. O. Grant, 'Meteorological satellites in support of weather modification', in *Bulletin of the AMS, 59*, pp. 271 and 276 (1978); Fig. 1.14, from Fig. 1 in the article by J. E. Stout, D. W. Martin, and D. N. Sikdar, 'Estimating GATE rainfall with geosynchronous satellite images', in *Monthly Weather Review, 107*, p. 587 (1979); Fig. 2.2 and 2.3, from Fig. 12 in the article by K-N Liou; 'Influence of cirrus clouds on weather and climate processes: a global perspective', in *Monthly Weather Review, 114*, p. 1185 (1986); Fig. 2.6, from Fig. 7 in the article by P. Minnis and E. F. Harrison, 'Diurnal variability of regional cloud and clear sky radiative parameters derived from GOES data. Part I: Analysis method, in *Journal of Climate and Applied Meteorology, 23*, p. 998 (1984); Fig. 3.1, from Fig. 20 in the article by P. Minnis and E. F. Harrison, 'Diurnal variability of regional cloud and clear sky radiative parameters derived from GOES data. Part II: November 1978 cloud distributions', in *Journal of Climate and Applied Meteorology, 23*, p. 1029 (1984); Fig. 2.7, from Fig. 1 in the

Acknowledgements

article by R. C. Savage, 'Radiative properties of hydrometeors at microwave frequencies', in *Journal of Applied Meteorology, 17*, p. 905 (1978); Figs. 2.8 and 2.9, from Figs. 2 and 6 in the aticle by T. T. Wilheit 'Some comments on passive microwave measurements of rain', in *Bulletin of the AMS, 67*, pp. 1227 and 1229 (1986); Fig. 2.10, from Fig. 1 in the article by R. C. Savage and J. A. Weinman 'Preliminary calculations of upwelling radiance from rainclouds at 37.0 and 19.35GHz', in *Bulletin of the AMS, 56*, p. 1273; Figs. 2.11 and 2.12a,b, from Figs. 1, 4 and 5 in the article by E. Rodgers, H. Siddalingaiah, A. T. C. Chang and T. Wilheit, 'A statistical technique for determining rainfall over land employing Nimbus 6 ESMR measurements', in *Journal of Applied Meteorology, 18*, pp. 979 and 983 (1979); Fig. 2.13, from Fig. 5 in the article by R. T. Pinker and J. A. Ewing 'Modeling surface solar radiation: model formulation and validation', in *Journal of Climate and Applied Meteorology, 24*, p. 393 (1985); Figs. 2.17 and 2.18, from Figs. 7 and 8 of the article by D. W. Reynolds and T. H. Vondar Haar 'A bispectral method for cloud parameter determination', in *Monthly Weather Review, 105*, p. 452 (1977); Fig. 2.19, from Fig. 2 in the article by S. K. Cox 'Cirrus clouds and climate', in *Journal of the Atmospheric Sciences, 28*, p. 1515 (1971); Fig. 2.21a,b,c, from Figs. 8, 9 and 10 in the article by C. Gautier 'Mesoscale insolation variability derived from satellite data', in *Journal of Applied Meteorology, 21*, pp. 55, 56 and 57 (1982); Fig. 3.2, from Fig. 1 in the article by R. M. Rabin, S. J. Stadler, P. J. Wetzel, D. J. Stenrud, M. Gregory, 'Observed effects of landscape variability on convective clouds', in *Bulletin of the AMS, 71*, p. 273 (1990); Fig. 3.5, from Fig. 7 in the article by E. M. Agee, 'Observations from space and thermal convection: a historical perspective', in *Bulletin of the AMS, 65*, p. 944 (1984); Fig. 3.9, from Fig. 3 in the article by R. DelBeato and S. L. Barrell, 'Rain estimation in extratropical cyclones using GMS imagery', in *Monthly Weather Review, 113*, p. 748 (1985); Figs. 3.14, 3.15 and 3.16, from Figs. 4, 8 and 15 in the article by B. A. Wielicki and R. M. Welch, 'Cumulus cloud properties derived using Landsat satellite data', in *Journal of Climate and Applied Meteorology, 25*, pp. 265, 268 and 272 (1986); Fig. 3.17, from Fig. 1 in the article by R. Koffler, G. DeCotiis and P. K. Rao, 'A procedure for estimating cloud amount and height from satellite infrared radiation data', in *Monthly Weather Review, 101*, p. 241 (1973); Fig. 3.18, from Fig. 9a,b in the article by G. Sèze and M. Desbois 'Cloud cover analysis from satellite imagery using spatial and temporal characteristics of the data', in *Journal of Climate and Applied Meteorology, 26*, p. 294 (1987); Figs. 3.19 and 3.20a,b, from Figs. 1 and 5 in the article by M. Desbois, G. Sèze and G. Szejwach, 'Automatic classification of clouds on METEOSAT imagery: application to high level clouds', in *Journal of Applied Meteorology, 21*, p. 402 (1982); Fig. 3.22, from Fig. 12 in the article by P. F. Clapp, 'Global cloud cover for seasons using TIROS nephanalysis', in *Monthly Weather Review, 92*, p. 506 (1964); Fig. 3.23a,b, from Fig. 11a,b in the article by W. B. Rossow, L. C. Garder and A. A. Lacis, 'Global seasonal cloud variations from satellite radiance measurements. Part II: Sensitivity of the analysis', in *Journal of Climate, 2*, p. 450 (1989); Fig. 3.24, from Fig. 15 in the article by L. L. Stowe, H. Y. M. Yeh, T. F. Eck, C. G. Wellemeyer, and H. L. Kyle, 'Nimbus 7 global cloud climatology. Part III: 'First year results', in *Journal of Climate, 2*, p. 700 (1989); Fig. 3.25, from Fig. 4 in the article by N. A. Hughes and A. Henderson-Sellers, 'Global 3D Nephanalysis of total cloud amount: climatology for 1979', in *Journal of Climate and Applied Meteorology, 24*, p. 679 (1985); Fig. 3.26, from Fig. 11 in the article by I. J. Barton, 'Upper level cloud climatology from an orbiting satellite', in *Journal of the Atmospheric Sciences, 40*, p. 445 (1983); Fig. 3.28a–d, from Fig. 12a–d, in the article by C. Prabhakara, R. S. Fraser, G. Dalu, M-LC Wu and R. J. Curran, 'Thin cirrus clouds: seasonal distribution over oceans deduced from Nimbus 4 IRIS', in *Journal of Applied Meteorology, 27*, pp. 395–8 (1988); Fig. 4.9 and Tables 9 and 10, from Fig. 3 and Tables 1 and 2 in the article by J. A. Augustine and K. W. Howard, 'Mesoscale convective complexes over the U.S. during 1985', in *Monthly Weather Review, 116*, pp. 690–92 and 686 (1988); Fig. 5.1, from Fig. 4 in the article by M. M. Poc, M. Roulleau, N. A. Scott and A. Chedin, 'Quanitative studies of METEOSAT water vapor channel data', in *Journal of Applied Meteorology, 19*, p. 873 (1980); Fig. 5.2, from Fig. 2 in the article by N. Eigenwillig and H. Fischer, 'Determination of mid-tropospheric wind vectors by tracking pure water vapor structures in METEOSAT water vapor image sequences', in *Bulletin of the AMS, 63*, p. 48 (1983); Figs. 5.3 and 5.4, from Figs. 1 and 2 in the article by C. Prabhakara, H. D. Chang, and A. T. C. Chang, 'Remote sensing of precipitable water over the oceans from Nimbus 7 microwave measurements', in *Journal of Applied Meteorology, 21*, pp. 61 and 62 (1982); Fig. 5.5, from Fig. 12 in the article by E. Raschke and W. R. Bandeen, 'A quasi-global analysis of tropospheric water vapor content from TIROS IV radiation data', in *Journal of Applied Meteorology, 6*, p. 476 (1967); Fig. 5.6a,b, from Figs. 7 and 9 in the article by C. Prabhakara, G. Dalu, R. C. Lo and N. R. Nath, 'Remote sensing of seasonal distribution of precipitable water vapor over the oceans and the inference of boundary-layer structure', in *Monthly Weather Review, 107*, pp.1396 and 1398 (1979); Figs. 5.8 and 5.9, from Figs. 12 and 13 in the article by E. G. Njoku and L. Swanson, 'Global measurements of sea surface temperature, wind speed and atmospheric water content from satellite microwave radiometry', in *Monthly Weather Review, 111*, pp. 1985 and 1986 (1983); Figs. 5.10 and 5.11, from Figs. 3 and 5 in the article by C. Prabhakara, D. A. Short and B. E. Vollmer, 'El Niño and atmospheric water vapor: observations from Nimbus 7 SMMR', in *Journal of Climate and Applied Meteorology, 24*, pp. 1316 and 1320 (1985); Fig. 5.12, from Figs. 4 and 5 in the article by D. P. Wylie 'An application of a geostationary satellite rain estimation technique to an extratropical area', in *Journal of Applied Meteorology, 18*, p. 1644 (1979); Fig. 5.13, from Fig. 15 in the article by C. G. Griffith, W. L. Woodley, P. G. Grube, D. W. Martin, J. Stout and D. N. Sikdar, 'Rain estimation from geosynchronous satellite imagery–visible and infrared studies', in *Monthly Weather Review, 106*, p. 1166 (1978); Fig. 5.16, from Fig. 2 in the article by L. J. Allison, E. B. Rodgers, T. T. Wilheit and R. W. Fett, 'Tropical cyclone rainfall as measured by the Nimbus 5 electrically scanning microwave radiometer', in *Bulletin of the AMS, 55*, p. 1076 (1974); Fig. 5.17a,b from Fig. 2a,b, in the article by C. Prabhakara, D. A. Short, W. Wiscombe and R. S. Fraser, 'Rainfall over oceans inferred from Nimbus 7 SMMR: applications to 1982–83 El Niño', in *Journal of Climate and Applied Meteorology, 25*, p. 1468 (1986); Fig. 5.18 from Fig. 12 in the article by P. A. Arkin and B. N. Meisner 'Relationship between large-scale convective rainfall and cold cloud over the western hemisphere during 1982–84', in *Monthly Weather Review, 115*, p. 70 (1987); Fig. 5.19a,b, from Fig. 1c,d, in the article by P. A. Arkin, A. V. R. Krishna Rao and R. R. Kelkar, 'Large scale precipitation and outgoing longwave radiation from INSAT-IB during the 1986 southwest monsoon season', in *Journal of Climate, 2*, pp. 622 and 624 (1989); Fig. 6.4, from Fig. 17a, in the article by W. B. Rossow, C. L. Brest and L. C. Gardner, 'Global, seasonal surface variations from satellite radiance measurements', in *Journal of Climate, 2*, p. 240 (1989); Fig. 6.10, from Fig. 13 in the article by C. C. Norton, F. R. Mosher and B. Hinton, 'An investigation of surface albedo variations during recent Sahel drought', in *Journal of Applied Meteorology, 18*, p. 1259 (1979); Fig. 6.15, from Fig. 12 in the article by W. L. Darnell, W. F. Staylor, S. K. Gupta and F. M. Denn, 'Estimation of surface insolation using sun-synchronous satellite data', in *Journal of Climate, 1*, p. 831 (1988); Table 4, from Table 1 in the article by L. Garand, 'Automated recognition of oceanic cloud patterns. Part I: Methodology and application to cloud climatology', in *Journal of Climate, 1*, p. 21 (1988); Table 7, from Table 1 in the article by H. Malberg, 'Comparison of mean cloud cover obtained by satellite photographs and ground-based observations over Europe and the Atlantic', in *Monthly Weather Review, 101*, p. 895 (1973); Table 12, from Table 1 in the article by E. C. Barrett 'Estimation of monthly rainfall from satellite data', in *Monthly Weather Review, 98*, p. 324 (1970); Table 13 from Table 4 in the article by F. Richards and P. Arkin 'On the relationship between satellite observed cloud cover and precipitation', in *Monthly Weather Review, 109*, p. 1090 (1981); Tables 15 and 17 from Tables 2 and 4 in the article by A. A. Rockwood and S. K. Cox 'Satellite inferred surface albedo over northwestern Africa', in *Journal of the Atmospheric Sciences, 35*, pp. 517 and 520 (1978); and the author, R. F. Cahalan, for Table 3, from Table 1 in the conference paper by R. F. Cahalan, 'Climatological statistics of cloudiness', in Preprints volume, *Fifth Conference on Atmospheric Radiation*, p. 206 (1983). The Australian Government Publishing Service (AGPS) for the following: Fig. 3.4, from Fig. 1 in the report by L. B. Guymer, 'Operational application of satellite imagery to synoptic analysis in the Southern Hemisphere', Tech. Rept. No. 29, Commonwealth Bureau of Meteorology, Melbourne, p. 4 (1978); Fig. 2.15a,b, from Fig. 8 in the article by N. A. Streten 'Some aspects of the satellite observation of frontal cloud over the Southern Hemisphere', in Conference Handbook of the Two-Day Workshop on Fronts, Bureau of Meteorology, Melbourne, pp. 3.6–3.7 (1977); Fig. 3.8, from Fig. 1 in the article by J. W. Zillman and P. G. Price 'On the thermal structure of mature Southern Ocean cyclones', in *Australian Meteorological Magazine, 20*, p. 36 (1972); Fig. 4.18, from Fig. 4a in the article by N. A. Streten and W. R. Kellas 'Some applications of simple image analysis techniques to very high resolution satellite data', in *Australian Meteorological Magazine, 23*, p. 11 (1975); Figs. 4.21, 4.22 and 4.23, from Figs. 1, 3 and 4 in the article by A. M. Carleton, 'Ice–ocean–atmosphere interactions at high southern latitudes in winter from satellite observation', in *Australian Meteorological Magazine, 29*, pp. 185, 187 and 189 (1981). Cambridge University Press for Fig. 1.2, from Fig. 1.12 in the chapter by H. E. Fleer 'Teleconnections of rainfall anomalies in the tropics and subtropics, from the book edited by Sir J. Lighthill and R. P. Pearce *Monsoon Dynamics*, p. 17 (1981). Canadian Meteorological and Oceanographic Society or Fig.

Acknowledgements

5.15, from Fig. 1a,b in the article by S. Lovejoy and G. L. Austin 'The delineation of rain areas from visible and IR satellite data for GATE and mid-latitudes', in *Atmosphere-Ocean, 17*, p. 82 (1979). Elsevier Science Publishing Co., Inc., for Figs. 4.13 and 4.20, from Figs. 1 and 7 of the article by A. M. Carleton 'Satellite-derived attributes of cloud vortex systems and their application to climate studies', in *Remote Sensing of Environment, 22*, pp. 272 and 282–3. Copyright 1987. W. H. Freeman and Company for Fig. 2.5, from Fig. 1.3 of the book by Floyd F. Sabins, Jr., *Remote Sensing Principles and Interpretation*, 2nd edition, p. 5 (1987). The International Association of Hydrological Sciences (IAHS) Press for Table 16, from Table 2 in the article by D. A. Robinson, G. Scharfen, R. G. Barry and G. Kukla, 'Analysis of interannual variations of snow melt on Arctic sea ice mapped from meteorological satellite imagery', In: *Large-Scale Effects of Seasonal Snow Cover*, IAHS Publication no. 166, p. 322 (1987). Kluwer Publishing for the following: Fig. 4.16, from Fig. 2 in the article by A. M. Carleton and D. E. Carpenter 'Intermediate-scale sea ice–atmosphere interactions over high southern latitudes in winter', *GeoJournal, 18*, p. 90 (1989); Fig. 2.16, from Fig. 1 in the chapter by S. Manable 'Cloudiness and the radiative convective equilibrium', from the book edited by S. F. Singer *The Changing Global Environment*, p. 176 (1975). Macmillan Magazines Ltd. and the respective authors for the following: Fig. 1.9, from Fig. 2 of the article by R. J. Charlson, J. E. Lovelock, M. O. Andreae and S. G. Warren, 'Oceanic phytoplankton, atmospheric sulphur, cloud albedo and climate', in *Nature, 326*, p. 659 (1987); Fig. 2.23, from Fig. 1 in the article by D. Robinson, K. Kukla, G. Kukla and H. Roth, 'Comparative utility of microwave and shortwave satellite data or all-weather charting of snow cover', in *Nature, 312*, p. 434 (1984); Fig. 6.14, from Fig. 2 in the article by R. W. Saunders an G. E. Hunt 'METEOSAT observations of diurnal variation of radiation budget parameters', in *Nature, 283*, p. 646 (1980). The above reprinted by permission from *Nature*. Copyright (C) 1980, 1984, 1987 Macmillan Magazines Ltd. Munksgaard International Publishers and the Swedish Geophysical Society for Fig. 3.13, from Fig. 3 of the article by K. J. Weston 'An observational study of convective cloud streets', in *Tellus, 32*, p. 435 (1980); and Table 8, from Table 2 of the article by G. S. Forbes and W. D. Lottes 'Classification of mesoscale vortices in polar airstreams and the influence of the large-scale environment on their evolutions', in *Tellus, 37A*, p. 137 (1985). Pergamon Press, Inc. for Fig. 1.3, from Fig. 6 of the article by J. S. Winston 'Climatic variations in the outgoing longwave radiation of the past decade as observed from NOAA polar orbiting satellite', in *Advances in Space Research, 5*, p. 124 (1985). The Royal Meteorological Society for the following: Fig. 1.12, from Fig. 2 of the article by A. M. Carleton 'Synoptic and satellite aspects of the southwestern U.S. summer "monsoon"', in *Journal of Climatology, 5*, p. 395 (1985); and Fig. 4.5, from Fig. 3b, of the article by A. M. Carleton and D. E. Carpenter 'Satellite climatology of "polar lows" and broadscale climatic associations for the Southern Hemisphere', in *International Journal of Climatology, 10*, p. 228 (1990). Springer-Verlag for the following: Fig. 3.11a,b, from Fig. 7a,b, of the article by N. A. Streten 'Cloud cell size and pattern evolution in Arctic air advection over the North Pacific', in *Archiv fur Meteorologie, Geophysik und Bioklimatologie, A24*, p. 223 (1975); and Fig. 6.11, from Fig. 1 of the article by H. J. Preuss and J. F. Geleyn, 'Surface albedos derived from satellite data and their impact on forecast models', in *Archiv fur Meteorologie, Geophysik und Bioklimatologie, A29*, p. 350 (1980). Taylor and Francis, Ltd., and the respective authors for the following: Fig. 1.13, from Fig. 2 of the article by R. G. Crane and M. R. Anderson 'Satellite discrimination of snow/cloud surfaces', in *International Journal of Remote Sensing, 5*, p. 215 (1984); and Figs. 3.3, 4.15 and 4.17, from Figs. 1, 5 and 9 in the article by A. M. Carleton 'Synoptic cryosphere–atmosphere interactions in the Northern Hemisphere from DMSP image analysis', in *International Journal of Remote Sensing, 6*, pp. 241, 245 and 249 (1985).

In addition, I acknowledge the following sources: the Meteorological Society of Japan for Fig. 4.12, from the article by T. Nakazawa 'Tropical super clusters within intraseasonal variations over the western Pacific', in *Journal of the Meteorological Society of Japan, 66* (1988). The U.S. Air Force Geophysics Laboratory, Meteorology Division report AFGL-TR-81-0218 'Short-Range Forecasting of Cloudiness and Precipitation through Extrapolation of GOES imagery' by H. Stuart Muench, for Fig. 5.14. The U.S. National Oceanic and Atmospheric Administration (NOAA) for the GOES images reproduced in Figs. 2.4, 3.6a,b,c, 3.7, 3.10, 3.12, 4.7, 4.8, and 4.14a,b,c. The U.S. National Snow and Ice Data Center (NSIDC) for Figs. 1.5, 1.8, 4.1, 4.2, 4.3, 4.4a,b., and 4.19a,b,c. Produced from USAF DMSP (Defense Meteorological Satellite Program) film transparencies archived from NOAA/NESDIS at the University of Colorado, CIRES/National Snow and Ice Data Center, Campus Box 449, Boulder, CO, USA 80309.

1 Introduction

1 Satellites, weather and climate

Earth-orbiting satellites and the data they provide have helped spawn a revolution in the sciences of meteorology and climatology over the past 30 years (Rao *et al.*, 1990). High-frequency surveillance of cloud patterns, which are tracers of atmospheric energy and moisture flows; the sensing of temperature and moisture profiles under a range of cloud conditions; remote wind determination over the oceans; and precipitation estimation, all comprise essential components of the synoptic analysis and weather forecasting problem provided by operational weather satellites (Smith *et al.*, 1986). These observational advances have occurred in tandem with developments in prediction modeling and computer compatibility (Dey, 1989).

Meteorological satellite information has enhanced our understanding of synoptic systems and provided insights into previously little known meso-scale circulations (Chapter 4). The Norwegian extratropical frontal cyclone model, as it had evolved through the first half of this century, has been largely confirmed using satellite data (see, e.g., Carlson, 1980). It has been further enhanced with the use of ground-based radar and space-borne passive microwave sensors. The former has permitted the resolving of meso-scale (approx. $0.5-5 \times 10^2$ km) details of the precipitation distribution in extratropical cyclones, while the latter reveals that fronts may be detected from the steep gradient in tropospheric water vapor and the precipitation that accompany them (Fig. 1.1). Similarly, more detailed 'models' of jet streams have been afforded by the use of satellite radiance data (see, e.g., Ramond *et al.*, 1981; Durran and Weber, 1988), and these expand upon those previously developed from conventional data and earlier satellite studies (e.g., Martin and Salomonson, 1970). Meteorological satellites have also provided insights into previously unknown circulation systems at the meso-scale as higher spatial and temporal resolutions have become possible (Shenk *et al.*, 1987). These include meso-cyclones that occur dominantly over the oceans in the cold season (Businger and Reed, 1989); warm season meso-scale convective systems (MCSs) over continental areas (see, e.g., Maddox, 1980; Velasco and Fritsch, 1987); and low-level convergence lines (Drosdowsky and Holland, 1987).

Aside from their obvious application to synoptic analysis and synoptic climatology, satellite data have broader application to climate dynamics, oceanography, and hydrology (Yates *et al.*, 1986; Ohring *et al.*, 1989). This is facilitated by the monitoring of climatically significant surface variables; particularly sea surface temperatures (SSTs), snow cover, sea ice, soil moisture, and biomass (both on land and in the ocean). The interannual variability of climate is largely the result of thermal and dynamic interactions between the atmosphere and Earth's surface that occur via non-linear feedbacks. Thus, the ability of satellites to monitor fluctuations and changes

1

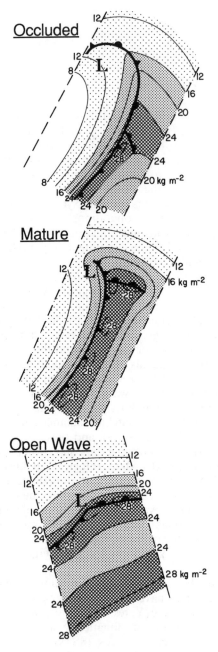

Fig. 1.1 Schematic of contours in Nimbus-7 SMMR integrated atmospheric water vapor in relation to surface fronts for Northern Hemisphere cyclones in three stages of development. Values are in kg m^{-2} and are representative of North Pacific winter storms. (From Katsaros *et al.*, 1989.)

in atmospheric and surface conditions regularly and over large space scales makes them an integral part of contemporary studies in climate dynamics (Nicholson, 1989). These studies are being increasingly viewed in the context of the Global Change initiative (Rasool, 1987). This program has, as a chief priority, the development of a dedicated satellite-based Earth Observing System (Eos) for extending the monitoring of climate variables into the twenty-first century, and the assessment of anthropogenic impacts on the global environmental system (Dziewonski, 1989). In this context must be included the following areas of contemporary climate research.

(1) Improved understanding of the dynamics of desertification and drought, both in the subtropics and middle latitudes, as they may result from remote forcing (teleconnections), enhanced by regional- or local-scale changes in land surface conditions involving vegetation and soil moisture (Gutman, 1990). These studies are largely descended from the initial work on Sahelian drought and the summer droughts of the central US and Western Europe conducted in the 1970s.

(2) The monitoring of deforestation, particularly in the tropics, for its influence on the atmospheric CO_2 and O_1 loading and regional-scale water balance, and potential long-range impact on the global-scale temperature, atmospheric circulation and climate (Woodwell *et al.*, 1987; Watson *et al.*, 1990; Helfert and Lulla, 1990).

(3) Monitoring of the destruction of stratospheric ozone by chlorofluoro-carbons (primarily in the Antarctic), and a resultant net thinning of the ozone layer with possible attendant changes in the atmospheric energy balance and circulation (Bowman, 1988, 1990).

(4) Assessing the impact of changes in cloud parameters as a result of pollution from urban areas, ships and jet aircraft. These may involve real increases in cloud amount (Henderson-Sellers, 1989) and the enhancement of cloud reflectance in solar wavelengths (Twomey *et al.*, 1984; Wigley, 1989), and interactions with energy budgets and atmospheric moisture regimes. These changes may compete with, and possibly counter, the perturbations induced in model-predicted scenarios of trace-gas induced global warming (Somerville and Remer, 1984).

2 Change detection

The anthropogenic impact on climate, as evaluated using satellite data, can only be clearly identified with the development of global-scale 'baseline' regimes for Earth surface and atmospheric variables (land cover, snow and ice, moisture, clouds, radiation and energy, dust and aerosols, gases); that is, with reliable satellite climatologies. For global cloud cover and Earth radiation budgets (ERBs), there probably now exists a sufficiently long record of satellite observations to permit the use of the term 'climatology', even where data may be derived from a range of different satellite systems having different orbital, spectral and resolution characteristics (see, e.g., Kandel, 1983; Stephens *et al.*, 1981; Preuss and Geleyn, 1980; Hughes, 1984).

However, for surface variables that have typically longer time scales of variability, such as sea ice and SSTs, it can be argued that the current satellite record, which may be the *only* record over large areas of the Earth's surface, is not yet sufficient to yield a reliable multi-year climatology suitable for the detection of trends. It may, however, be adequate to define the seasonal cycle and general patterns of variability (see Parkinson *et al.*, 1987; Carleton, 1989; Alfultis and Martin, 1987; Reynolds *et al.*, 1989; Chang *et al.*, 1990). Table 1 gives required accuracies for certain satellite-derived variables of importance to climate. These requirements are, for the most part, not yet being met, except perhaps in the area of surface solar irradiance estimation. The latter is important for predictions of crop yield and for site selection for solar energy development. Accurate monitoring of the solar constant (Table 1) is essential for albedo determination from satellite altitudes and for assessing the net radiative effects of clouds in the climate system.

The detection of trends is very sensitive to the period of record. For example, Strong (1989) has reported an upward trend in the operational SST product from the NOAA polar orbiting satellites for the period 1982–8. The warming amounts to about 0.1 °C yr^{-1} for the global ocean, but is double that identified from conventional SST analyses for the same time period. The greatest contribution to the warming has occurred in the October–December

Table 1 Accuracy requirements of satellite sensed variables important in climate

Variable	Required accuracy	Reference
Albedo	±0.05	Henderson-Sellers and Wilson (1983)
SSTs	±0.05°C	Njoku *et al.* (1985)
Surface solar irradiance	±5–10%	Gautier *et al.* (1980)
Solar constant	±0.1%	Rasool (1987)
Cloud: 30-day averages (fraction)		Schiffer and Rossow (1983)
total cloud amount	±0.03	
cirrus cloud amount	±0.05	
middle cloud amount	±0.05	
low cloud amount	±0.05	
deep convect. amount	±0.05	
cirrus height (km)	±1.00	
middle cloud height (km)	±1.00	
low cloud height (km)	±0.50	
deep convect. height (km)	±1.00	
CTT (all types) (K)	±1.00	
Precipitation	±10%	Chen (1985)
Vegetation cover	±5%	
Snow cover	±3%	
Soil moisture	±0.05 g cm^{-3}	
Sea ice cover	±3%	

period for both hemispheres. Trends are reported as significant over the Atlantic and Pacific basins but not over the Indian Ocean. The reality of this upward trend in global SST has been seriously questioned by Reynolds *et al*. (1989) using conventional and blended (conventional plus constrained satellite) SSTs. These authors report no significant trend in SST for the same period. They suggest that the SST retrievals do not show the strong El Niño Southern Oscillation (ENSO)-related warming early in the 1982–3 period owing to underestimates of the satellite SSTs by atmospheric absorption due to aerosols from the El Chichon eruption. These could give rise to spuriously low SSTs and the appearance of an upward trend. Moreover, marine air temperatures, which correlate very highly with SSTs, show no trend for the 1982–8 period. Similarly, the nine-year period of record from the Nimbus-7 Scanning Multichannel Microwave Radiometer (SMMR: 1978–87) has been examined for trends in sea ice parameters for the Arctic, Antarctic and global (combined) regions by Gloersen and Campbell (1988). Although they find no significant trends in the ice areal extent of either polar region, or in the maxima or minima of the annual extents, significant negative trends were found in the annual global ice extent maxima and in the annual maxima of the global open-water area within the pack. Adding the earlier period of record for the Nimbus-5 Electrically Scanning Microwave Radiometer (ESMR: 1973–6) to that for the SMMR confirms a longer 15-year negative trend in the global ice extent maxima of about 6%. For the Arctic, regional-scale trends are apparent for the 1973–87 period even though the entire polar basin shows no significant trend.

3 Monitoring interannual variability

Notwithstanding the relative shortness of the record, satellite information is an essential tool in climate analysis and prediction, particularly for monitoring surface–atmosphere interaction over large space scales (see, e.g., Streten, 1978a) and associated climate teleconnections. Most notable among these is the El Niño Southern Oscillation (ENSO) phenomenon. The ENSO is a coupled oceanic and atmospheric oscillation which dominates in the tropical Pacific and has associations with global-scale climate variations (Graham and White, 1988). The oscillation has poles in the western and eastern tropical Pacific linked by the zonal Walker Circulation, and is characterized by reversals in oceanic and atmospheric conditions on time scales of between three and seven years. The set of anomalies accompanying an El Niño ('warm') event (in, e.g., 1972–3, 1976–7, 1982–3, 1986–7) involves an increase in the SSTs of the eastern and central equatorial Pacific, reduced oceanic upwelling near coastal Peru, lowered surface pressures in the South Pacific anticyclone and enhanced precipitation in the tropical and subtropical dry zones. At the same time, anomalies tend to be of opposite sign in the western Pacific, such that reduced rainfall (and even drought) occurs over the 'maritime continent' and Australia. At certain other times, the eastern equatorial Pacific may experience a 'supernormal' mode of ENSO: the La Niña or 'cold' SST event (e.g., 1987–8). Because of the usually gradual evolution of the ENSO phenomenon over many months, and its source over the data-sparse tropical Pacific, satellites are an ideal means of monitoring its progress (Ardanuy *et al.*, 1987). These data can provide input to coupled ocean–atmosphere general circulation models (GCMs) that are used to predict ENSO (Barnett *et al.*, 1988).

Variables that are routinely monitored from satellites, and which are relevant to ENSO, include the tropical SSTs and the locations of greatest convective cloud cover (Hastenrath, 1990). The latter are identified as highly reflective clouds (HRCs) in visible (solar) wavelength data (Ramage, 1975; Kilonsky and Ramage, 1976) and as regions of low outgoing longwave radiation (OLR) in thermal (infrared) wavelengths. These areas represent the sites of heaviest precipitation (Kidder and Vonder Haar, 1977) and, accordingly, are major sources of heat and moisture to the tropical atmosphere, and of momentum to the extratropics (Fig. 1.2). They are involved in the maintenance of the subtropical jet streams of both hemispheres (Nogues-Paegle and Mo, 1988). Since the cloud, radiation and precipitation anomalies also tend to overlie the regions of positive (warm) SST anomalies, they are an indicator of the state of the Southern Oscillation (Kousky, 1987; Strong, 1986). Figure 1.3 shows the February 1983 satellite-derived OLR expressed as anomalies from the 1974–83 baseline period. This month corresponds to the peak of a warm ENSO, and shows particularly the strong anomalies (increased cloud cover and precipitation) over the central Pacific. Additional variables important in the monitoring of ENSO over the Pacific Ocean, which have been obtained from satellite microwave sensors, include the atmospheric water vapor and cloud liquid water content; surface winds; and instantaneous rain rates (Chapter 5). The need to monitor climate teleconnections from satellites is expressed in large-scale observational and modeling programs, such as TOGA (Tropical Ocean Global Atmosphere) and the World Ocean Circulation Experiment (WOCE) (McBean, 1989).

4 Climate theory and modeling

Satellite information may provide essential input for climate modeling. Key areas of contemporary research include the improved parameterization of surface conditions (vegetation, snow and ice extent, soil moisture, and the associated fluxes of latent and sensible heat); and atmospheric distributions, particularly clouds. Table 2 summarizes the important methods for satellite surveillance of climate variables. For climate modeling purposes, the data provide realistic boundary conditions for the models; for example, in regard to the surface albedo (the reflectance integrated across all solar wavelengths) (Robock, 1983). Satellite data on clouds act as a check on the model parameterizaton of important atmospheric processes, such as the convective cloud scheme (Puri and Miller, 1990). It turns out that two areas most in need of high-resolution global-scale climatologies are probably the surface albedo and clouds. Both variables strongly influence the Earth–atmosphere radiation budgets (Chapters 2 and 6). Cloud and land surface conditions, as well as

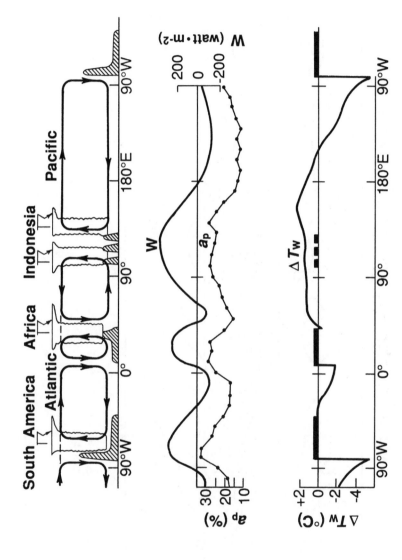

Fig. 1.2 Schematic of the zonal (west–east) Walker Circulation and its tropical teleconnections (top panel); longitude departures of the zonal mean SST (bottom panel); and the associated net heat input of an atmospheric column and planetary albedo (a_p) (a function of cloud cover), in middle panel. (From Lighthill and Pearce, 1981.)

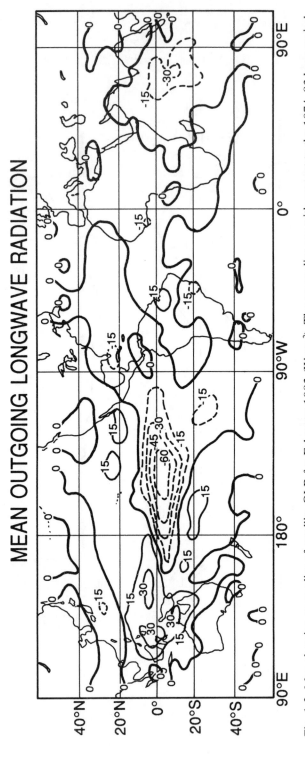

Fig. 1.3 Map showing anomalies of satellite OLR for February 1983 (W m^{-2}). The anomalies are with respect to the 1974–83 base period. Negative anomalies are shown dashed. (From Winston, 1985.)

ERBs, are currently the subject of intensive satellite monitoring programs (Table 2). The ultimate objective is to establish sufficiently long-term climatologies and to apply these to problems in climate modeling.

(a) Clouds in climate and the cirrus question

Clouds modulate the shortwave (from the sun) and longwave thermal (from the Earth) radiation streams and, in association with the underlying surface characteristics, help determine the surface temperature (see, e.g., Crane and Barry, 1984). Clouds effectively remove the strong differences in annual-averaged planetary albedo and net radiation that would be expected between Northern and Southern Hemispheres on the basis of the very different land–sea distributions (Henderson-Sellers, 1978; Stephens *et al.*, 1981). These competing effects of clouds can be expressed by the following equation (after Ohring and Gruber, 1983):

$$\delta = \frac{\partial Q}{\partial A_{ovc}} + \frac{\partial F}{\partial A_{ovc}} \tag{1}$$

where Q is the absorbed solar radiation, F is the emitted longwave flux, and A_{ovc} is the cloud amount.

The first expression on the right-hand side of equation (1) is the albedo effect of clouds; the second expression is the greenhouse enhancement effect. The cloud–climate sensitivity is a function of many cloud attributes on scales ranging from the microphysical to the macro. An understanding of the synoptic-scale variations (climatology) in cloud microphysical properties, such as cloud particle radii and ice-water content, is crucial to the accurate determination of the cloud–climate forcing and could be achieved using satellites (Liou and Ou, 1989). It has been argued from model studies that the sign of the feedback for the cloud forcing in a CO_2-enriched atmosphere could be changed according to the perturbed cloud particle radii (Slingo, 1990). Figure 1.4 shows that, for cloud particle radii smaller than the climatological mean values, the cloud albedo effect dominates over the greenhouse enhancement (i.e., the feedback is negative). However, the reverse is apparently the case for cloud droplet sizes that are larger than climatology. These forcings arise from changes in cloud liquid water content; smaller (larger) radii lead to reduced (enhanced) precipitation and, accordingly, an increase (decrease) in the cloud liquid water content. Somerville and Remer (1984) have argued, from simulation with a radiative-convective model, that the increase in cloud optical thickness arising from a CO_2-induced warming, could reduce the change in surface temperature by about one-half. They suggest that the albedo enhancement may operate for most clouds, but not for thin cirrus.

Satellites are beginning to help researchers assess the net effect of different cloud types, both as a global average and in specific regions (Kiehl and Ramanthan, 1990). Such information is critical for the generation of reliable climate scenarios in numerical modeling experiments, especially those concerned with trace-gas induced global warming (Mitchell *et al.*, 1989). Much of

Table 2 Sensors and satellite systems for monitoring weather and climate variables

	Variable	Spectral region(s)		Representative sensors (nominal spatial resolution)
Clouds:	coverage; type	VIS	0.4–0.7 μm	SR (4 km); AVHRR (1 km); VISSR (1 km); VHVIS (0.6 km); HVIS (2.7 km)
	coverage; type; CTT	TIR	8–14 μm	SR (8 km); AVHRR (1 km); VISSR (8 km); VHIR (0.6 km); HIR (2.7 km)
	cirrus	NEAR IR NEAR IR CO_2	2.6, 2.7 μm 1.0 μm 15 μm	SCR (25 km) SAGE (100 km) VAS (14 km)
Water vapor:	mid/upper tropospheric	water vapor	6.3–6.7 μm	VAS (14 km); HIRS/TOVS (30 km) THIR (22 km)
	column total;	water vapor	8–9 μm	VAS (14 km)
	cloud liquid water	passive microwave	8, 21, 22, 31, 37GHz	SMMR (55–27 km) SSM/I (12–25 km)
		passive microwave	90, 183 GHz	AMSU (planned)
Precipitation:	cumulative	VIS/TIR		VISSR (1 km; 7 km)
	instantaneous	passive microwave	19GHz	ESMR (25 km): ocean only
			18, 21, 37GHz	SMMR (55–27 km) = ocean and convective over land
		passive microwave	+85.5GHz	SSM/I (12–25 km)
Winds:	low-level	VIS		VISSR
		p. microwave; active microwave	13.4–14GHz	SMMR (55–27 km) SASS (25 km)
	high-level	TIR water vapor	8–14 μm 6.3 μm	VISSR; AVHRR; HIR THIR (23 km); VAS
Sea surface temps:	cloud-free scenes	TIR	8–14 μm	AVHRR (1 km)
	cloudy scenes	p. microwave	6.6GHz T_V	SMMR (55–27 km)

Satellite platform(s)	Technique	Corroborating data	Operational products	Special programs	Reference
NOAA; GOES; DMSP	threshold; bi-spectral HBTM; pattern recognition	surface observer; all-sky cameras	USAF 3-D/RT Nephanalysis	ISCCP; FIRE	Schiffer and Rossow (1985)
NOAA; GOES; DMSP		radiosondes			Cox et al. (1987)
Nimbus-5 Explorer GOES	limb sounding vertical sounding	surface obs; aircraft		Cirrus IFO	Barton (1983b) Woodbury and McCormick (1986) Wylie and Menzel (1989)
GOES; METEOSAT; TIROS-N; Nimbus-4; 5; 6	w.v. absorption	radiosondes	UTH		Eyre (1981)
GOES	multi-frequency sounding				Chesters et al. (1987)
Nimbus-7; Seasat; DMSP	T_B Differencing	aircraft			Wilheit and Chang (1980)
Eos					Wang et al. (1989)
GOES; METEOSAT; GMS	cloud indexing; bispectral; life history	rain gauges; radar	GPI; EPI	Winter MONEX TRMM	Arkin and Ardanuy (1989)
Nimbus-5; -6	absorption based	radar			Wilheit (1986)
Nimbus-7	scattering based				Spencer and Santek (1985)
DMSP					Petty and Katsaros (1990)
GOES; METEOSAT; GMS;	cloud tracking	ships; buoys	NMC analyses		Gruber et al. (1971)
Nimbus-7; Seasat Seasat; ERS-1	scatterometry	synoptic data			McMurdie et al. (1987)
GOES; NOAA; DMSP Nimbus; METEOSAT	cloud and water vapor feature displacement	radiosondes; aircraft	UTH		Turner and Warren (1989)
NOAA	split-window IR	ships; buoys	NOAA MCSST	TOGA	Kastner et al. (1980) McClain et al. (1983)
Nimbus-7; Seasat	vertical polarization	ships; buoys			Gloersen (1984)

Table 2 (*cont.*)

	Variable	Spectral region(s)		Representative sensors (nominal spatial resolution)
Snow-cover:	extent; reflect.	VIS, NEAR-IR; TIR		VISSR; AVHRR; HVIS
	depth; volume	p. microwave	18, 37GHz T_H	SMMR (55; 27 km)
	snow-cloud discrimination	Near IR Near IR	3.55–3.9 μm 1.51–1.63 μm	AVHRR ch. 3 (1 km) HVIS, VHVIS (2.7, 0.6 km)
Sea ice:	extent	VIS, TIR (cloud-free) p. microwave (cloudy)		AVHRR (1 km); HVIS (0.6, 2.7 km) ESMR (25 km); SMMR (55 km)
	type; concentration; surface melt	p. microwave a. microwave (23 cm)		SMMR (55 km); SSM/I (12–25 km) SAR (25 m); SIR-B (25 m)
Vegetation:	biomass	VIS/NEAR-IR (0.6; 0.9 μm) p. microwave (37GHz)		AVHRR bands 1 and 2 (1 km); MSS bands 5, 7 (80 m) SMMR (27 km)
Soil moisture:	amount; depth	TIR (8–9 μm) p. microwave (8, 21, 37GHz)		VISSR (8 km) SMMR (55 km)
ERB:	surface albedo	VIS; NEAR-IR (0.4–1.0 μm)		SR; AVHRR; VISSR
	surface solar irradiance	VIS		SR: VISSR
	OLR	TIR		SR; NFOVR/WFOVR (150 km)
Other:	ocean color/ chlorophyll	VIS	0.4–0.7 μm	CZCS (825 m)
	jet contrails	TIR	11, 12 μm	AVHRR: channels 4, 5
	aerosols	NEAR-IR	1.0 μm	SAGE (100 km)
	stratospheric ozone	MID-UV	0.2–0.3 μm	SBUV/TOMS (200 km)

Satellite platform(s)	Technique	Corroborating data	Operational products	Special programs	Reference
GOES; NOAA; DMSP	subjective classif.; band differencing	sfc. obs; aircraft	NOAA/NESS	SNOTEL	Matson and Wiesnet (1981) Strong et al. (1971)
Nimbus-7	T_B differencing	ground; aircraft			Chang et al. (1990)
NOAA DMSP (experimental)	narrow band sensing	aircraft			Kidder and Wu (1984) Crane and Anderson (1984)
NOAA; DSMP Nimbus-5; -7	pattern recognition T_B spatial gradient	shore obs; ships; aircraft	Joint Ice Center charts (Navy-NOAA)	MIZEX	Swift and Cavalieri (1985)
Nimbus-7, DMSP Seasat; Shuttle	T_B differencing T_H in L-band	surface; aircraft surface; aircraft			Weaver et al. (1987)
NOAA; Landsat	band ratioing	ground; aircraft	NOAA GVI/ NDVI	ISLSCP; FIFE	Tarpley et al. (1984)
Nimbus-7	T_H–T_V differencing	ground; aircraft			Choudhury et al. (1987)
GOES Nimbus-7	AM/PM T_S difference T_B polarization	rainfall (gauge) ground			Wetzel et al. (1984) Schmugge (1978)
NOAA; GOES ERBS; METEOSAT	minimum brightness	pyranometer; aircraft		ERBE; Nimbus-6; -7	Barkstrom et al. (1989) Ohring and Gruber (1983) Bess et al. (1989)
NOAA; GOES; METEOSAT	statistical; physical (models)	pyranometer	NOAA AgRISTARS		Tarpley et al. (1979) Gautier et al. (1980)
NOAA; ERBS; Nimbus-6; -7	window exitance	rainfall	CAC tropical analyses	HCMM;	Stephens et al. (1981)
Nimbus-7	band differencing	ships	CZCS (NASA)	GOFS	Feldman et al. (1989)
NOAA	band differencing	surface obs.			Lee (1989)
Explorer	limb sounding	surface obs; other satellite (VIS)		SAGE	Woodbury and McCormick (1983)
NOAA; Nimbus-7	backscattered u.v.	Dobson profiles	SBUV (NOAA/ NASA)		Ohring et al. (1989)

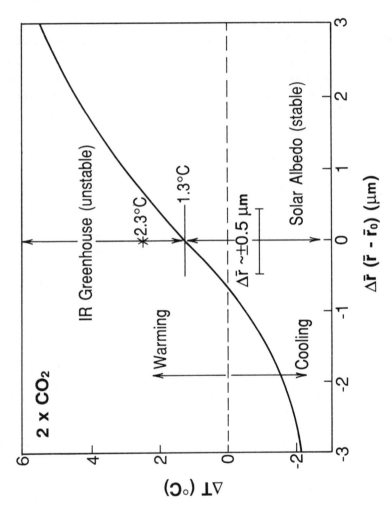

Fig. 1.4 The influence of cloud particle radius, r, on the surface temperature change for a modeled double CO_2 run. The change in the mean particle radius $\Delta\bar{r}$ is defined as the deviation of the mean radius for high, middle and low clouds, from the climatological means ($\bar{r}_0$). The horizontal bar in the center of the diagram denotes the standard temperature change. (From Liou and Ou, 1989.)

the variation in GCM responses to increased atmospheric CO_2 concentrations results from differences in the parameterization of clouds and cloud processes (Cess *et al.*, 1989). This is particularly true of the high cloud, or cirrus, fraction (Liou, 1986; Platt, 1989; also Chapter 3). While considerable efforts have been made in recent years to estimate the amount, characteristic optical thickness, microphysical properties, and spectral variations of cirrus clouds from satellite altitudes, they are a particularly difficult cloud layer to retrieve because of their wide range of radiative properties (Dowling and Radke, 1990). Moreover, there can be major differences between surface-observed and satellite-retrieved estimates of cirrus, especially over the oceans and high albedo snow and ice surfaces. For the latter in particular, cirrus clouds may radically alter the surface net radiation budgets (Cogley and Henderson-Sellers, 1984), so it is crucial that their presence be determined. They also influence the satellite retrieval of surface parameters, particularly SSTs. The cirrus cloud problem is a focus of this monograph on satellite climatology.

A closely related question to that of cirrus clouds involves the impacts on radiative and moisture regimes of anthropogenerated cirrus in the form of jet contrails (Fig. 1.5; see also Manabe, 1975; Carleton and Lamb, 1986). There is some speculation that documented increases in cloud cover during this century may be derived, at least in part, from increases in jet air traffic and the occurrence of contrails in preferred regions (Machta and Carpenter, 1971; Changnon, 1981; Lee and Johnson, 1985; Seaver and Lee, 1987; Angell *et al.*, 1984; Henderson-Sellers, 1989). As in the case of natural cirrus, the precise impact of contrail cirrus on climate is unclear. Kuhn (1970) showed from aircraft measurements that thick persistent contrails may result in a depletion of about 12% in the net radiation just below cirrus level and a 7–8% decrease at the Earth's surface, depending on latitude. This translates into a surface temperature decrease of $-5.3\,°C$ but which reduces to about $-0.15\,°C$ when assuming a 5% persistence of contrails. Figure 1.6 shows this effect for a mid-latitude contrail. These results, along with those of Reinking (1968), argue that the albedo effect of 'false' cirrus and cirrostratus exceeds the greenhouse enhancement in both middle and tropical latitudes (see also Fleming and Cox, 1974; Griffith *et al.*, 1980). Other studies favor the greenhouse enhancement effect of thin cirrus and contrails (Stephens and Webster, 1981; Detwiler and Pratt, 1984; Cogley and Henderson-Sellers, 1984; Manabe, 1975). Obviously, a fuller appraisal of the contrail cirrus—climate question requires an accurate determination of the frequency of occurrence of contrails; and their characteristic thickness, persistence and frequency of lateral spreading to form cirrostratus. Bryson and Wendland (1975) estimated an increase in cirrus cloudiness of 5–10% for North America due to contrails, but only about 1/20 of this percentage for the globe as a whole. They assume that 50% of aircraft in the air at one time make contrails lasting an average of two hours. Using Landsat data, Detwiler and Pratt find contrail widths over the north-eastern US to be highly skewed towards the lower values, and with a median width of 2–3 km.

A step towards the derivation of a large-scale (North America and adjacent ocean areas) contrail climatology using high-resolution DMSP imagery (e.g., Fig. 1.5) is currently underway (DeGrand *et al.*, 1990). Figure 1.7

Fig. 1.5 Portion of a DMSP TIR orbital swath (1.0 km resolution) showing multiple jet contrails over the Midwest US (1621Z April 19, 1979). The contrails are the bright lines in the cirriform cloud mass located in the upper center of the image.

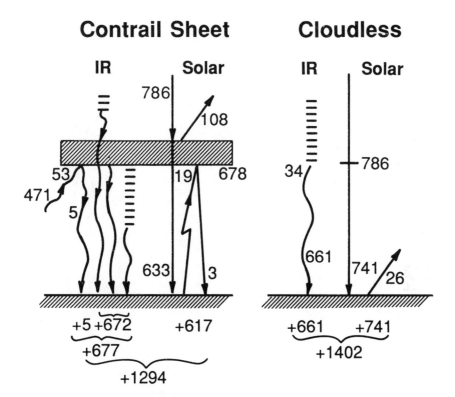

Surface Total Power Depletion : 8 %
IR Increase 2.4 %
Solar Decrease 16.7 %

Fig. 1.6 Radiation budget of mid-latitude cirrostratus derived from jet contrails, in comparison with clear-sky conditions. Note the net cooling effect of this case. (From Kuhn, 1970.)

shows the normalized frequencies of contrails for three years' (1977–9) mid-season months, and illustrates the high concentrations in certain regions; notably, the south-west Pacific coast, the west Canada coast, the Midwest US and south-east US. These distributions and their seasonal variations appear to be more a function of synoptic circulation controls than aircraft flight frequencies.

(b) Gaia and cloud–climate feedback

Satellite information is proving invaluable in the development of a holistic theory of climate and climate change that is compatible with the growing

Mean Monthly Distribution of Jet Condensation Trails

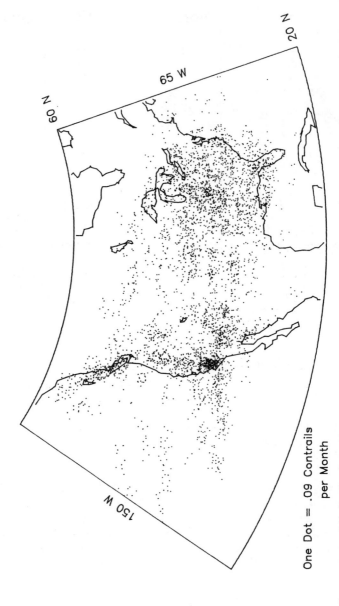

One Dot = .09 Contrails
per Month

Fig. 1.7 The mean monthly distribution of jet contrails for North America, determined from subjective interpretation of high-resolution DMSP visible and IR direct readout imagery (0.6, 1.0 km resolution). The map shows contrail frequencies in 1° × 1° grid cells for three years' mid-season months of 1977–9, excluding January 1979. (From DeGrand *et al.*, 1990.)

emphasis on Global Change and Earth System Science. Perhaps central to this approach is the so-called Gaia Hypothesis, or the theory that the biota on land and in the oceans helps modulate the Earth's climate and keeps it generally amenable to the continued proliferation of life (Lovelock, 1989). The distribution and status of the biomass can be monitored effectively from satellites (Goward, 1990) and estimates made of the primary productivity (Platt and Sathyendranath, 1988). This knowledge is an essential component of the CO_2-climate question because of the role of the biota as a CO_2 sink. It is expressed, for example, in the Global Ocean Flux Study (GOFS), and which has a major satellite component to monitor 'ocean color' (Brewer et al., 1986). Ocean color refers to the preferential reflectance of certain solar wavelengths as the chlorophyll concentrations associated with phytoplankton increase, and which can be detected from satellite altitudes (see, e.g., Holligan et al., 1983). Evidence supporting the idea of some degree of climate interaction by the phytoplankton blooms has come from satellite observations of reflectivity in marine stratus clouds over the eastern North Pacific (Coakley et al., 1986). The cloud reflectivity was found to be increased along shipping lanes, where the release of pollutants from ship stacks act as cloud condensation nuclei (CCN) to both reduce the drop sizes and increase the total concentrations of cloud droplets (Albrecht, 1990; Radke et al., 1989) (Fig. 1.8). These observations are important since it is known that phytoplankton emit a form of CCN as dimethylsulphide (DMS), and that the rate of CCN production is high in regions of high DMS flux and solar radiation receipt (Bates et al., 1987). The implication is that, in areas of high primary productivity in the ocean, the phytoplankton may help modulate the surface temperature (T_s) via the enhancement of cloud reflectivity. The possible feedback loop involved in the phytoplankton–DMS–CCN–cloud reflectivity, which may be negative, is shown in Fig. 1.9. Note that there is still uncertainty as to the details of the feedback loop, and these await further research. In the broader context, these observations point strongly to the still incompletely understood role of clouds in climate, and which constitutes a key focus of ongoing satellite cloud research and of this monograph.

5 Satellite sampling strategies for climate research

Spatial and temporal scales of climate processes are closely interconnected (Salby, 1989): the smaller the spatial scale of the phenomenon, the generally shorter-lived it is (cf. cumulus clouds, thunderstorm cells, squall lines, and mid-latitude cyclones). It has been noted from satellite data that, at both large (e.g., 32 km) and small (e.g., 1 km) spatial resolutions, there is a tendency towards uniformity of cloud conditions: either cloudy or clear skies occur (Rossow, 1989). However, at intermediate scales (e.g., 4 km) the frequency of occurrence of partly cloudy scenes increases and the texturing associated with different types becomes apparent. Similarly, as the period of temporal averaging increases there is a tendency towards the majority of regions being characterized by partly cloudy conditions, rather than the dominantly clear

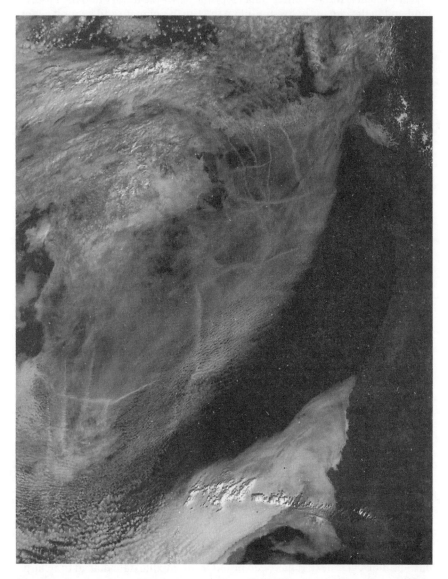

Fig. 1.8 Section of a DMSP visible orbital swath (2.7 km resolution) showing multiple ship tracks embedded in the stratiform clouds off the west coast of North America. The tracks are the linear features that are brigher than the surrounding cloud (1705Z, August 22, 1988).

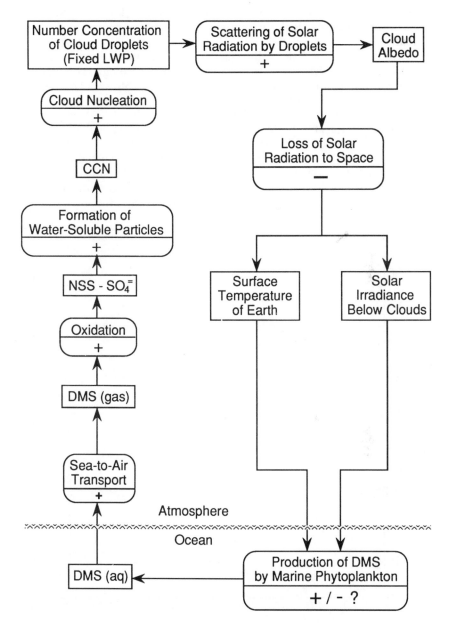

Fig. 1.9 Conceptual diagram of a possible DMS-climate feedback loop. The rectangles enclose measured quantities; the ovals represent processes linking the rectangles. LWP = liquid water path of the cloud. This is influenced by the cloud droplet size and cloud thickness; it influences the cloud albedo. (From Charlson *et al.*, 1987.)

or cloudy conditions characteristic of short time scales (Fig. 1.10). Surface-based observations, particularly for the oceans, tend to confirm these cloud attributes (Henderson-Sellers, 1978; Henderson-Sellers *et al.*, 1981). They give dominantly U- or J-shaped frequency distributions of cloud amounts, although there is evidence of a strong stratification according to cloud group (cf. cumuliform, stratiform) (Barrett and Grant, 1979; Stowe *et al.*, 1988): Fig. 1.11. Advantage may be taken of the statistical characteristics of cloud-cover for different regions (e.g., Bean and Somerville, 1981) to provide boundary conditions for modeling the global radiation balance (Yang *et al.*, 1988). The resolution influences the probability of the co-occurrence of overlapping cloud layers (Tian and Curry, 1989). At larger spatial scales, the probability of overlap is reduced, particularly for cloudier situations. Considerations such as these are important in the development of one or more global-scale cloud climatologies, as under the umbrella of the International

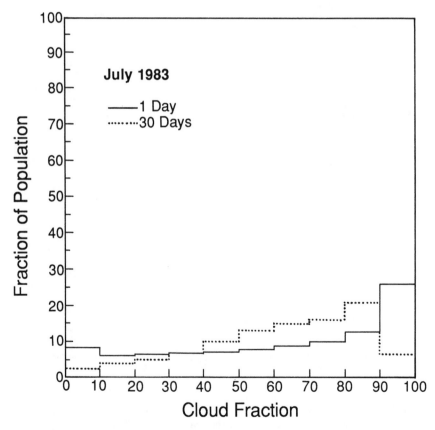

Fig. 1.10 The influence of temporal averaging period on the regional-scale (2.5° equal-area) cloud fraction. The cloud fraction is determined from the number of cloudy pixels in that area, from July 1983 ISCCP data. The solid line represents the distribution averaged over 1 day; the dotted line, over 30 days. (From Rossow, 1989.)

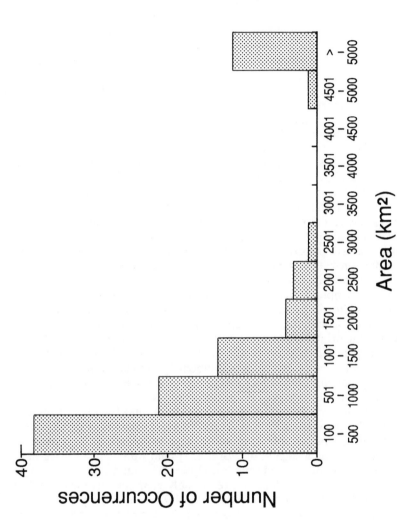

Fig. 1.11 Histogram of cloud-size distribution for the Miles City, Montana, area as observed by ATS-3. The satellite resolved area is 113 km². (From Reynolds *et al.*, 1978.)

Satellite Cloud Climatology Project (ISCCP), and which may involve sensors from different satellites having different resolutions (Schiffer and Rossow, 1983, 1985).

Of considerable relevance in the formulation of spatial sampling strategies involving satellite data is the observation that cloud and rain areas may be fractals over at least six orders of magnitude (Lovejoy, 1982). That is, they have structures at all scales but no characteristic length. This is especially true in the tropics where convection dominates. This scale invariance, which has strong implications for fundamental physical processes of turbulence and energy cascades, implies that satellite observations of clouds and rain are independent of sensor resolution and may be characterized statistically (Lovejoy and Schertzer, 1990). Fractal dimensions derived from establishing perimeter–area relationships over a range of satellite resolutions, have been investigated (Kuo *et al.*, 1988; Welch *et al.*, 1988; Cahalan and Joseph, 1989). Of the types studied using this technique, only cirrus clouds approach true fractal character. Other clouds, particularly fairweather cumulus, cumulonimbi in the ITCZ, and marine stratus, appear to be bi-fractal; there is a change in the fractal dimension at a characteristic size. These results have suggested very different dynamical and possibly microphysical processes occurring between cirrus and boundary layer clouds (Joseph and Cahalan, 1990). Artificially produced clouds, such as contrail cirrus, have a higher perimeter–area dimension compared with 'natural' cirrus (Kuo *et al.*, 1988), and this may provide a useful means of detecting these features.

A key question in satellite climatology research is: What are the most appropriate temporal frequencies and spatial resolutions for sampling weather and climate processes? Formulation of an appropriate sampling strategy involves the characteristic patterns of variability of the processes concerned and the orbital characteristics of the satellite system (see e.g., Brown, 1970). It is particularly problematic in the case of cloud cover conditions, surface solar irradiance estimates and albedo, precipitation rates and ERB components. Satellite climatologies of these variables that do not adequately sample the marked diurnal cycle in the tropics or over mid-latitude land areas in summer, will be unrepresentative (Minnis and Harrison, 1984a, b; North, 1987; Ramanathan, 1987). Accordingly, geostationary meteorological platforms are better able to monitor these variations, at least over the lower latitudes, than a single-polar-orbiter system with its fixed twice-daily (ascending, descending) orbital nodes. However, even a one-polar-orbiter system is capable of sampling in a reasonably representative fashion. For example, the 0900 local time overpass of the NOAA satellites is considered a good proxy for the diurnally averaged planetary (or system) albedo (Briegleb and Ramanathan, 1982), and which represents the contributions to the upwelling radiation from the Earth's surface, clouds and the atmosphere. For surface solar irradiance estimates, a two-polar-orbiter system with overpasses at around 0900 and 1500 hours has considerable reliability when compared with higher temporal sampling obtained using a geostationary platform (Tarpley, 1979).

To optimize the monitoring of some variables, such as the diurnal variation of tropical precipitation rates, and Earth surface thermal properties, requires dedicated platforms in orbits that are neither geostationary (and high altitude) nor polar orbiting. NASA's Heat Capacity Mapping Mission (HCMM) involved a satellite that was capable of sensing the diurnal variations of the thermal inertia of heterogenous surfaces for use in ERB studies (e.g., Carlson *et al.*, 1981; Price, 1982). For monitoring the strongly diurnal rainfall regime of the tropics, a relatively low-altitude low-inclination orbit has been proposed for the TRMMS (Tropical Rainfall Measuring Mission Satellite), which will be equipped with both passive and active (radar) microwave sensors (Simpson *et al.*, 1988). Kedem *et al.* (1990) have argued, from sampling considerations, that such a satellite should be able to yield reasonable estimates of the area-averaged three-week mean rain rate for regions on the size of GATE (about 350×359 km^2). Kidder and Vander Haar (1990) advocate the use of satellites in Molniya-type orbits for high frequency monitoring of higher-latitude processes, particularly 'polar lows'.

6 Satellite and conventional data

Conventional meteorological and geophysical data are essential for validating the satellite observations ('truth'). However, it is important to realize that the satellite sensor is not necessarily 'seeing' the same thing as the corroborating conventional data. The radiometric signals must be transformed into meaningful information using statistically or physically based retrieval algorithms that consider the characteristic spectral responses of the variables. Information on typical variables include surface observations of cloud (both visually and with all-sky cameras); solar radiation using pyranometers; the vertical profiles of temperature and humidity from radiosonde ascents; surface-based radar observations of precipitation, and SSTs from ocean drifting buoys (see Table 2). This distinction between the satellite retrieval and 'ground truth' may be obvious, as in the case of clouds, which are viewed 'top down' and 'bottom up', respectively. This difference influences the determination of cloud amount for co-occurring cloud levels. Thus, as the high cloud fraction increases due to the spreading out of convective clouds, the low cloud fraction appears to decrease when viewed from satellite altitudes, especially for large amounts of cloud (e.g., Fig. 1.12; also Ackerman and Cox, 1981; Minnis and Harrison, 1984a; Henderson-Sellers *et al.*, 1987; Sèze *et al.*, 1986). It is important to know the co-occurrence probabilities of different cloud layers for radiative considerations and also for improving the modeling of cloud effects in GCMs. Less obviously, satellite thermal infrared sensing of SSTs tends to provide a surface 'skin' temperature that correlates better with observations from drifting buoys than with the bulk temperatures sensed by ships (Strong and McClain, 1984); water vapor sensed in the absorption band around 6.3 μm is generally applicable to a *layer* rather than a distinct atmospheric level sensed by a radiosonde; and a true surface albedo is not usually detected or derived from satellite sensors. Rather, it is the *reflectance* in a relatively narrow wavelength band, typically 0.5–0.7 μm (Matthews and Rossow, 1987), to which a weighted narrow-band to broad-band (albedo) conversion

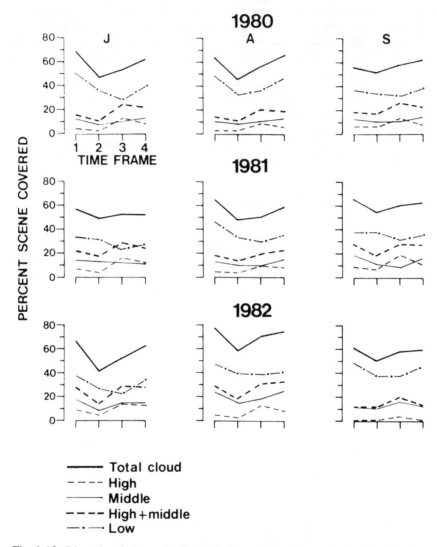

Fig. 1.12 Diurnal variations of GOES-W-observed cloud cover by level and for the total cloud, over the south-west US: three summers (July, August, September 1–15), 1980–2. The time frames are: (1) 845-1245Z; (2) 1345-2045Z; (3) 2145-245Z; (4) 345-745Z. Statistics are grouped according to month. Note the out-of-phase relationship between lower and upper cloud amounts. (From Carleton, 1985a.)

would need to be applied. In addition, the influence of changing satellite–target–sun geometry, solar elevation, atmospheric attenuation by water vapor and aerosols, and other effects need also to be incorporated in order to approximate a true surface albedo (Chapter 2).

By extension, satellite remote sensing can also yield information on weather and climate variables indirectly. For example, all-weather microwave sensing of surface winds over the open ocean (e.g., by SASS and Nimbus-7 SMMR) actually senses the effect of the wind on the ocean surface in the form of increased roughening and wave formation, rather than the airflow directly (Davison and Harrison, 1990). Of course, more direct methods are available, as in the automated tracking of cloud displacements at short time increments to provide winds at particular cloud levels (see e.g., Gruber *et al.*, 1971; Wylie *et al.*, 1981). The deployment of night-time lightning detectors on satellites (e.g., DMSP) is a more obvious surrogate for thunderstorm activity (see, e.g., Christian *et al.*, 1989).

Satellite and conventional data are not directly comparable in many circumstances, for the following reasons.

(1) There are differences in the spatial coverage of the sensor and 'ground truth' systems. Conventional data are typically for a point whereas satellite observations are usually aggregated over an area that is a function of the sensor resolution. Satellite estimates of surface solar irradiance, cloud cover, precipitation and SSTs are particularly sensitive to this point–area differential, so it is often necessary to average the conventional data spatially to make them comparable with the satellite retrievals (see, e.g., Griffith, 1987; Emery and Schluessel, 1989). In general, the satellite retrievals tend to correspond fairly closely with the surface observations of these variables over space scales of about 50 km^2 (Henderson-Sellers *et al.*, 1981; Barrett and Grant, 1979; Njoku, 1985). However, for processes occurring on sub-grid scales, such as convective precipitation in coarse resolution microwave data, significant underestimation of their magnitude can occur due to area-averaging.

(2) The asynopticity of the satellite and conventional meteorological data may lead to insufficient sampling of processes at smaller scales (e.g., convection) and their subsequent underrepresentation when averaged into larger scales (Salby, 1989). Such a problem necessitates a careful consideration of sampling strategies for weather and climate monitoring, according to the characteristic time and space scales of the variables concerned.

(3) Different physical properties may be sensed; for example, the surface radiant temperature sensed by a satellite is typically different from the physical (thermometric) surface temperature. Also, the surface shortwave reflectance sensed by a scanner contrasts with the sky-hemisphere averaged surface albedo measured using a pyranometer.

(4) Satellite sensor degradation may occur through time ('drift'), and it may not be possible to determine the necessary correction in the absence of on-board calibration. This problem is particularly pertinent to reliable satellite estimates of ERB components, such as the albedo (Chapter 6). In that case, the ability of the satellite to monitor the solar constant helps constrain the

irradiance estimates for albedo determination, and acts as a check on the sensor performance over time.

(5) The 'truth' data may themseves be in error. For example, humidities sensed by radiosondes are considered unreliable in the generally drier mid- and upper-troposphere. Also, calibration of satellite rain rates using ground-based radar presupposes that the radar itself is correctly calibrated against gauge measurements (but see Barnston and Thomas, 1983; Petty and Katsaros, 1990).

(6) Inaccuracies in satellite–Earth colocation. This influences particularly small time- and space-scale variations of a variable, such as convective precipitation, sea ice concentration, or the spatial variability of the solar irradiance (Raphael and Hay, 1984).

7 Retrieval algorithms

Algorithms to retrieve climatically significant information on atmospheric and Earth surface variables tend to be either statistical (empirical) or physically based. In the former, linear regression analysis is often used to derive associations between the satellite sensor response in a particular wavelength or group of wavelengths and a conventionally measured parameter under a wide range of conditions. For example, the satellite determination of surface solar radiation receipts by Tarpley's (1979) method compares satellite-derived brightness and surface pyranometer estimates separately for cloud-free, partly cloudy and overcast scenes for the US Great Plains. Similarly, algorithms have been developed to derive satellite SST estimates for clear-sky conditions by comparison with buoys, but without explicitly modeling the effects of the intervening atmosphere (see, e.g., Strong and McClain, 1984). Cloud detection algorithms, necessary both for study of the 'uncontaminated' Earth scene (Chapter 6), and for the clouds themselves (Chapter 3), tend to utilize the different spectral responses of targets. The task is easier in visible wavelengths over the ocean, but far more complex over high albedo snow and ice surfaces or when the overlying clouds are optically thin. In the broad-band visible and thermal infrared (TIR) wavelengths characteristic of most meteorological satellite sensors, clouds and snow have broadly similar radiances. However, it is possible to utilize certain narrow wavelength bands where the differences between snow and clouds, and even between water and ice clouds, are at a maximum and can be used for cloud detection over high albedo surfaces (Table 2). These include the 3.7 μm or the 1.5–1.6 μm channels (Fig. 1.13). More complex statistical algorithms, such as that used to estimate soil moisture from satellite IR data (Wetzel and Woodward, 1987) rely on the inclusion of several contributory variables such as surface wind speed, vegetation cover, low-level temperature advection, and an antecedent precipitation index.

A problem with such empirical algorithms is that they may not be transferrable without modification to other climatic regimes and latitudes (see, e.g., Raphael and Hay, 1984). Thus, a satellite precipitation estimation scheme utilizing visible and TIR data and developed for the tropical Atlantic,

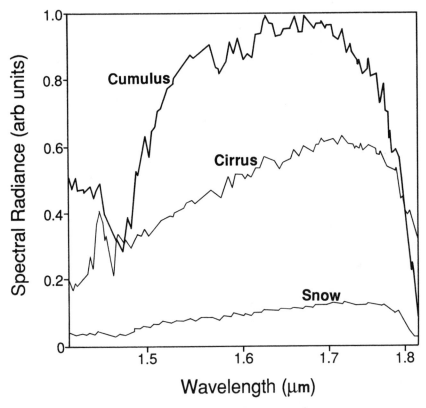

Fig. 1.13 Spectral irradiance of snow compared with water clouds (cumulus) and ice clouds (cirrus) in the near-IR. Strong reflectances at around 1.6 μm are useful for satellite differentiation; as with the DMSP snow-cloud discriminator. (From Crane and Anderson, 1984).

will not be directly applicable to semi-arid or arid land areas, where the sub-cloud layer is much drier; or to middle latitudes, where most precipitation derives from widespread ascent associated with fronts and cyclones (Griffith *et al.*, 1981). Similarly, satellite estimates of surface albedo and solar irradiance are particularly sensitive to variations in aerosol loading, and this can vary quite strongly spatially. The variation of atmospheric water vapor content with latitude over the oceans needs to be considered when estimating SSTs from IR data (Barton, 1983a), and estimates of sea ice—water concentration need to consider also the presence of surface meltwater (Crane and Anderson, 1989). A further problem with such statistical algorithms occurs when the relationship between satellite-derived and conventional data is not linear over the full range of values. In such cases, values on the dependent variable may be transformed mathematically (e.g. logarithmically). A prime example of this problem involves the indirect estimation of convective precipitation from visible and infrared radiance variations, which exhibits a strong life cycle (Fig. 1.14). In the early to mature stages of a convective cloud the increasing cloud brightness in visible wavelengths and decreasing cloud top temperature (CCT) in IIR wavelengths tends to correlate well with increasing precipitation intensity. However, the association breaks down as the maximum cloud area is reached, and which arises largely from the detection of cold cirriform clouds as the convective cell dissipates.

Physically based retrieval algorithms attempt to model mathematically the net effects on the satellite sensor response of a host of subsidiary contributions from surface and atmospheric variables, and also the characteristics of the satellite sensor. The determination of a reliable surface albedo from a satellite-sensed planetary albedo, or top of the atmosphere (TOA) brightness, needs to account for all of the following important effects: surface-cover type and its change through time; the atmospheric effects of absorption, scattering and reflection by atmospheric gases, especially O_3 and water vapor, and by liquid water in the form of clouds; attenuation by dust and other aerosols; changes in solar elevation according to time of day and season; changes in the sun–Earth–satellite geometry; and the representativeness of the sensor spectral resolution (see, e.g., Barker and Davies, 1989).

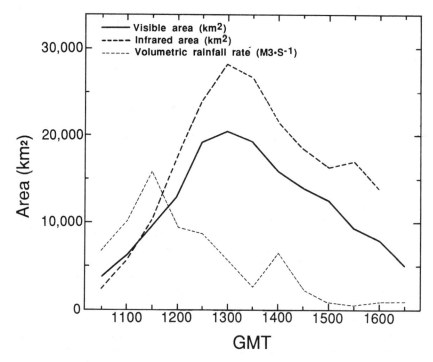

Fig. 1.14 Temporal evolution of a typical tropical convective rain cloud in terms of satellite-observed cloud area (km^2) and volumetric rain rate, as measured by radar ($m^3 s^{-1}$). Note that the peak in rainfall rate precedes the maximal coverage of the cloud area. (From Stout *et al.*, 1979.)

Although a non-trivial task, good results have been obtained from the use of radiative transfer models, even though some assumptions are still often required (Diak and Gautier, 1983). It is necessary to validate the physically derived algorithm on independent satellite data sets and this often takes the form of a regression-type analysis (see, e.g., Darnell *et al.*, 1983; Justus *et al.*, 1986). Similarly, retrieval of cloud properties may make use of physical modeling by comparing results for cloudy atmospheres with the satellite-observed radiances (Rossow *et al.*, 1989a).

Ideally, once relationships have been established between satellite and 'truth' data, it is possible to capitalize on the strong wavelength-dependent sensitivities of targets in two or more spectral bands to retrieve information from satellite altitudes. A prime example is the determination of land cover types by difference-ratioing reflected solar radiance information in narrow-band visible (0.5–0.7 μm) and the near-IR (0.7–1.1 μm). (Table 2). This procedure capitalizes on the very different spectral responses of actively growing from stressed vegetation, bare soil and snow-covered surfaces (e.g., Figure 1.15), and is used operationally by NOAA to derive a normalized difference vegetation index (NDVI) on a weekly basis (Ohring *et al.*, 1989; also Chapter 6). Similarly, the determination of chlorophyll concentration in the upper ocean layers ('ocean color') from the Nimbus-7 Coastal Zone Color Scanner (CZCS) relies upon changes in the band reflectance between about 500 and 550 nm as chlorophyll concentrations increase (Fig. 1.16). In cloud-free daytime scenes this results in a shift from the blue to the green portion of the visible spectrum (Hovis *et al.*, 1980). Satellite ocean color data sets are available for climatological studies at a range of spatial and temporal resolutions (Feldman *et al.*, 1989).

8 Developments in satellite climatology through the 1980s

A characteristic of climatological studies through the 1980s was the increased use of satellite information, either as the key data source on an atmospheric phenomenon (see, e.g., Carleton and Carpenter, 1990; Kuhnel, 1989) or to supplement information from other sources (Table 2). These other sources, both standard meteorological and non-standard, include:

(1) ground-based radar, particularly for rainfall estimation studies (e.g., Woodley *et al.*, 1980);
(2) aircraft, for corroborating satellite estimates of surface albedo (Rockwood and Cox, 1978) and high-latitude meso-cyclone studies (Shapiro *et al.*, 1987);
(3) all-sky cameras for observations of clouds (McGuffie and Henderson-Sellers, 1988, 1989);
(4) surface observer data, particularly for clouds; either on land or at sea (Henderson-Sellers *et al.*, 1987);
(5) other surface-measured data, such as rain gauges for corroboration with satellite rainfall estimates (Griffith, 1987) and pyranometers for use in satellite estimation of surface solar radiation receipts (Tarpley, 1979; Justus *et al.*, 1986);

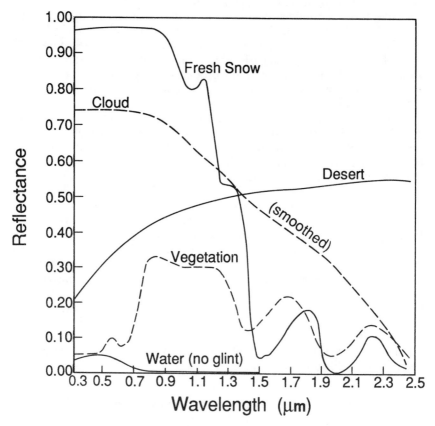

Fig. 1.15 Generalized spectral reflectance curves of five different targets in the 0.3–2.5 μm (visible, near-IR) region. The curve for clouds does not consider particular types. (From Davis *et al.*, 1984.)

(6) radiosonde data for corroboration of satellite vertical temperature and moisture profiles ('sounder') data (see, e.g., Prabhakara *et al.*, 1982);

(7) drifting buoys in the open ocean and in the pack ice regions to verify SSTs, surface temperatures, and ocean current speeds (Strong and McClain, 1984; Barry and Maslanik, 1989).

There has also been a tendency in recent years to use information from another satellite or sensor system as truth. This is seen in the efforts to verify the retrieval algorithms for sea ice concentration in the relatively coarse resolution microwave data of the Nimbus-7 SMMR; comparisons of cloud detection in meteorological satellite data with high-resolution Landsat MSS (Multispectral Scanner System); and comparisons of land biomass characteristics in NOAA Advanced Very High Resolution Radiometer (AVHRR) and SMMR.

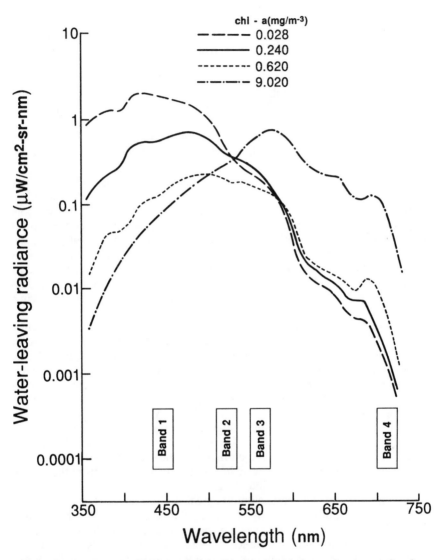

Fig. 1.16 Spectral distribution of light backscattered from the ocean for four different concentrations of chlorophyll (mg m⁻⁶) measured at the water surface ('ocean color'). Note the reversal that occurs between about 500 and 550 nm (0.5–0.55 μm) for low and high pigment concentrations. Also shown are Nimbus-7 CZCS spectral bands 1 through 4. (From Hovis *et al.*, 1980, and Yates *et al.*, 1986.)

This tendency towards the use of multiple observing platforms has been most evident in special monitoring programs. The Marginal Ice Zone Experiment (MIZEX) examined the meteorological, glaciological, oceanographic and biological characteristics of the seasonal sea ice zone in the Bering Sea (MIZEX West) and Fram Strait/Greenland Sea (MIZEX East). The marginal ice zone marks the transition from open water to polar ice regimes, and the interactions between water, ice and the atmospheric boundary layer are therefore highly complex (MIZEX Group, 1986). At the same time, these regional differences in interactions at the MIZ are compounded by seasonal differences in the physical processes near the ice edge (Johannessen, 1987). MIZEX used combinations of satellite, meteorological and remote sensing aircraft, ships and satellite-tracked buoys. The satellite component consisted chiefly of passive microwave sensing using the Nimbus-7 SMMR.

Satellite data on vegetation status can be used along with other remotely and conventionally measured variables to estimate surface evapotranspiration and its effect on surface temperature (Nemani and Running, 1989). Note also the attempts to improve satellite rainfall estimation using combinations of microwave, visible and TIR data (Barrett *et al.*, 1988, 1990), and the classification of clouds over high albedo Arctic surfaces using different spectral bands (Key *et al.*, 1989; Ebert, 1987, 1989). These developments have largely resulted from improvements in satellite sensor systems and data handling capabilities. In this regard we should note particularly the following.

(1) Improvements in the coverage of the operational environmental satellites and their sensing capabilities, particularly of the USA, but also of other countries and consortia (e.g., the European Space Agency). These major improvements may be considered to date back to the mid- to late-1970s with the deployment of the Japanese GMS and European METEO-SAT geostationary platforms, and the latest series of US NOAA polar orbiting satellites commencing with TIROS-N. An important feature of the last for meteorological and climatological research is the five-channel high-resolution AVHRR and the TOVS (TIROS-N Operational Vertical Sounder). Other improvements have included the wide access for researchers to high-resolution hard-copy cloud image sets, such as those from the DMSP (Defense Meteorological Satellite Program).

A characteristic of satellite climatological studies in recent years has been the ability to optimize the utility of a particular data set by using it to obtain information on more than one weather or climate variable. For example, the retrieval of the surface reflectance of solar radiation yields not only the surface albedo but, in conjunction with cloud information, the surface solar irradiance (see, e.g., Tarpley, 1979; Gautier *et al.*, 1980); satellite limb sounding of tropospheric ozone and aerosols can improve the estimates of surface solar irradiance by improving the modeling of the atmospheric transmittance (Darnell *et al.*, 1988); knowledge of the atmospheric water vapor distribution from satellite soundings can greatly improve the retrieval of SSTs from satellites (Barton, 1983a), as can a knowledge of the incidence of optically thin cirrus clouds (Prabhakara *et al.*, 1982, 1988) and the distribution of volcanic aerosols and dust (Strong, 1986; Lee, 1989a). An interesting feature of satellite climatological studies of the 1980s has been the tendency to go back to data sets collected on Research and Development (R&D) platforms, such as the earlier Nimbus satellites, and reanalyze them in the context of contemporary climate issues. For example, Prabhakara *et al.* (1979, 1988) have used data from the Nimbus-4 IRIS (Infrared Interfero-meter Spectrometer) instrument to assess, respectively, the total precipitable water and the occurrence of thin cirrus over the oceans for the April 1970–January 1971 period.

By the early 1980s, a five-geostationary satellite configuration was in place to permit more or less continuous observation of the Earth disc, especially of the low, subtropical and middle latitudes. Weather and climate processes benefiting particularly from such platforms, owing to their strongly diurnal or short-lived nature, are rain estimation; surface albedo and estimates of the surface receipt of solar radiation; cloud cover; and detection of meso-scale phenomena such as shower bands and thunderstorms, and monitoring of meso-scale convective systems (MCSs).

(2) Advances in satellite systems. The deployment of operational satellite sounders, particularly the TOVS and Microwave Sounding Unit (MSU) on the NOAA series, and the VAS (VISSR Atmospheric Sounder: VISSR Visible and Infrared Spin-Scan Radiometer) on the GOES series have made major contributions to climatological studies. These include cloud climatologies (Wylie and Menzel, 1989) and assessments of climatic change in tropospheric temperatures (Spencer and Christy, 1990). There has also been an increase in what might be considered non-standard satellite monitoring systems: those having orbital inclinations designed for a specific purpose (e.g., NASA's HCMM, and the proposed TRMM), or with sensors that do not scan in a more-or-less vertical plane. In the latter category must be included the limb-sounding sensors, such as that on the SAGE (Stratospheric Aerosol and Gas Experiment) satellite. Limb sounding involves sideways sensing through the atmosphere, and is particularly useful for concentrating atmospheric con-stituents that are present in trace amounts (e.g., stratospheric ozone, aerosols). Data from these missions have made contributions to the problem of detecting high-altitude (cirrus) clouds (Woodbury and McCormick, 1983, 1986), particularly where they are optically thin since they would be very difficult to detect using vertical sensing of the atmosphere.

(3) Greater diversity in spectral sensitivity of satellite sensors. Satellite climatological studies through the mid- to late-1970s were dominated by data in the shortwave visible (0.4–0.7 μm) and thermal infrared (8–14 μm) regions of the electromagnetic spectrum (EMS) since these were the standard sensors on operational meteorological satellites. Through the 1980s, the use of visible and IR data continued to be important in synoptic circulation studies, rain estimation and ERB analyses. At the same time, other wavelength regions of the EMS were being increasingly applied to satellite climatological problems, most notably the microwave frequencies (approx. 1–500 GHz). Passive microwave sensing and the development of appropriate retrieval algorithms was a prime concern of the NASA R&D satellites which began with the single-channel Nimbus-5 ESMR and

continued with the long-lived multi-channel Nimbus-7 SMMR. These developments are continuing with the DMSP SSM/I (Special Sensor Microwave/Imager) (Weaver *et al.*, 1987). The key areas of climate monitoring in which passive microwave radiometry has made major contributions in recent years (Table 2) include the following:

(a) sea ice extent, concentration and ice type, and detecting the presence of surficial meltwater;
(b) continental snow-cover extent, snow depth and snow volume;
(c) instantaneous rain rates, particularly over the oceans but increasingly also for convection situations over land;
(d) vertical temperature and moisture profiling;
(e) spatial distributions of tropospheric water vapor (precipitable water) and cloud liquid water content;
(f) estimates of cloud vertical thickness over the ocean;
(g) soil moisture and vegetation status;
(h) surface winds over the open ocean;
(i) sea surface temperatures in cloudy scenes;
(j) estimation of meteorological parameters (temperature, pressure) of tropical cyclones.

The trend towards using passive microwave for geophysical and climate studies will continue through the 1990s, with sensors deployed on the next generation of GOES satellites. These developments will, moreover, be complemented by the deployment on polar orbiters of imaging and non-imaging active microwave (radar) sensors that continue the advances made by the short-lived Seasat mission of 1978.

The application to climate studies of data in the microwave frequencies has been the major, but not the only, development through the 1980s. Sensors having spectral sensitivities in certain narrow bands of the shorter solar and thermal wavelengths have proven useful, particularly with regard to the following.

(a) The separation of clouds from snow in daylit scenes using the 3.7 μm channel of the AVHRR (Kidder and Wu, 1984) or the 1.5–1.6 μm band on the DMSP (Crane and Anderson, 1984). The former is also particularly useful for the detection of ship tracks in oceanic stratus clouds (Coakley *et al.*, 1987).
(b) Determination of land biomass in terms of photosynthetic activity, moisture stress and leaf area index by difference ratioing of the near (reflective) infrared and shortwave visible channels (Tarpley *et al.*, 1984). These same channels can be used to monitor large-scale air pollution, such as that generated by forest fires (Ferrare *et al.*, 1990) and, in combination with the thermal bands, the extent of biomass burning (Kaufman *et al.*, 1990).
(c) Determination of oceanic primary productivity ('ocean color') in the green portion of the visible spectrum, and its association with sea surface temperatures in the thermal IR (Feldman *et al.*, 1989).
(d) Water vapor in the absorption band centered on 6.3 μm and the tracking

of water vapor 'features' as indicators of the mid- to upper-tropospheric airflow (Eigenwillig and Fischer, 1982).

(e) Detection of cirrus clouds associated with jet aircraft contrails by the differencing of two narrow bands in the thermal IR (Lee, 1989b).

9 Contemporary issues in satellite climatology

The above discussion suggests the following current and ongoing themes in satellite applications to climatological research. Many of these are closely intertwined (e.g., clouds in climate and the ERB). They are the chief foci of this monograph.

(1) The role of clouds and cloud processes in climate over a wide range of scales (Rossow, 1989). Particular thrusts involve the development of one or more global-scale cloud climatologies, most notably under the auspices of NASA's International Satellite Cloud Climatology Project (ISCCP) (Schiffer and Rossow, 1983) and related intensive field programs (Cox *et al.*, 1987; Starr, 1987); and the determination of the net effects of clouds on global temperatures and climate stability (the *cloud–climate sensitivity*: Ramanathan, 1987; Ramanathan *et al.*, 1989a,b).

(2) Analyses of tropospheric circulation systems revealed by their characteristic vortical or banded cloud structures, both as visual indicators of the transports of moisture, energy and momentum and for their interactions with surface conditions (e.g., SSTs, sea ice extent). These particularly involve synoptic-scale frontal cyclones, tropical cyclones and tropical-extratropical cloud band connections (TECBs: Kuhnel, 1989) but also, increasingly, are emphasizing meso-scale circulations (boundary layer flows, cold air meso-cyclones, thunderstorms and MCSs).

(3) Patterns and processes of atmospheric moisture, especially water vapor and precipitation estimation, and as revealed increasingly by passive microwave techniques (Arkin and Ardanuy, 1989). A key emphasis involves the associations between satellite-derived water vapor, cloud liquid water, precipitation, surface winds, and SSTs, and their signal in hemispheric and global-scale climate variations, particularly ENSO.

(4) Global-scale radiation and energy budgets of the Earth–atmosphere system (ERBs), particularly under the umbrella of NASA's Earth Radiation Budget Experiment (ERBE): see, e.g., Bess *et al.*, 1989; Barkstrom *et al.*, 1989). Principal avenues of satellite-based research include: determination of unbiased surface albedos for different land cover types; accurate estimation of the surface solar irradiance and net radiation budgets; and the impacts of cirrus clouds (natural, contrails) in particular, and clouds in general, on the planetary albedo and ERB.

(5) The characterization of variations in the Earth's surface properties and their influence on climate and climate variations. These range from the micro- to the meso-scale effects of land cover types, and as examined in NASA's International Satellite Land Surface Climatology Project (ISLSCP) and associated field programs (Rasool, 1984; Sellers *et al.*, 1988, 1990). In

addition, the regional-scale climatic impacts of sea ice (extent, ice-water concentration) and snowcover (extent, depth, water equivalent), are important emphases; as are the hemispheric- and global-scale associations of prominent SST variations with circulation changes, especially those connected with ENSO.

Suggested further reading

Browning, K. A., 1982: *Nowcasting*. Academic Press. 256 pp.
Cracknell, A. P. (ed.), 1981: *Remote Sensing in Meteorology, Oceanography and Hydrology*. Wiley and Sons, UK. 542 pp.
Deepak, A. (ed.), 1980: *Remote Sensing of Atmosphere and Oceans*. Academic Press, New York. 641 pp.
Houghton, J. T., 1984: *The Global Climate*. Cambridge University Press. 233 pp.
Ohring, G. and Bolle, H-J., 1985: Space Observations for Climate Studies (special issue). *Adv. Space Res.*, *5*: 6. 396 pp.
Rao, P. K., Holmes, S. J., Anderson, R. K., Winston, J. S., and Lehr, P. E., 1990: *Weather Satellites: Systems, Data and Environmental Applications*. American Meteorological Society, Boston, Mass. 503 pp.
Szekieldei, K-H., 1988: *Satellite Monitoring of the Earth*. Wiley and Sons. 326 pp.

2 Principles of satellite remote sensing applied to climate studies

The climate system involves solar forcing, the atmospheric variables of clouds, water vapor and aerosols; Earth surface variables (oceans, land surface, snow and ice cover), and their interactions via energy and moisture fluxes (Chapter 1). These 'targets' interact differently with electromagnetic radiation and can, therefore, be detected using remote sensing techniques. Useful information can be derived if one knows the spectral sensitivity of different targets; or the wavelength regions in which the amount of target information is at a maximum; and has some idea of the physical properties of the target giving rise to the observed electromagnetic signature. An important consideration in satellite remote sensing for climate is the fact that, for assessment of atmospheric targets, the Earth's surface is a source of contamination that may interfere with retrieval of an atmospheric signal. Similarly, where desiring to obtain accurate information on the Earth's surface, the attenuating effects on radiation by the atmosphere must be minimized. Given the range of variables to be remotely sensed, satellite-based monitoring of climate and climate processes requires:

(1) a large range of sensor types that are sensitive to different portions of the electromagnetic spectrum (EMS) of energy, and include a range of imagers, scanners and sounders;
(2) a variety of satellite platforms and orbital configurations in an effort to capture the appropriate time and space scales of target variation in different regions and latitude zones (e.g., tropical cloud and rainfall versus polar sea ice extent).

1 The electromagnetic spectrum and radiation laws

Electromagnetic energy is radiated across space by bodies that possess a temperature above 0 K (absolute zero). Whether envisaged as travelling oscillations or photons, this energy possesses certain properties. These include the wavelength (in μm, $= 10^{-6}$m; or nm $= 10^{-9}$m); the frequency (cycles per unit time: GHz $= 10^9$ sec^{-1}); and the intensity. The intensity of the radiation (flux per unit area: $E =$ W m^{-2}) is dependent on the temperature of the radiating body, as described by Stefan–Boltzmann's Law:

$$E = \sigma T^4 \tag{2}$$

where σ is the Stefan–Boltzmann constant ($= 5.6677 \times 10^{-8}$ W m^{-2} K^{-4}).

For most purposes in remote sensing, the sensor is calibrated according to blackbody conditions. The concept of a blackbody carries no connotation of actual color. It simply refers to the efficiency of a body to radiate the maximum amount of energy for a given wavelength or group of wavelengths at that temperature. A 'grey body', in contrast, is less efficient in re-emitting the incident energy relative to its temperature; some of which is attenuated by processes of reflection and transmission. These processes are important in the atmosphere, and the radiation intensity of the Earth-atmosphere system is better described by the following equation:

$$E = \varepsilon \sigma T^4 \qquad (3)$$

where ε is the emissivity.

The emissivity is described by Kirchoff's Law, which can be reduced to the following expression:

$$E_{emit}/E_{abs} = f(\lambda, T) \qquad (4)$$

where E_{emit} is the intensity of the emitted energy,
E_{abs} is the intensity of the absorbed energy,
λ is the wavelength.

Thus, by definition, a blackbody has an emissivity of 1.0, and a grey body $0 < \varepsilon < 1$. The sun approaches blackbody status over most visible wavelengths; however, targets in the Earth–atmosphere system may approximate blackbodies only in longer wavelengths. For example, snowcover dominates the emission and absorption of thermal infrared (TIR) radiation since it approximates a blackbody in those wavelengths (Foster $et\ al.$, 1987). The spectral distribution of the radiation according to wavelength for a blackbody radiator is given by Planck's Law, which relates the emitted radiation intensity to the temperature, T, and spectral emissivity of the surface:

$$B_{(\lambda, T)} = \frac{c_1}{\lambda^5 [\exp(c_2/\lambda T) - 1]} \qquad (5)$$

where $B_{(\lambda, T)}$ is the Planck function ($Wm^{-2}\mu m^{-1}$), c_1 and c_2 are constants: $3.7413 \times 10^8 W$, $(\mu m)^4 m^{-2}$ and $1.4388 \times 10^4 \mu m\ K$, respectively.

The wavelength of the maximum emission (λ_{max}), which is inversely related to T, is described by Wien's displacement law:

$$\lambda_{max} = 2897/T \qquad (6)$$

where 2897 is a constant ($\mu m\ K$).

Thus, not only is the intensity of the solar radiation very different from that emitted by the Earth (Stefan–Boltzmann's Law), so too is the λ_{max}. Substituting the appropriate radiative temperatures of the sun (6000 K) and the Earth (300 K) into equation (6) give λ_{max} of about 0.5 μm and 10 μm, respectively. Thus, solar radiation peaks in the short wavelengths (higher frequencies) of visible light (0.4–0.7 μm) while radiation emitted by the Earth–atmosphere system peaks in the TIR (8–14 μm) (lower frequency) band.

Knowledge of the target emissivity is extremely important for remote sensing in the thermal IR and longer (microwave) wavelengths. The radiant temperature (T_{rad}) of a body is related to its kinetic temperature (T_{kin}) thus:

$$T_{rad} = \varepsilon^{1/4} T_{kin} \qquad (7)$$

For a blackbody, with negligible reflection or transmission of energy $T_{rad} = T_{kin}$. Targets having similar kinetic temperatures but different emissivities have very different radiant temperatures when sensed by a radiometer. In the absence of reliable information on the target emissivity, as over heterogeneous land surfaces, an equivalent blackbody temperature (T_{BB}) may be derived. The T_{BB} assumes that the emissivity is equal to 1.0 and that T_{rad} equals T_{kin}. This is reasonable for most Earth materials in the 8–14 μm band, and which have emissivities generally between about 0.8 to 0.95. The emissivity of water is close to 0.99 in the TIR. For targets in which the transmission of energy is significant (e.g., for thin cold cirrus clouds overlying warmer surfaces or lower clouds) T_{rad} will tend to be significantly higher than that expected from its consideration as a blackbody. Accordingly, the determination of sea surface temperatures using TIR remote sensing will be in error where thin cirrus overlies the ocean but goes undetected. Similarly, an accurate retrieval of the total cirrus cloud fraction will be difficult where there is an assumed cloud emissivity of 1.0.

The microwave frequencies, which extend from about 3 to 300 GHz (10–0.1 cm), are a major area for satellite remote sensing applications to climate (Chapter 1, Section 8). The satellite-measured upwelling microwave emission can be expressed as a brightness temperature (T_B: K). This is a function of the emissivity, the physical temperature, and the wavelength, and has varying contributions from the Earth's surface and subsurface; as well as from the atmospheric gases, cloud liquid water, and precipitation. Thus:

$$\varepsilon = T_B / T \tag{8}$$

where

$$T_B = \varepsilon T = \frac{E\lambda^4}{2\pi c\sigma} \tag{9}$$

E is the intensity of thermal radiation from a blackbody and c is the speed of light.

For a given microwave frequency, the emissivity is influenced by a host of variables. These include the type of surface (ocean, land, ice); the surface roughness, which at micro-scales is related to the particle or grain size; and dielectric properties, such as soil moisture, or the salinity in the upper layers of sea ice. In the case of sea ice at the 19 GHz (ESMR) frequency, melt cycles accompanying increasing age of the ice reduce the salinities in the upper layers and, consequently, the emissivity and T_B relative to new ice. At the 19 GHz frequency, microwave retrieval algorithms for sea ice extent and concentration and also precipitation rates over the oceans, have been quite successful since these targets show increased T_B against the radiometrically cold ocean background. There is also a lack of attenuation by the atmospheric gases and by non-raining clouds, especially at higher latitudes. The emissivity is higher and also much more variable over land. The latter complicates the assessment of precipitation over land using single channel microwave data.

The location of the sea ice–ocean boundary can be determined with relative ease from the ESMR data, given the strong contrasts in T_B between the relatively cold ocean (emissivity approx. 0.4 at 19 GHz), newly formed sea ice ($\varepsilon = 0.95$) and multi-year ice ($\varepsilon = 0.84$). Figure 2.1 shows the ESMR T_B changes associated with a period of strong cold air advection in the western Arctic (October 11–17, 1976). The large positive values are associated with the formation of new sea ice and the advection southward of older ice, giving an accompanying advance of about 200 km in the Chukchi Sea ice-edge location during this six-day period (Carleton, 1984a). Regional interactions between ESMR T_B data on sea ice and the synoptic-scale atmospheric circulation are documented for the Arctic by Crane (1983) and Crane *et al.* (1982), and for the Antarctic by Cavalieri and Parkinson (1981). The presence of meltwater on the ice surface can radically alter the T_B of sea ice. Thus, determination of the ice-water concentration, which is important for estimates of surface–atmosphere energy fluxes, can only be made in the absence of surface melt (typically in winter) and in the presence of one ice type. At the same time, the ability to determine ice type (new, second-year,

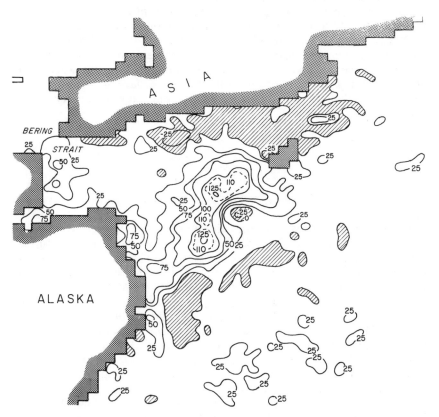

Fig. 2.1 Nimbus-5 ESMR T_B difference map (K) for the western Arctic; October 17 minus 11, 1976. T_B increases of the order of 110–125 K are associated with rapid 'advance' of the sea ice. The feature off the Arctic coast of Asia in the pack ice zone is Wrangel Island. (From Carleton, 1984a, b.)

multi-year), which is a function of the ice thickness, its salinity and surface roughness, can only be reliably made for a 100% ice concentration in the ESMR. The presence of open water, with its lower T_B, increases the risk of misinterpreting newly formed ice for older ice in the relatively coarse resolution (about 30 km) of the ESMR. For example, a T_B corresponding to about 50% ice-water concentration in first-year ice may be closer to about 65% for a field of multi-year sea ice, and these differences are enhanced at higher ice concentrations (Parkinson, 1989).

There are major limitations with the use of a single microwave sensor that is nadir (vertically) pointing. These include, particularly, the inability to determine unequivocally the sea ice type and concentration, snow depth on land, or precipitation over land, or to separate surface moisture from precipitation. These result from the high and variable emissivities of land surfaces, which swamp the weak microwave signal from these other sources. The deployment of the five-channel SMMR was a major improvement towards resolving these and other deficiencies, mainly by:

(1) the addition of higher frequencies;
(2) the ability to detect differences in the polarization (horizontal, T_H; and vertical T_V) of the microwave emission for a given wavelength; and
(3) the ability to separate the scattered from the absorbed contributions to the upwelling microwave radiation, which is particularly important in precipitation monitoring (below).

Polarization refers to the tendency for the radiation emission to not be equal in all directions. The polarization increases with look angle away from nadir and for smooth or specular (e.g., ocean) surfaces. Polarization is small or negligible for atmospheric gases, cloud droplets and rain droplets. Accordingly, the presence of heavy rain over the ocean tends to depolarize the emission from the ocean surface, and the difference in the horizontally and vertically polarized T_B at the 37 GHz channel of the SMMR is an indicator of the obscuration of the ocean surface by rain clouds (Petty and Katsaros, 1990). Many satellite retrieval algorithms for SMMR data, such as snow depth, cloud liquid water, and sea ice concentration, rely on difference ratioing of the vertically and horizontally polarized components of the microwave emission at a given frequency. An alternative retrieval technique involves the exploitation of two microwave frequencies but like polarizations. This holds particular promise for rain estimation over land (Barrett *et al.*, 1990).

2 Radiation interactions with the atmosphere and Earth surface

In satellite remote sensing, much information has been obtained from sensors that detect:

(1) the reflection of shortwave solar radiation, or the albedo (reflectance integrated across all visible wavelengths);
(2) the thermal infrared (TIR) emission of Earth targets as an indicator of their physical temperature (T); and

(3) the microwave emission from the Earth's surface and as modulated by atmospheric constituents. The interpretation of target signatures in these wavelength bands is dependent on the radiation interactions with the target and with the intervening atmosphere.

(a) Attenuation due to atmospheric gases and hydrometeors

Figure 2.2 shows the shortwave radiation spectrum for the top of the atmosphere (TOA), and as depleted by passing through the atmosphere in the absence of clouds. Unlike the situation for the thermal IR wavelengths, where the only significant atmospheric attenuation is by absorption, attenuation of short wavelength radiation occurs by reflection due to clouds, water vapor, aerosols and air; scattering due to air molecules smaller than the radiation wavelengths (Rayleigh scatter) and by particles that are comparable in size to the radiation wavelengths, such as aerosols and cloud liquid water (Mie scatter); and absorption by ozone in shorter wavelengths (<0.3 μm) and water vapor in the longer visible wavelengths (>1.0 μm).

Figure 2.3 shows the thermal IR absorption spectra and the major contributor gases to that absorption and re-emission. The smooth curves are for blackbody radiators at the specified temperatures (from Planck's Law). Note how the wavelength peaks shift towards longer wavelengths (smaller wavenumbers) as temperature decreases (Wien's Law). Around 15 μm there is a major atmospheric absorption band due to CO_2 and water vapor; there is significant attenuation by O_3 around 9.6 μm. While these absorption bands cannot be used to retrieve useful information about the Earth's surface and targets within the atmosphere (clouds), satellite sensors that are sensitive to these wavelengths can provide many data on atmospheric properties; principally, the patterns and amount of water vapor (Chapter 5). For those spectral regions on either side of the absorption bands (the atmospheric windows), the atmosphere is largely transparent to upwelling radiation from the Earth's surface and from clouds. The GOES satellite image (Fig. 2.4) is taken in the water vapor absorption band at 6.3 μm, and is built up from many individual vertical soundings made by the VAS instrument. Emittances in this band are a function of the vertical variations of relative humidity, and which can be estimated if the temperature profile is known (Chapter 5, Section 1a). Decreased column radiances indicate increased mid-tropospheric humidities associated with ascent. The brighter signatures in Fig. 2.4 are mostly associated with clouds (e.g., the deep convection in the ITCZ and the tropical vortex in the lower part of the image); however the greyer tones may not be associated with clouds. They are water vapor 'features' (Allison et al., 1972). The darkest tones (increased radiances) are associated with drying and subsidence through much of the troposphere. This occurs particularly behind frontal systems and on the cold (poleward) side of jet streams, and results in steep horizontal water vapor gradients. Radiances in the water vapor band are also very useful for detecting thin cirrus clouds.

Figure 2.5 depicts the atmospheric transmittance of electromagnetic radiation according to wavelength. Note that, for visible wavelengths, the

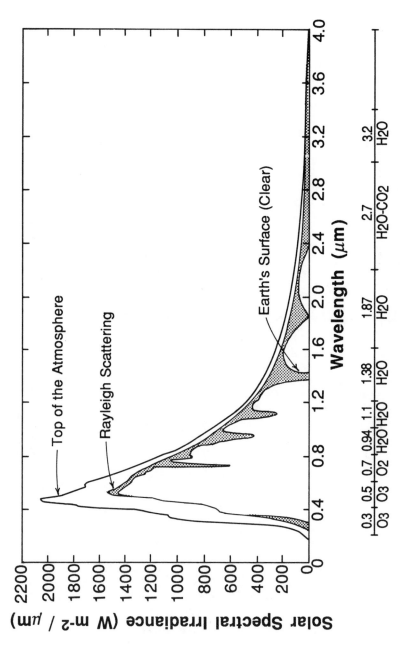

Fig. 2.2 The wavelength dependence of the solar irradiance for the top of the atmosphere (TOA) and clear-sky conditions at the Earth's surface. Note the dominance of Rayleigh scatter for the attenuation in visible wavelengths. Absorption bands also shown. (From Liou, 1986.)

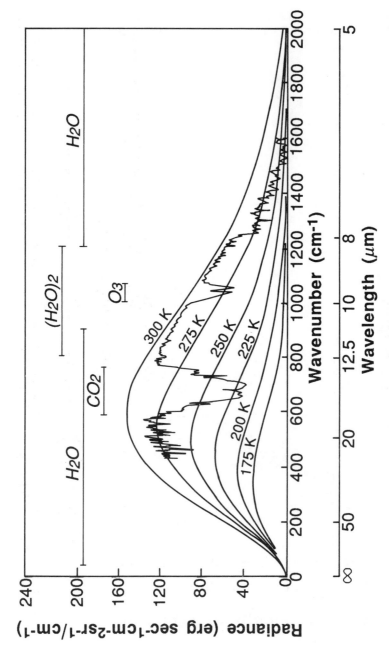

Fig. 2.3 Solar and TIR absorption spectra, and the major absorbing gases. This plot shows the shift of the wavelength$_{max}$ towards longer wavelengths as temperature decreases, and the main window regions (high radiance) and absorption bands (reduced radiance). Note the inverse relationship of wavelength and wavenumber. (From Liou, 1986.)

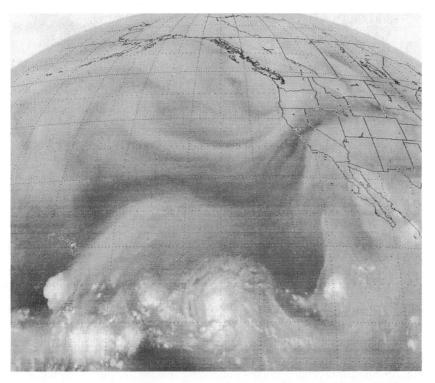

Fig. 2.4 Portion of a GOES-W water vapor image for the north-east Pacific Ocean (1215Z, August 10, 1982). Refer to the text for explanation of the radiometric features.

attenuation due to the gases is mostly small, and is due dominantly to scattering. The atmospheric absorption bands in the near (reflective) IR and TIR are indicated by low values of the transmission. They are mainly due to water vapor and CO_2. The window regions, especially that between about 8 and 14 μm, contain narrow absorption bands. It is observed that, beyond about the 1 cm wavelength, atmospheric attenuation is negligible for these microwave frequencies.

Since the emissivity of the Earth's surface is close to unity in the TIR window, a satellite-measured radiance can be entered into Planck's Law (equation 5) and solved for T_s as an inversion problem. However, satellite-retrieved clear-sky surface temperatures are not the same as those measured by thermometers (Fig. 2.6). The diurnal variability is usually larger in the former, and this is a function of variables such as atmospheric moisture content, wind speed and surface type influencing the shelter temperature (Minnis and Harrison, 1984a). Moreover, large differences in diurnal variability of the clear-sky temperatures occur for different surface (and climate) types. Note also in Fig. 2.6 the lag in maximum surface air

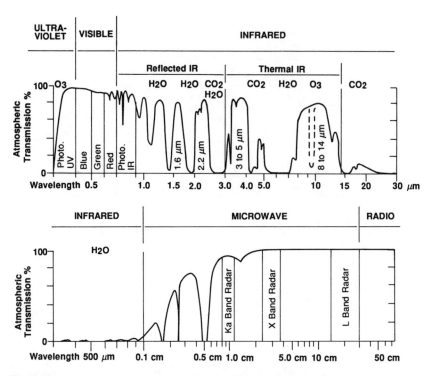

Fig. 2.5 Spectral dependence of the atmospheric transmission (inverse: atmospheric absorption) and the dominant absorbing gases. (From Sabins, 1987.)

temperature compared with the clear-sky radiative temperature, which arises from the transfer of sensible heat from the Earth's surface to the air at shelter level. The retrieval of IR surface skin temperatures from satellites, when used in the appropriate statistical or physical (model) algorithm, can provide information on a range of Earth surface ERB parameters (Price, 1982). These include the latent (evaporation) and sensible heat fluxes over land, the soil moisture, and surface thermal inertia (Carlson *et al.*, 1981; Wetzel and Woodward, 1987; Nemani and Running, 1989; Klassen and VanDenberg, 1985).

Given the presence of absorption bands in TIR wavelengths, it is possible to reduce the atmospheric effect on the satellite-sensed T_{BB} by:

(1) using a sensor with a spectral sensitivity that selects for a confined portion of the IR wherein the atmospheric attenuation is known to be minimal (This is the principle of narrow-band sensing.);

(2) using a sensor that gives broad-band radiances but in which the temperature-dependent absorption spectra of the various atmospheric gases can be subtracted out;

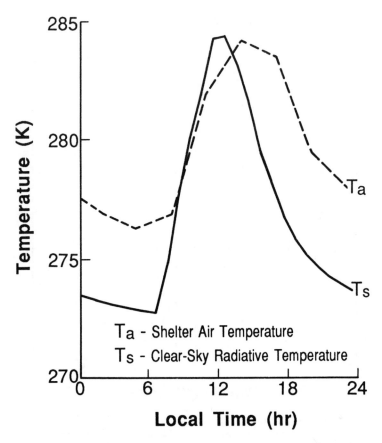

Fig. 2.6 Comparisons between the GOES-observed clear-sky radiative temperatures (T_s) and surface-observed (shelter) air temperature (T_a) for November 1978 at a point: lat. 34.8° N; long. 111.5° W. (From Minnis and Harrison, 1984b.)

(3) using a combination of non-adjacent narrow bands in which atmospheric absorption is weak to arrive at a weighted spectral irradiance.

The degree of atmospheric attenuation of microwave radiation by absorption due to atmospheric gases and scattering due to liquid water (cloud droplets, precipitation) is wavelength-dependent. At certain shorter wavelengths and around 22 GH_3 absorption by water vapor and oxygen occurs. This forms the basis of vertical temperature profiling from satellites when compared with regions of low absorption (e.g., 31 GH_3). The ESMR systems sense in a region of generally low gaseous absorption (Fig. 2.7). Accordingly, between 22 and 60 GHz, attenuation by absorption and scattering occurs in a raining atmosphere but will be much smaller in clear or cloudy (but non-precipitating) conditions because of differences in the liquid water content

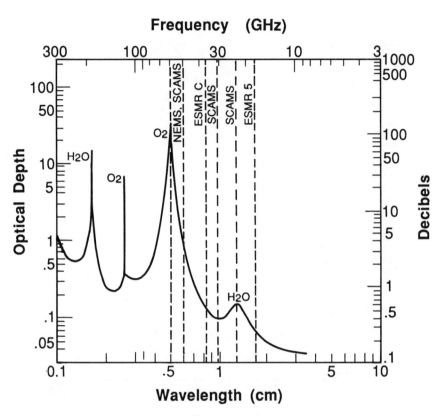

Fig. 2.7 Attenuation due to atmospheric gases in the microwave spectrum along a zenith path (decibels, right-hand scale; optical depth, left-hand scale). The assumed atmosphere is maritime polar. Note the frequencies that have been used extensively for atmospheric observation (dashed lines). (From Savage, 1978.)

and drop-size distribution of the clouds. The effect is minimal in higher latitudes where the atmospheric water vapor content is generally low. At frequencies greater than about 60 GHz, scattering of the microwave radiation by hydrometeors dominates over absorption and re-emission. For very small particles with respect to the wavelength considered, Rayleigh scatter dominates; for larger particles (cloud droplets, precipitation), Mie scatter is important. The latter increases as drop radius (and rain rate) increases for a given wavelength. Only at low frequencies (long wavelengths) does heavy precipitation (large particles) extinguish microwave radiation to a significant degree; at higher frequencies, or shorter wavelengths (e.g., 300 GHz), all rain clouds extinguish microwave radiation and drop size becomes important to the Rayleigh regime. For frequencies below 60 GHz, scattering can always be neglected for clouds, and also for rain rates less than 10 mm h⁻¹ and frequencies below 31 GHz (Lovejoy and Austin, 1980). Thus, cumulus clouds can be distinguished from cumulonimbus in the microwave.

At ESMR wavelengths (19 GHz), attenuation of microwave radiation comes about dominantly through *absorption* by rain drops, and is a function of the assumed freezing-level altitude, as shown in Fig. 2.8. When viewed against a more-or-less constant low emissivity background (ocean) the increase in T_B relates directly to rain rate, at least up to about 20 mm h^{-1}. Note that, at higher rates, T_B falls owing to reflection of the cold cosmic background radiation back to the satellite (Fig. 2.8). Over surfaces of variable emissivity (especially land), there is ambiguity in the T_B signal that necessitates use of higher microwave frequencies and dual polarization, as effected by non-nadir viewing sensors (e.g., SMMR). However, at those higher frequencies *scattering* of the microwave radiation within the rain column becomes dominant, and is influenced particularly by the presence of ice. Figure 2.9 shows the effect of even a relatively thin (0.5 km) ice layer on the decrease of T_B at 92 GHz associated with increasing rain rate. At the 37 GHz channel of the ESMR-6 and Nimbus-7 SMMR, both scattering and absorption of microwave radiation by hydrometeors occurs. Thus, multispectral and/or polarization observations are clearly desirable to extract the signal of precipitation over land and to separate the scattered from the absorbed contribution (Weinman and Guetter, 1977). The very low T_B associated with thunderstorm precipitation derives from scattering by the frozen hydrometeors. This permits the determination of heavy convective precipitation over land by the SMMR (Spencer *et al.*, 1983a).

Using a theoretical rain-cloud model, Savage and Weinman (1975) computed, for 19.35 and 37.0 GHz, the relationship of T_B with rainfall rates separately over land and ocean (Fig. 2.10). These confirm the inability of ESMR-5 (19.35 GHz) to distinguish precipitation over land. However, at 37.0 GHz, the brightness temperature decrease associated with precipitation over land is much greater (to about 20 mm h^{-1}). Clearly, ESMR-5 is better for precipitation estimation over ocean and SMMR over land (note the reversal in T_B change over ocean for low rain rates at 37 GHz: Fig. 2.10). Since the emissivity of a land surface is lowered by the addition of moisture (its dielectric constant increases), the surface emissivity becomes highly polarized when viewed obliquely (Rodgers *et al.*, 1979). Radiation transfer model results (Fig. 2.11) demonstrate the change from highly polarized radiation emitted from wet soils $(T_V > T_H)$ to more-or-less unpolarized radiation for heavy rainfall areas. Although radiation over dry land is essentially unpolarized, brightness temperatures (T_B) are higher compared with wet ground or heavy rainfall areas. Rodgers *et al.* (1979) demonstrated the utility of horizontally and vertically polarized T_B pairs in differentiating dry ground, wet ground and rain over land for the Nimbus-6 ESMR. However, they also found a strong thermodynamic influence. The three categories can only be statistically discriminated where surface temperatures exceed 15 °C; they are not well differentiated for T_s between 5 °C and 15 °C (Fig. 2.12a,b). The influence of dew was also shown to affect the classification of wet ground and rain.

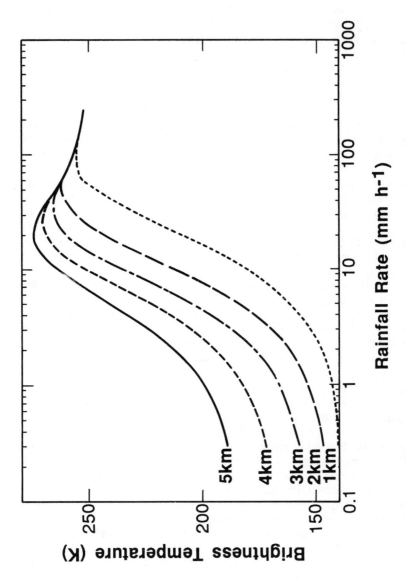

Fig. 2.8 ESMR (19 GHz) brightness temperatures (T_B) for nadir viewing over an ocean surface and for various assumed freezing levels. Note the monotonic increase in T_B as rain rate increases, up to about 20–25 mm h^{-1}. (From Wilheit, 1986.)

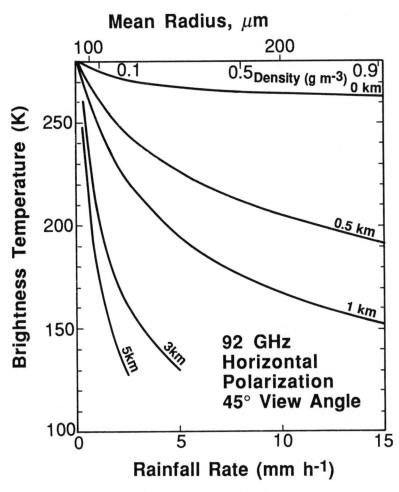

Fig. 2.9 Brightness temperature calculations at 92 GHz and with a 45° viewing angle for various thicknesses of the cloud ice layer as a function of rain rate (horizontal axis). The two main abscissae give the mean particle radius and the particle density (for both the ice and liquid phases). (From Wilheit, 1986)

(b) Cloud effects

Clouds exert a significant modulating influence on the solar and terrestrial (longwave) radiation streams by two competing processes: the albedo effect and the greenhouse enhancement (or longwave absorption and re-emission). Differences in the relative values of each, and hence on the net effect for the Earth's surface temperature (T_s), are functions of the following parameters: cloud type, cloud height, cloud thickness and the vertical water or ice

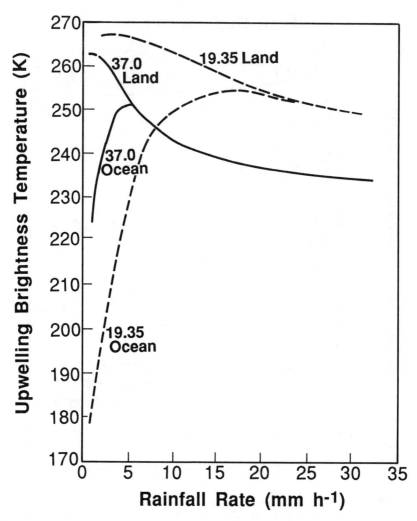

Fig. 2.10 Comparisons between 19.35 GHz (ESMR) and 37.0 GHz (SMMR) upwelling brightness temperatures (zenith angle = 48.6°) of a 4.47 km rain cloud over land ($\varepsilon = 0.90$) and ocean surfaces ($\varepsilon = 0.42$ at 19.35 GHz; 0.47 at 37.0 GHz). (From Savage and Weinman, 1975.)

content. The last two influence the cloud optical thickness (τ), its emissivity (ε) and, accordingly, the radiant temperature (T_{rad}) sensed by a space-borne radiometer. Figure 2.13 shows model-generated relationships between the cloud optical thickness and the shortwave radiation fluxes at the surface and TOA. Note that, as cloud optical thickness increases, the surface global flux decreases strongly, the diffuse solar flux increases up to a point and then decreases, and the TOA flux increases strongly at greater values of τ. The last

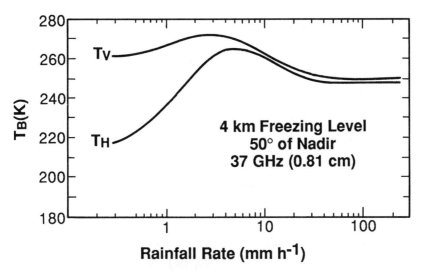

Fig. 2.11 Horizontally and vertically polarized T_B at 37.0 GHz computed as a function of rain rate. The plot assumes (1) a 40° incidence angle with the Earth's surface; (2) a 4 km freezing level. (From Rodgers *et al.*, 1979.)

is an indicator of the albedo enhancement effect of clouds. In the infrared wavelengths, it is necessary to consider the emissivity of clouds.

Over most wavelengths in the 8–14 μm band, the majority of clouds (except cirrus) approach blackbody status; that is, they have emissivities close to 1.0. This is important for satellite interpretation of clouds because it means that the CTT obtained by the sensor corresponds closely with the physical temperature at cloud top altitude sensed by a radiosonde (Fig. 2.14). Figure 2.15a,b shows the retrieved CTTs along a section of an atmospheric front in the Southern Hemisphere on two days, as sensed in the thermal IR. The reduction in the area of the coldest (highest) cloud tops between the two days and the obvious fragmentation of the cloud band, is an indicator of changes in cloud level (low, middle, high) since the cold clouds have the lowest outgoing IR flux densities compared with lower (and warmer clouds). Wexler (1983) used Nimbus-5 THIR (Temperature Humidity Infrared Radiometer) data in the window region to investigate the distribution and diurnal variation of high cold clouds over the tropical Atlantic and Pacific between 0° and 15° N for late July and early August of 1973. Two threshold values of T_{BB} were selected: less than or equal to 220 K and less than or equal to 240 K, and the frequencies of occurrence in 1° × 1° latitude/longitude grid cells were tabulated. It was found that daytime cloud cover exceeds that at night for the entire tropical Atlantic, but is more variable for the Pacific belt.

In normal lapse rate conditions, the Earth's surface is the warmest target. Important exceptions to this association occur in inversion conditions, when radiatively 'warm clouds' overlie colder surfaces (e.g., in the Arctic and

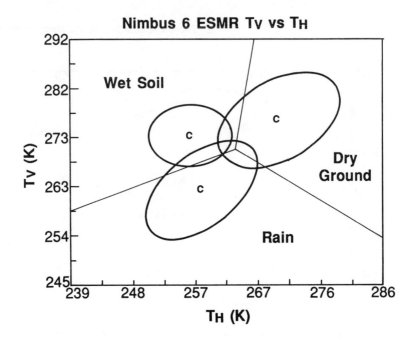

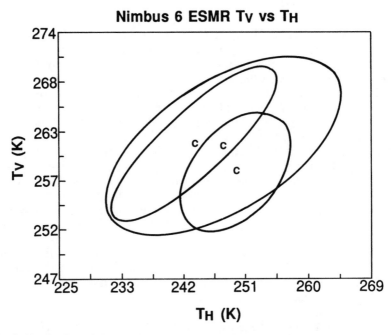

Fig. 2.12a, b Comparisons of the horizontally (T_H) and vertically (T_V) polarized T_B (K) for separation of rain from dry ground and wet soil using the Nimbus-6 ESMR: (a) for surfaces with thermodynamic temperatures >15°C; (b) for surfaces with thermodynamic temperatures of 5–15°C. 'C' is cluster center. (From Rodgers *et al.*, 1979.)

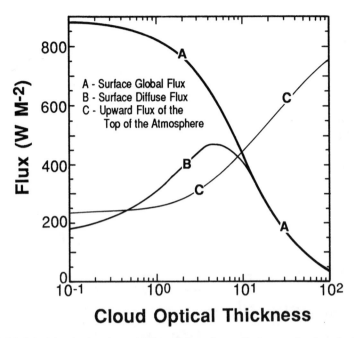

Fig. 2.13 Model-calculated sensitivity of the solar radiation to cloud optical thickness. Curves A, B and C refer, respectively, to the surface global flux, surface diffuse flux, the upward flux at the top of the atmosphere (TOA). The plotted values are for a surface albedo of 0.22; solar zenith angle of 30°; precipitable water of 1.4 cm; and Julian day 140. (From Pinker and Ewing, 1985.)

Antarctic), and in the presence of clouds that are semi-transparent to upwelling IR radiation from the Earth's surface and from clouds at lower altitudes (Saunders, 1988). The latter is typical of thin cirrus clouds. For those clouds, large departures of the T_{BB} from the CTT indicate a cloud emissivity that is significantly lower than unity. The climatological mean cirrus IR emissivity may be closer to 0.47 (Liou, 1986).

Figure 2.16 shows model-generated surface temperature effects of different clouds. It should be noted that the 'full-black' emissivities of low and middle clouds tend to result in a net surface cooling. However, the variable cirrus cloud emissivities suggest a surface warming, even where the emissivity approaches unity. The implications for satellite remote sensing of 'non-black' cirrus are major and include the following.

(1) Cloud retrieval algorithms, such as that of the United States Air Force (USAF) three-dimensional (3-D) or real time (RT) nephanalysis, which assume cloud emissivities of 1.0, will tend to underestimate the amount of cirrus cloudiness (Henderson-Sellers, 1986).
(2) If the retrieved fractional amount of cirrus cloudiness is inaccurate, then the cloud–climate sensitivity (net radiative cooling or warming) cannot

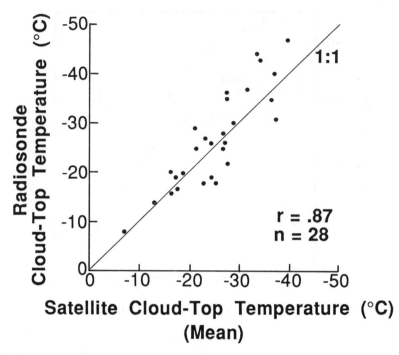

Fig. 2.14 Relationship between mean satellite-observed CTTs and radiosonde-determined CTTs for the winter season 1973–4 for the San Juan Mountain area of Colorado. (From Reynolds *et al.*, 1978.)

be accurately determined. Figure 2.17 shows the measured relationship between cirrus cloud albedo and T_{BB} over the Gulf of California (low surface albedo). Note that, as the cloud albedo increases with increased optical thickness, less of the emission from the Earth's surface is received at the satellite and, accordingly, the T_{BB} decreases. Towards the upper portion of the curve, the satellite is sensing the effective CTT and the cloud is 'black'.

Figure 2.18 gives the generalized relationship between the emissivity and albedo of cirrus clouds (Reynolds and Vonder Haar, 1977), and shows the large range of values possible between thin and thick cirrus; hence, also the large possible variation in the cloud–climate forcing in cirrus cloud regions (see also Liou *et al.*, 1990). These perturbations may be comparable to those associated with a doubling of CO_2 and a 2% decrease in the solar constant (Liou, 1986). There may also be latitude differences in the climatic influence of cirrus clouds (Cox, 1971; Platt *et al.*, 1987). Figure 2.19 shows mean emissivity variations with height (pressure) of cirrus in tropical and mid-latitude regions compared with the 'critical blackness' curve. Values to the right of this curve would result in net warming at the Earth's surface, especially in the tropics; values to the left, a net cooling effect, mainly in

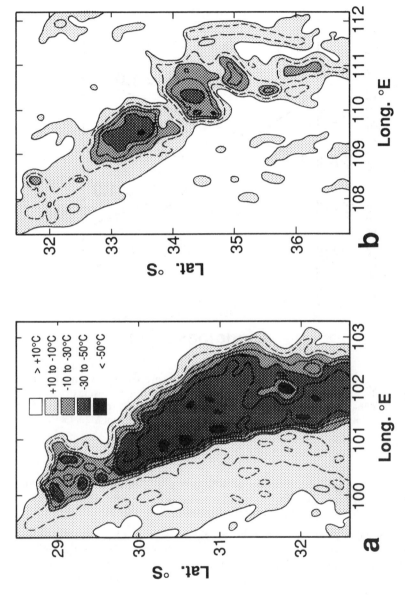

Fig. 2.15a, b Details of a NOAA SR map of the CTTs over a front in the Western Australian region on two consecutive days (about 00Z): (a) June 15, 1976; (b) June 16, 1976. Note the south-eastward displacement of the sector of interest. (From Streten, 1977.)

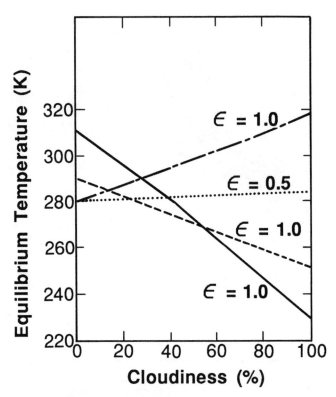

Fig. 2.16 Dependence of the Earth's surface equilibrium temperature on the cloudiness at different levels, as calculated using a radiative-convective model. Two regression lines appear for cirrus (CI) clouds (dot-dashed, dotted) owing to their wide range of possible emissivities. Note that cirrus mostly tends to warm the surface; other clouds to cool it. (Altostratus: dashed; low cloud: solid line). (From Manabe, 1975).

middle latitudes. This question is still being evaluated from modeling, surface, aircraft and satellite studies, particularly in the context of the underlying surface albedo (Cogley and Henderson-Sellers, 1984).

(3) Retrieval algorithms that rely on clear-scene brightnesses to obtain information on Earth surface parameters—notably SSTs in the IR, and the surface solar irradiance and surface albedo in visible wavelengths—will give inaccurate results when thin cirrus occurs but is not detected (see, e.g., Wu, 1984).

(4) Radiative transfer modeling of satellite-derived data; for example the surface irradiance of longwave fluxes, will show large errors if cloud emissivity is assumed equal to unity and the presence of thin cirrus not accounted for (see, e.g., Darnell *et al.*, 1983).

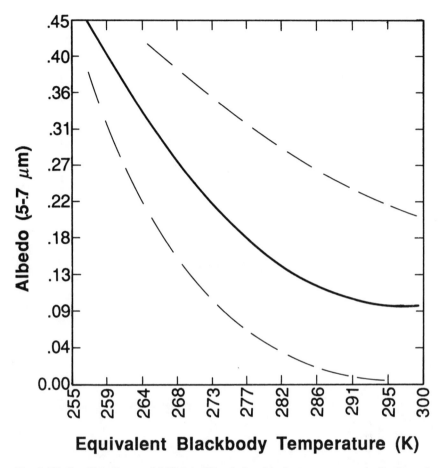

Fig. 2.17 Satellite-observed NOAA SR relationship between spectral albedo and blackbody temperatures of cirrus clouds over the Gulf of California. The dashed lines enclose about 95% of all the observations. (After Reynolds and Vonder Haar, 1977.)

There are various satellite-based methods developed in recent years that have been used to derive the cirrus fraction (refer to Chapter 3, Section 3e). However, there is still considerable uncertainty as to its frequency of occurrence relative to optically thick cirrus. The Selective Chopper Radiometer (SCR) on the Nimbus-5 had a pair of channels centered at 2.7 μm and 2.63 μm. These are both located in atmospheric absorption bands (e.g., Fig. 2.5); the former with respect to both water vapor and CO_2, and the latter mainly for absorption by water vapor. Owing to the strong absorption of radiation emitted from the Earth's surface and lower- and middle-level clouds, the SCR data provide a climatology of 'upper-level' clouds occurring at heights exceeding abut 6 km (i.e., mainly cirrus).

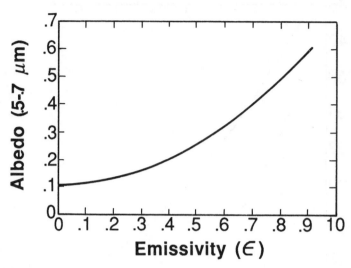

Fig. 2.18 Generalized empirical relationship between the spectral albedo and IR emissivity of cirrus clouds. Note the wide range of possible values. (From Reynolds and Vonder Haar, 1977.)

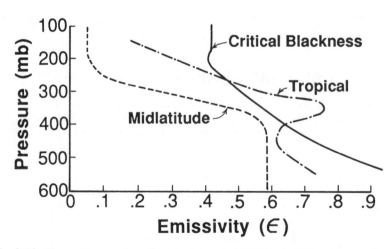

Fig. 2.19 Comparisons of measured cirrus cloud IR emissivities in two climatic zones and the model-calculated critical blackness curve, plotted against height. The latter separates the net greenhouse and albedo enhancement effects of cirrus (cooling on the left; warming on the right of the curve). (From Cox, 1971.)

3 Retrieval of parameters at the Earth's surface

Satellite studies that involve the sensing of TOA incident and upwelling (reflected) solar radiation include the following.

(1) Estimates of the Earth's surface *albedo* and the surface plus atmosphere reflectance (the planetary, or system, albedo). Knowledge of the surface albedo is essential for the satellite determination of the energy budget, while an understanding of the planetary albedo is necessary for assessing the stability of the climate, particularly in regard to high albedo surfaces such as snow and ice.
(2) Estimates of the surface receipts of solar radiation (the irradiance) for agricultural and energy development, using a knowledge of the surface albedo in conjunction with cloud cover (both from satellites).

(a) Earth reflectance and albedo

A critical component of the Earth–atmosphere energy budget is the surface albedo (α_s), or the ratio of the amount of shortwave radiation reflected to the total incident. The reciprocal of this quantity $(1 - \alpha_s)$ gives the amount of shortwave energy absorbed which, along with the infrared emittance, comprises the surface net radiation. The last is partitioned into the sensible and latent heat fluxes, thus (after Henderson-Sellers and Wilson, 1983):

$$(1 - \alpha_s)R_s = \varepsilon(\sigma T_s^4 - R_L) + SH + LH \tag{10}$$

where ε is the infrared surface emissivity,
    T_s is the surface temperature,
    R_L is the downward thermal radiation,
    R_s is the amount and nature of incident solar flux at surface,
    SH and LH are the sensible and latent heat fluxes.

The determination of *reliable* surface albedos from satellites is anything but straightforward (Chapter 6, Section 4). Difficulties attending the derivation of accurate surface albedo involve the following.

(1) *The surface characteristics and their interaction with the shortwave radiation*   Surface effects range from the broad-scale, such as the distinctions between land and sea at similar latitudes (see, e.g., Saunders and Hunt, 1980), down to meso-scale variations in land-use type (vegetated, bare soil, urban), and the role of surface moisture. In addition, micro-scale influences of surface roughness, soil particle size and texture, leaf size and orientation, and others, become important.

(2) *The characteristics of the intervening atmosphere*   Aside from the attenuating effects of Rayleigh scatter and ozone absorption, and of water vapor in the near IR; all of which can be modeled, the optical depth (transmissivity) of the atmosphere is influenced particularly by aerosols. These

vary spatially and temporally. Accordingly, concurrent observations of the atmospheric state are normally required, or assumptions made about the spatial gradient of the optical depth (see, e.g., Pinty and Szejwach, 1985). In addition, there are problems with ensuring a truly uncontaminated (by clouds) 'clear-sky' albedo, particularly as the spatial resolution of the sensor degrades and for low clouds over the ocean, as the sun's elevation above the horizon becomes low (Taylor and Stowe, 1984).

The amount of reflected solar energy depends on the sun's elevation and the direction of reflectance (Pinker, 1985). When sensed by a scanner, which views only a portion of a hemisphere, the amount of reflected energy is found to be non-Lambertian. That is, the radiation field is not diffused equally in all directions nor is it of uniform intensity (isotropic). Instead, it becomes more specular (mirror-like) and contains an anisotropic (forward to backward scattering) component. It is influenced by the solar zenith angle; the zenith angle of reflection; and the relative azimuth angle, or difference in the satellite–sun azimuth angle (Fig. 2.20). Norton *et al.* (1978) found increasingly large errors introduced by the isotropic assumption as solar zenith angle increases away from local noon. From Fig. 2.20 it can be seen that differences in the sun–target–sensor geometry will occur from orbit to orbit (for a polar orbiter) and also with latitude, time of day and season. The anisotropic reflectance factor, which tends to increase with increasing solar zenith angle (Halpern, 1984), can be defined thus:

$$f(\theta, \phi) = \frac{\pi \bar{R}(\theta, \phi)}{\int_0^{2\pi} \int_0^{\pi/2} \bar{R}(\theta, \phi) \cos\theta \sin\theta \, d\theta \, d\phi} \tag{11}$$

where the numerator represents the reflected radiation computed on the basis of the isotropic assumption (equal scattering in all directions), and the denominator represents the reflected radiation computed from observations in different directions (after Pinker, 1985).

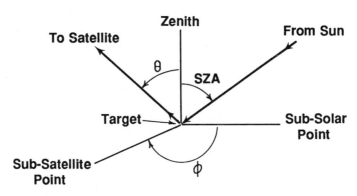

Fig. 2.20 Terminology and geometry of a reflected (scattered) ray of solar radiation. Note particularly the solar zenith angle (SZA) and the relative azimuth angle (ϕ). (From Taylor and Stowe, 1984.)

In order to interpret satellite observations of surface reflectance reliably it is generally necessary to develop models of the bidirectional reflectance. Taylor and Stowe (1984) used the Nimbus-7 ERB sensor to develop such models for several homogeneous surfaces (water, land, snow-covered) and low, middle and high clouds over a range of solar zenith angles and relative azimuths. They showed that marked anisotropy can exist, especially for water surfaces as solar zenith angles become large. Snow and high ice clouds appear to be the most isotropic surfaces. At moderate angles, the albedo of land and water may be quite similar, and at large angles the ocean becomes highly reflective. Taylor and Stowe note the necessity of separating the single land surface type into several sub-types. Snow albedo, calculated for higher latitudes where solar zenith angles are larger, shows a decrease at very large angles due to increased atmospheric attenuation of the incoming solar radiation. Sea ice and snow surfaces appear broadly similar in angular dependence of the reflectance, except that the sea ice has an albedo that is lower by about 10–20%. This may result from melt puddling and possible shadowing because of a rougher surface. For cloud types, the albedo change with SZA is much greater for low than for higher clouds, and low clouds are also darker. The latter effect may result, at least in part, from the reduced attenuation by water vapor associated with high clouds in the near-IR. The role of albedo studies in climate is discussed more fully in the context of satellite monitoring of surface types in Chapter 6 (Section 4).

(b) Surface solar irradiance

Data on surface solar radiation receipts at local- to meso-scales and on temporal scales approaching real time (hourly to daily) are important in agriculture as input to various crop yield models (e.g., NOAA AgRISTARS). These enable the assessment of crop growth rates (Tarpley, 1978). Satellite determination of the surface solar irradiance requires information on the surface albedo and the cloud conditions (Tarpley, 1979). The latter includes particularly the amount of cloud and its thickness since these influence the fraction of the TOA incoming flux (Q_0) that is reflected (Q_R) and absorbed (Q_A), or:

$$Q_0 = Q_R + Q_A \qquad (12)$$

The solar irradiance under cloud-free conditions is dominated by the solar zenith angle. However, cloud cover competes with SZA since it tends to reduce the surface solar radiation receipts from the clear-sky values by increasing the scattering and absorption (e.g., Fig. 2.13). Satellite-based methods for obtaining the sky (cloud) conditions differ in terms of translating the satellite brightness measurements into surface solar irradiance values. Tarpley (1979) found that the inclusion of information on standard meteorological variables did not significantly improve the satellite estimates of solar irradiance beyond those determined using the satellite brightness values. In most studies, the cloud amount and thickness are obtained from reflected radiances that increase above the calculated or observed background (minimum brightness) values for snow-free surfaces (Tarpley, 1979; Gautier *et al.*, 1980; Brakke and Kanemasu, 1981; Moser and Raschke, 1984). In certain cases, the surface albedo may be determined from ground-test sites. Gautier and Katsaros (1984) demonstrate the utility of such a technique over the ocean, and indicate possible inaccuracies related to shipboard measurements of solar irradiance. Methods used to retrieve the surface irradiance from satellite data may be classified as statistically (regression) based (see, e.g., Tarpley, 1979; Brakke and Kanemasu, 1981); physically based (see, e.g., Gautier *et al.*, 1980; Halpern, 1984; Moser and Raschke, 1984); or having attributes common to both (see, e.g., Powell *et al.*, 1984; Justus *et al.*, 1986).

The pioneering statistical study of Tarpley (1979) for the US Great Plains used GOES visible reflectances for 50 km target areas calibrated with available surface-based pyranometer data. Pixel brightnesses were assigned to three classes of sky conditions: clear, partly cloudy and cloudy. The satellite hourly irradiance estimates gave the closest correspondence for clear-sky cases (standard error of estimate <10%), which is a characteristic of several methods (Raphael and Hay, 1984). Accuracies decline under partly cloudy and cloudy conditions, with a tendency to overestimate insolation at low surface light levels and to slightly underestimate it at high levels. This feature is also characteristic of several methods applied to a variety of environments (e.g., Brakke and Kanemasu, 1981; Powell *et al.*, 1984). There is a non-linear association between cloud brightness and cloud thickness: clouds of different thickness may appear similarly bright. In addition, cloud geometric effects (cumuliform compared with stratiform) and cloud shadowing become accentuated at low SZA, and the satellite estimates are influenced by the cloud cover thresholding scheme. Refinements to the Tarpley regression-based model have tended to center on the cloud detection method and the treatment of cloud fraction (Brakke and Kanemasu, 1981; Justus *et al.*, 1986). For all sky states, the standard errors of estimates from the satellite data decrease to about 10% as the period of integration is lengthened to the *daily* time scale (Tarpley, 1979); again, this is a feature of several methods (see, e.g., Darnell *et al.*, 1988). When applied to a mid-latitude region of varied topography and with a high density of coincident pyranometer data (Raphael and Hay, 1984), the Tarpley model performance for clear days in summer shows a dependence on distance from the coast. Over-prediction of surface solar radiation occurs at inland stations, and this is ascribed to the increased aerosol loading that is not explicitly handled by this method.

The second group of satellite surface solar irradiance estimation techniques are those that are physically based. These attempt to account explicitly for atmospheric processes of scattering, absorption and reflection that contribute to the satellite brightness variations. These can be measured directly (Halpern, 1984). Information on precipitable water, ozone and aerosols (Pinker and Ewing, 1985) as well as snowcover detection (Darnell *et al.*, 1988), might also be included. The Gautier *et al.* (1980) model and its subsequent refinements (see, e.g., Diak and Gautier, 1983) has shown the value of a basically simple physical approach. The method is based upon the energy conservation model of Hanson, as used in the Rockwood and Cox

(1978) method to estimate surface albedo over north-west Africa (Chapter 6, Section 4). For the clear-air (cloud-less) model of Gautier *et al.*, the surface solar irradiance is given thus:

$$SW\uparrow = F_0\alpha + F_0(1 - \alpha)[1 - a(u_1)][1 - a(u_2)](1 - \alpha_1)A \qquad (13)$$

where F_0 is the instantaneous shortwave solar radiant flux at TOA ($= I_0\cos\theta$ in W m^{-2}); or reduction of the solar constant value as a function of the solar zenith angle,

α, α_1 is the reflection coefficient for beam, diffuse radiation (dimensionless),

$a(u_1), a(u_2)$ are the absorption coefficients for slant water vapor, u_1 for sun angle and u_2 for satellite angle, is the surface albedo;

and the *net shortwave energy budget at the surface* can be written:

$$SW_{net} = F_0(1 - \alpha)[1 - a(u_1)][1 - A(1 - \alpha_1) - A^2\alpha_1] \qquad (14)$$

Cloud effects on upwelling radiances received at the satellite (brightnesses) are also modeled to include scattering and absorption above and below cloud height using assigned fractions of precipitable water. Thus, at satellite level,

$$\begin{aligned}
SW\uparrow = {}& F_0\alpha + F_0(1 - \alpha)[1 - a(u_1)_t](1 - \alpha_1) \\
& \times A_c[1 - a(u_2)_t] + F_0(1 - \alpha)[1 - a(u_1)_t] \\
& \times (1 - A_c)^2[1 - a(u_1)_b]A(1 - \alpha_1) \\
& \times [1 - a(u_2)_t](1 - abs)^2[1 - a(u_2)_b]
\end{aligned} \qquad (15)$$

where $a(u_1)_t$ is the absorption of SW$\downarrow$ above cloud (dimensionless),
$a(u_1)_b$ is the absorption of SW$\downarrow$ below cloud (dimensionless),
$a(u_2)_t$ is the absorption of SW$\uparrow$ above cloud (dimensionless)
$a(u_2)_b$ is the absorption of SW$\uparrow$ below cloud (dimensionless),
abs   is the cloud absorption (dimensionless),
A_c is the cloud albedo,

and the three terms of equation (15) represent energy scattered from the atmosphere to the satellite, energy reflected from the cloud to the satellite, and energy passing through the cloud, reflected from the ground back through the cloud to the satellite (important for high albedo surfaces).

With these cloud effects included, the incident shortwave at the surface is written:

$$SW\downarrow = F_0(1 - \alpha)[1 - a(u_1)_t](1 - A_c) \times (1 - abs)[1 - a(u_1)_b] \qquad (16)$$

and the net shortwave received at the surface is

$$\begin{aligned}
SW_{net} = {}& F_0(1 - \alpha)[1 - a(u_1)_t](1 - A_c) \\
& \times (1 - abs)[1 - a(u_1)_b](1 - A) \\
= {}& SW\downarrow(1 - A).
\end{aligned} \qquad (17)$$

Unlike the Tarpley model, the Gautier *et al.* (1980) method permits bulk cloud effects to be modeled continuously, rather than assigning sky conditions to one of three discrete classes. The application of the clear or cloudy model is determined using a threshold technique with respect to a minimum brightness (surface albedo) background at a fixed time of day, and in which the effects of SZA and atmospheric attenuation are accounted for. Tests with the model for southern Canada under two different cloud regimes showed good results ($\pm 9\%$ of daily mean insolation measured by pyranometers) (Gautier, 1982). Subsequent application of this model to the Vancouver–Fraser Valley region (Raphael and Hay) showed that reliable *hourly* estimates of insolation under partly cloudy and cloudy conditions could be achieved. The high spatial resolution of the GOES sensor and use of the individual pixel brightnesses in the Gautier *et al.* model permit the derivation of maps of meso-scale variability of surface solar irradiance on a range of time scales (Gautier, 1982). Figure 2.21a–c are maps of, respectively, the satellite-derived mean insolation, surface albedo and 'natural variability' of the insolation (or ratio of the standard deviation to the mean) for eastern Canada in May 1978. They indicate the large spatial variability noted for the springtime and the influence of the Lakes on surface solar irradiance. The natural variability (Fig. 2.21c) enables direct comparisons between different months of the year.

Refinements to the basic solar radiation estimation model of Gautier *et al.* (1980) include parameterizations of Rayleigh scatter and ozone absorption specific to the VISSR sensor rather than the previous broad-band assumptions, and attempts at including the effects of absorption for clouds that are below the sensor resolution (Diak and Gautier, 1983). These authors report an improvement in the standard error of estimate of solar insolation of only about 1% compared with the earlier model.

Methods that could be considered to represent a combination of the regression and physical techniques of satellite surface solar irradiance include those of Powell *et al.* (1984) and also Justus *et al.* (1986). The latter extends the original study by Tarpley (1979). These methods incorporate some degree of parameterization of atmospheric effects (including clouds) but there is a strong statistical component to the estimates. Powell *et al.* (1984) used large numbers of GOES hard copy (photographic) images to determine the spatial variability of solar radiation in Arizona on a seasonal basis. Cloud cover for 64 km diameter circles was extracted manually and found to correlate highly with surface-observed cloud at 20 sites. The attenuation of the clear-sky radiation by clouds was determined by means of an empirically derived 'cloud cover factor' developed using ground-based solar radiation data. The model was validated by comparisons with surface-observed irradiance. Its performance compared favorably with other methods, and seasonal maps of solar radiation variability in Arizona were derived. These show a somewhat more complex pattern in summer compared with conventionally derived solar radiation maps, and indicate the role of orography in convective cloud development in that season.

Tarpley's (1979) method was refined (Justus *et al.*, 1986) to include modeled transmissivities for clear and cloudy skies in the operational NOAA

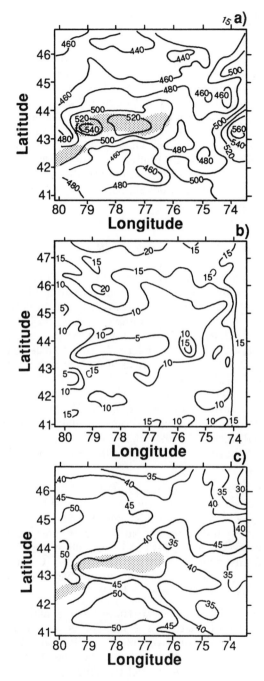

Fig. 2.21 (a) Satellite (GOES)-derived mean isolation map (ly per day) for eastern Canada; (b) surface albedo map (%); and (c) 'natural variability' map (%), for May 1978. The natural variability is defined as the ratio of the standard deviation of the variable to the mean. (From Gautier, 1982.)

daily solar radiation maps for the US. Accuracies to within 5% of mean values of the satellite-estimated monthly averaged insolation were reported with this modified technique. Maps of surface solar radiation receipts were developed from daily analyses for the US, Mexico and parts of South America. In general, the lowest daily variability about the monthly means of solar radiation is observed for those regions having highest radiation receipts (deserts). However, there can be marked meso-scale variability, particularly in association with topography. These authors also note some marked regional-scale differences in the annual mean of the daily total insolation for the US in 1983 compared with a longer-term mean derived from surface data, which they postulate may be related, at least in part, to the effects of the El Chichon volcanic eruption.

(c) Determination of SSTs in the thermal IR

The remote detection of sea surface temperature (SST) regimes for the global ocean is a prime example of the utility of satellites to combine timely information with synoptic-scale coverage. Moreover, since regional SST variations of only about 1–1.5 °C may have major impacts on atmospheric circulation and climate, a satellite-derived data set of global SSTs must be able to attain accuracies to within ±0.5 °C over space scales of about 2–10 km (Njoku *et al.*, 1985). It must also be able to resolve important thermal features such as oceanic fronts and eddies (Strong and McClain, 1984). An assumed ocean emissivity of 1.0 in TIR wavelengths means that satellite detection of SSTs can utilize the window region (Rao *et al.*, 1972; McClain *et al.*, 1983). A satellite SST operational product has been available using the multichannel NOAA AVHRR since 1978 (the MCSST).

There are certain problems attending the derivation of SSTs in the TIR from satellites. These include:

(1) the need to obtain clear column radiances owing to significant absorption by clouds and atmospheric water vapor within the 8–14 μm band (Rao *et al.*, 1972; Strong and McClain, 1984); and

(2) the need for accurate intercomparisons between satellite-retrieved and surface-observed (ship, buoy) SSTs (i.e., algorithm determination and validation). Given the required accuracy of ±0.5 °C, satellite-*in situ* root-mean-square (RMS) errors of no more than about 0.25–0.5 °C are desirable (Emery and Schluessel, 1989).

The broad-band TIR radiances are subject to attenuation in the absorption bands for water vapor (Barton, 1983a). The effects of this gas vary with viewing angle of the sensor, from place to place and from time to time (Chapter 5). Since the water vapor content of the atmosphere is typically greatest in the tropics, single-channel IR estimates of SST may be most in error for those regions when compared with ship-measured SST. Moreover, the errors tend to increase as the satellite-measured equivalent temperature increases above about 290 K (Barton, 1983a). Improved accuracy of the satellite retrievals is possible with dual channels that have contrasting water vapor absorption properties. Barton (1983a) describes an algorithm for use with two-window channels (11.4, 3.7 μm) of the Nimbus-5 SCR for night-time determination of tropical SSTs, and for which there is a close linear association. Corrections for angular viewing can also be determined by linear regression in a given channel over selected viewing angles.

Clear column radiances are not possible in regions of persistent oceanic cloud, particularly at higher latitudes (Emery and Schluessel, 1989) or where an apparently uncontaminated scene has present optically thin cirrus or volcanic dust or other haze (see, e.g., Strong and McClain, 1984; Njoku, 1985; Strong, 1986). Thus, cloud detection is an essential component of satellite SST determination in the TIR. Cloud filtering may involve time-compositing over a few days using minimum brightness (visible wavelengths) or maximum temperature (TIR) techniques (Rao *et al.*, 1972). However, where high temporal resolution is required (Njoku *et al.*, 1985), threshold techniques need to be applied to different channels of the AVHRR to eliminate cloudy or partly cloudy fields of view (Saunders, 1986). For the MCSST product, various spatial and spectral tests of uniformity are employed on the daytime visible and IR data, and SSTs derived using IR channels 4 and 5 for cloud-free scenes. At night, the 3.7 μm channel facilitates cloud detection by comparison with the two IR channels.

In the MCSST product, the SSTs are based upon equations specifying the statistical relationship between channel radiances and buoy-derived temperatures. However, there are major differences in the nature of the satellite and *in situ* methods of SST retrieval. These are:

(1) the satellite sensor resolution gives an area-averaged SST whereas a ship or buoy observation is for a point (Rao *et al.*, 1972; Njoku, 1985); and
(2) the satellite senses a surface 'skin' temperature owing to the strong IR absorption by liquid water. An observation made from a ship or buoy (fixed or drifting) is a bulk temperature that includes the subsurface.

Scatter plots of colocated ship and satellite SST observations show that the mean differences may be quite small but the scatter can be significant. However, a reduction in the RMS errors and improved calibration of satellite SST observations occur if the surface measurements made over longer time scales (>12 hours) and over space scales exceeding 50 km, are averaged and correlated with the satellite measurements (Emery and Schluessel, 1989; Njoku, 1985). This is also shown to hold true for satellite microwave (e.g., SMMR) estimates of SST (Gloersen *et al.*, 1984).

In situ observations of SST are typically subsurface, being of the order of 0.5–1.0 m for buoys and 5 m or more in the case of ships. Accordingly, satellite retrievals tend to correlate better with drifting buoy observations (Strong and McClain, 1984). Root-mean-square differences of 0.6 °C are reported for satellite-drifting buoy observations, and these contrast with up to 1.8 °C for satellite-ship differences. The skin temperature may undergo greater fluctuations than corresponding bulk SST observations because of the generally opposing influences of solar radiation receipts in cloud-free conditions and evaporative cooling. The evaporation rate is positively related to the wind speed and the vapor pressure gradient at and above the ocean surface. Thus, a lower skin temperature relative to the bulk SST dominates globally, especially in the subtropics (Emery and Schluessel, 1989; see also Strong and Pritchard, 1980). Conversely, skin temperatures tend to exceed bulk SSTs over the tropical oceans, since large solar radiation receipts and a persistently high atmospheric water vapor content occur there. The latter reduces longwave loss and evaporative cooling. When plotted by latitude zone (Fig. 2.22), the frequency distribution of the bulk SST–skin SST differences tends to be Gaussian outside the tropics (Emery and Schluessel, 1989). The greatest mean differences (0.8 °C) occur in the southern tropical zone (0–20° S). These differences argue for improved calibration of satellite

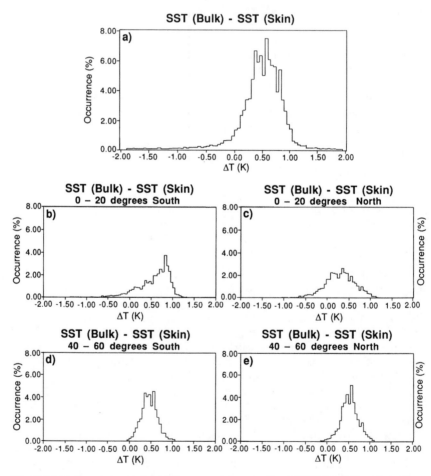

Fig. 2.22 Histograms of the differences between the MCSST 'skin' and surface-measured 'bulk' SSTs for (a) the global ocean; (b) latitude zone 0–20° S; (c) 0–20° N; (d) 40–60° S; and (e) 40–60° N; for the period October 9–22, 1982. (From Emery and Schluessel, 1989.)

retrievals of SST by incorporation of skin effects, especially in the tropics, since heat flux estimates so derived may be subject to considerable error.

Intercomparisons have been made of satellite SST 'anomalies' derived from multi-channel IR window and microwave measurements (Njoku, 1985). On a monthly basis over open ocean regions and for comparable grid cells, some interesting differences are observed. These include a warm bias in the GOES VAS retrievals and the bigger effects of El Chichon volcanic dust contamination on the AVHRR SSTs compared with the HIRS/MSU estimates of SST. In general, although the correlations between absolute SST measurements may not be particularly high, they tend to reproduce the broad-scale patterns of SST anomalies and gradients in regions significantly lacking in conventional observations; the tropics and Southern Hemisphere (Njoku, 1985; Gloersen, 1984).

(d) Snow cover

Snow cover exhibits a strong wavelength dependence (Pinker and Ewing, 1986). It is highly reflective over most visible wavelengths (Fig. 1.15), where it may approach 90% for freshly fallen snow, but is much more absorptive in the near-IR (Warren 1982). Thus, the absorbed fraction of the incident solar radiation is small, and extensive snowfields in winter tend to be regions of net radiation deficit and lowered T_s. Accordingly, snow-cover anomalies have been shown to influence significantly the 30-day forecasts of temperature generated by GCMs (Walsh and Ross, 1988). There is some suggestion that the thermal anomalies induced by snow-cover extent anomalies are manifest in synoptic circulation changes involving cyclone tracks, cyclonic intensity, the persistence of continental anticyclones, and climate teleconnections (see, e.g., Namias, 1962; Walsh et al., 1982). Also, model simulations of the climatic response to increases in CO_2 show that the temperature increases are generally greatest at higher latitudes in the vicinity of the snow/ice limit (the ice-albedo feedback). The high reflectance of snow complicates considerably the detection of clouds over snowfields; especially thin cirrus, and leads to unreliable cloud climatologies for those regions (Chapter 3, Section 2). Moreover, an increase in SZA and cloud cover tends to increase the all-wave albedo (Warren, 1982).

Satellite sensing lends itself to the regular monitoring of snow cover since the conventional data network tends to be sparse, irregularly spaced and inhomogeneous in terms of quality (Foster et al.,1987). The retrieval of snow cover parameters utilizes several portions of the EMS that detect specific snow properties. These include the liquid water content (from visible and near-IR reflectances; TIR emittances and microwave T_B); grain size (visible/near-IR, TIR, microwave); particulate concentrations, or impurities (visible, near-IR); the surface roughness (visible, near-IR, microwave); the temperature at the surface and with depth (TIR, microwave); and dielectric properties of the underlying surface (microwave) (Foster et al., 1987). The determination of areal extent using visible and near-IR wavelengths is optimized during spring melt owing to the strong differences in absorption and reflectance between snow-covered and snow-free land (Strong *et al.*, 1971). Accurate detection of the snow in these wavelengths is only possible in the absence or near-absence of clouds, although snow–cloud discrimination is possible in the 1.6 μm and 3.7 μm wavelengths (e.g., Fig. 1.13).

In the visible and near-IR, the albedo is lowered substantially with increasing age of the snow pack since melting and refreezing lead to increased grain size and the incorporation of impurities, such as soot and dust (Foster *et al.*, 1987). Calculations have been made of the change in spectrally averaged albedo compared with that in visible wavelengths, resulting from impurities in Arctic snow samples (Warren and Clarke, 1986). An average reduction of spectrally averaged albedo of 0.02, when input to the solar radiation climatology for February 1 to July 15, yields a net gain by the Earth–atmosphere system of $+4.6 \times 10^7$ J m^{-2}. This is in the same direction, and is of comparable magnitude to, the value of $+8.2 \times 10^7$ J m^{-2} established for the net effect of the Arctic haze on absorption of solar radiation at 75° N (Warren and Clarke, 1986). It is postulated that the resulting increased rates of snow melt in the spring and summer are part of a positive feedback; the lower albedo of the underlying ground or sea ice surface is exposed earlier with potential far-reaching effects on the surface–atmosphere heating rates and general circulation (see, e.g., D. A. Robinson *et al.*, 1986).

The retrieval of snowcover depth using the two SMMR frequencies at 18 GHz and 37 GHz relies on the differences in emissivity and T_B between snowfree and snowcovered land surfaces (Fig. 2.23), and the reduction of microwave emission by the surface as snow depth increases (Foster *et al.*, 1987). The emissivity of snowfree land is observed to change little over a range of frequencies compared with dry snow where the emissivity decreases with increasing frequency (Robinson *et al.*, 1984). The latter is itself related to the water equivalent of the snowpack, which further reduces the emissivity. An appropriate SMMR algorithm for snow depth (SD, in cm) is given by Chang *et al.* (1990):

$$SD = 1.59 \left(T_{18H} - T_{37H} \right) \qquad (18)$$

where T_{18H} and T_{37H} are the horizontally polarized brightness temperatures at 18 and 37 GHz respectively; 1.59 is a constant derived by using the linear portion of the 37 and 18 GHz responses to obtain a linear fit of the differences between the two channels. This method limits snow depth retrieval to depths of 1 m or less. The algorithm assumes a snow density of 0.3 g/cm^3 and a snow crystal radius of 0.3 mm for the entire snowpack. Comparisons of SMMR-derived snow cover over Eurasia with interpretation of visible DMSP imagery for March 1979 show reasonable agreement for various classes of snow cover (Robinson *et al.*, 1984). Differences between the analyses are ascribed mainly to the differences in sensor and temporal resolutions of the SMMR and DMSP. Nocturnal microwave signatures from dry sand are also similar to those of thin dry snow, owing to the existence of strong vertical temperature gradients. Thus, some snowfree ground can be confused with the snow cover. There are also problems with snow depth retrieval in coniferous vegetation areas and at the margins of the

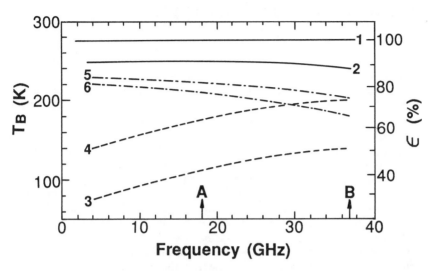

Fig. 2.23 Graph of brightness temperature variations and emissivity in microwave frequencies. The plots are for (1) an ideal blackbody with assumed physical temperature of 273 K; (2) snowfree ground; (3) and (4) a flat sea surface; (3) horizontal and (4) vertical polarizations for a physical temperature of approx. 283 K and incident angle approx. 50°; (5) snow-covered ground or old sea ice; and (6) increasing snow depth (horizontal polarization). A and B mark the SMMR frequencies for snow retrieval. (From Robinson *et al.*, 1984.)

snow cover where the snow may be very thin. However, snow volume estimates using the passive microwave are reported to be far superior to those based on local or regional surface measurements (Foster *et al.*, 1987).

Suggested further reading

Barrett, E. C. and Curtis, L. F., 1976: *Introduction to Environmental Remote Sensing*. Wiley and Sons. 336 pp.
Campbell, J. B., 1987: *Introduction to Remote Sensing*. Guilford Press. 551 pp.
Henderson-Sellers, A. (ed.), 1984: *Satellite Sensing of a Cloudy Atmosphere: Observing the Third Planet*. Taylor and Francis, London and Philadelphia. 340 pp.
Lillesand, T. M. and Kiefer, R. W., 1979: *Remote Sensing and Image Interpretation*. Wiley and Sons. 612 pp.
Lo, C. P. 1986. *Applied Remote Sensing*. Longman: Harlow, Essex, UK. 393 pp.
Szekieldei, K-H., 1988. *Satellite Monitoring of the Earth*. Wiley and Sons. 326 pp.

3 Clouds and cloud climatologies

Clouds are the dominant feature of satellite data in the visible (e.g., Fig. 1.8) and IR wavelengths, and their fluctuations occur on all spatial and temporal scales (Table 3). A space–time analysis of satellite OLR data for the Pacific in the extreme seasons (Cahalan *et al.*, 1982) reveals the dominance of relatively short time (1–2 days) scales, even over large space scales. There is a latitude dependence to the persistence of the IR cloud fluctuations, being generally more persistent at a given point where cloudiness is low (the sub-tropics) and less persistent in regions of greater cloud (the tropics and middle latitudes). The persistence is a function both of the speed of movement of cloud features and the development and dissipation of new clouds. There is also a significant inter-diurnal component to cloud variability in the tropics and also middle latitudes in the warm season, which is best captured by higher frequency imaging from geostationary satellites.

Minnis *et al.* (1987) have investigated the characteristic cloud cover distributions for the equatorial eastern Pacific for 15 days in July 1983 using the GOES. The authors identified four different cloud regimes, some with characteristic diurnal dependencies. These are:

(1) trade cumulus regions, characterized by low albedo, low altitude and low cloud amount;
(2) stratocumulus regions, with slightly higher albedo, but larger total cloud amount;
(3) convective clouds in the ITCZ, with relatively high albedos, low CTTs and substantial cloud amounts; and
(4) tropical storms with similar visible and infrared characteristics as in (3) but larger cloud coverage.

Minnis and Harrison (1984a,b) find, for the region imaged by GOES-E (±45° latitude, 30°–125° W) in November 1978, substantial differences in the timing of maximum and minimum cloudiness between land and ocean

Table 3 Scales of cloud fluctuations

Time	Space	Process
Hourly	Sub-grid scale < 1 km	Turbulent convection
Daily	Synoptic scale > 1000 km	Atmospheric dynamics
Monthly and interannually	Planetary scale > 5000 km	Volcanos, Southern Osc., sea ice, etc.

Source: Cahalan, 1983.

(Fig. 3.1), and also with latitude. Low cloud in the eastern Pacific exhibits a morning maximum, but this is later (near local noon) in the western Atlantic and over most of South America. Middle clouds are greatest in the evening over oceans, but in early morning over land. Over tropical and subtropical land areas the diurnal range in CTT is greater than over the oceans, owing to the effects of daytime convection by surface heating. This is enhanced over topography. Moreover, the diurnal change of the low cloud and high cloud fraction between the eastern Pacific and South America is out of phase (Fig. 3.1), and suggestive of a physical connection (Minnis and Harrison, 1984a,b).

The role played by clouds in the Earth's radiation and energy budgets is being increasingly recognized as crucial for an understanding of climate and its variations (Chapter 1). This is inextricably linked with the need to develop a global climatology of clouds that incorporates understanding of the means and variances of cloud amount (total, by layer and type); and cloud optical thicknesses and microphysical properties as input to climate models (Hughes, 1984). This is the major thrust behind ISCCP and its related field programs (Schiffer and Rossow, 1983, 1985; Cox *et al.*, 1987; Starr, 1987).

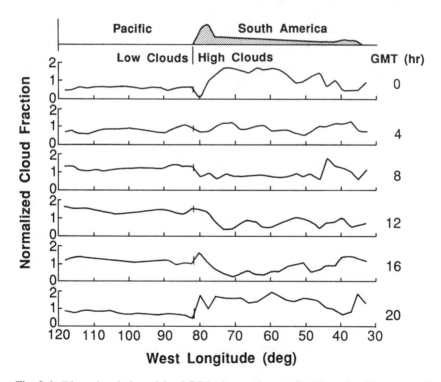

Fig. 3.1 Diurnal variation of the GOES-observed normalized low cloudiness west of South America and normalized high cloudiness over South America at 5.6° N during November 1978. Note the out-of-phase diurnal change. (From Minnis and Harrison, 1984a.)

1 Meteorological interpretations of cloud imagery

From the earliest days of the meteorological satellite program that began with TIROS, the primary application of the remotely sensed cloud cover data has been to provide synoptic cloud cover information over extensive data-sparse regions; particularly the oceans and low latitudes (see, e.g., Anderson *et al.*, 1969, 1973; van Loon and Thompson, 1966; Oliver and Anderson, 1969). In this section we examine examples of imagery from polar-orbiting and geostationary meteorological satellites. These examples are in no way intended to substitute for the excellent cloud image atlases of, for example, Anderson *et al.* (1969) or Scorer (1986). Rather, they are presented in order to illustrate some of the different types of spectral information that are routinely available from meteorological satellites; to act as an introduction to basic image interpretation of cloud features; and to assist in the understanding of the dominant physical processes at work to produce these features.

(a) Satellite cloud classification

Satellite observations of cloud cover confirm the impression gained from ground level that the atmosphere exhibits considerable linearity on a variety of scales (Kuettner, 1959). These range from the roll clouds associated with local topographic or thermal forcing to the very long (about $3-15 \times 10^3$ km) quasi-meridional cloud bands that connect circulations in the tropics with synoptic systems in middle and high latitudes, and which are frontal along at least part of their length.

Cloud classification is best achieved using combinations of visible and IR digital data. These provide the maximum range of spectral characteristics, notably the cloud brightness (reflectance), which is a function of cloud height, optical thickness, the SZA, and spectral interval; cloud morphological characteristics (texture, form); and cloud level (from CTT).

The determination of individual cloud types from satellites is complicated for co-occurring cloud layers, especially given the spatial resolutions typical of meteorological satellites. Accordingly, automated techniques, such as that of Garand (1989), tend to group clouds that share certain morphological characteristics and can be discriminated spectrally. Examples of such distinctive groups include low clouds, middle clouds, highly reflective (cumulonimbus) clouds, and cirriform clouds. Increasingly, cloud retrieval algorithms are employing a human–computer format: humans can best detect patterns and make certain decisions about the cloud presence or absence over, say, a high albedo surface. The computer, once trained (supervised), can process the vast amounts of pixel information required of a large-scale cloud climatology (see, e.g., Ebert, 1987, 1989; Key *et al.*, 1989).

The physical processes at work to produce the characteristic morphology of cloud groups and sub-types (e.g., Table 4) may be summarized as follows.

(1) The altitude and depth of the moist air. This determines the characteristic height (low, middle, high) and cloud thickness.

Table 4 List of the 20 classes of the
Garand cloud classification scheme

1.	Clear
2.	Stratus
3.	Scattered cumulus
4.	Broken cumulus
5.	Scattered stratocumulus
6.	Broken/overcast stratocumulus
7.	Cloud streets
8.	Bright rolls
9.	Polygonal open cells
10.	Strongly convective open cells
11.	Bright closed cells
12.	Nimbostratus
13.	Altocumulus
14.	Altocumulus with cumulus
15.	Altocumulus with stratocumulus
16.	Thin cirrus
17.	Multilayers with cirrus
18.	Multilayers with cumulonimbus
19.	Dense cirrostratus
20.	Overcast cumulonimbus

Source: Garand, 1988, 1989.

(2) The atmospheric stability, as given by the vertical temperature and moisture profiles. This parameter influences cloud character (stratiform, cumuliform) and the transformation of one kind to another, say, under conditions of nocturnal radiation loss from stratiform cloud tops.

(3) The cloud microphysical properties; most notably the dominant sizes of condensation nuclei and cloud droplets, and the degree of supersaturation within the cloud. This is one of the least understood areas of satellite cloud analysis, yet it is an extremely important consideration for the cloud-climate sensitivity (Chapter 1, Section 4a).

(4) The influence of local-scale forcing, such as orography, lakes, coastlines, and land cover heterogeneities (see, e.g., Anthes, 1984; Segal *et al.*, 1989).

(5) The synoptic-scale atmospheric circulation. For example, the streaky appearance of cirrus owes much to its occurrence at jet stream level (see, e.g., Durran and Weber, 1988).

An illustration of the importance of at least some of these processes is the satellite study of Rabin *et al.* (1990). These authors examined the influence of land surface heterogeneities in Oklahoma in the development of warm-season convective clouds. Under conditions of weak (generally anticyclonic) synoptic-scale flow, the timing of daytime convective cloud development differs according to surface type. As a general principle, clouds form first over dry ground when the ambient air is dry, since it requires less of the total energy (latent plus sensible heat) to initiate convection (Fig. 3.2). On days when the ambient air is moist, clouds tend to form later over dry ground since it requires more energy to initiate convection in that setting. At those times, convective clouds tend to develop first over moist (frequently forested) surfaces. The importance of this principle in the context of Earth surface–climate interactions and drought is discussed in Chapter 6.

In the context of satellite cloud analysis, it is appropriate to note the work of the pioneering satellite meteorologists (e.g., Conover, 1962; Hopkins, 1967; Anderson *et al.*, 1969; Lee and Taggart, 1969), and on which the subsequent automated techniques using high-resolution data have been based. These early classifications made much use of the following cloud descriptors (Barrett, 1974), some of which are discussed in the context of the single visible DMSP mosaic for the Europe–northern North Atlantic sector shown in Fig. 3.3.

(1) *Cloud texture* This refers to the physical appearance of the cloud: it is generally amorphous in the case of stratus; mottled for cumulus, owing to the gaps between the cloud elements; and striated for cirriform clouds (Fig. 3.3). Diffuse or fuzzy outlines to cumulus clouds, especially if asymmetrical in appearance, tend to indicate cumulonimbus and glaciated tops. Moreover, the differences are enhanced at relatively low sun angles, as in higher latitudes or in early morning or late afternoon satellite passes.

(2) *Cloud brightness* Although the cloud reflectance is partly a function of the sun–target–satellite geometry (bidirectional reflectance) and the spectral interval of the sensor, it is dominated by characteristics such as cloud thickness, the dominant particle size, and the phase (ice, liquid water) of the cloud. Thus, deep cumulonimbus clouds (highly reflective clouds: HRCs) and thick frontal systems tend to be the brightest of clouds. Cirriform clouds, being thinner, are characteristically less bright (Fig. 3.3). However, the albedo of the underlying surface may also be important for the cloud albedo, particularly for cirrus clouds (Henderson-Sellers and Wilson, 1982).

(3) *Cloud form* The form of the cloud elements may vary within a given cloud field, according to changes in the static stability, the strength of the wind and the vertical wind shear. This is particularly true of cumuliform clouds, which show a close association with surface thermal forcing and the wind speed. Rogers (1965) identified typical ranges of low-level wind speed for cumuliform cloud configurations over the ocean. However, he noted considerable ambiguity in the direction of the wind for certain broad ranges of wind speed. A more reliable determination of wind direction, other than that obtained from microwave sensing of the ocean surface, is made by frequent (half-hourly or hourly) satellite images of the same region, so that individual cloud elements may be tracked (see, e.g., Gruber *et al.*, 1971; Sadler and Kilonsky, 1985; Turner and Warren, 1989), or by reference to the larger-scale patterns of the clouds.

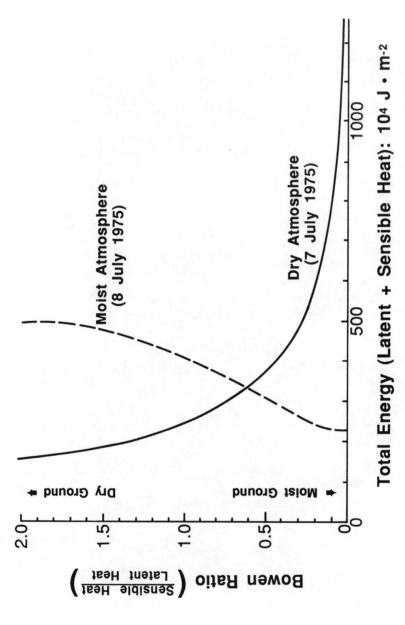

Fig. 3.2 Relationships between convective cloud development and the Bowen ratio at the ground, as modeled for two different initial atmospheric profiles of humidity and temperature. (From Rabin *et al.*, 1990.)

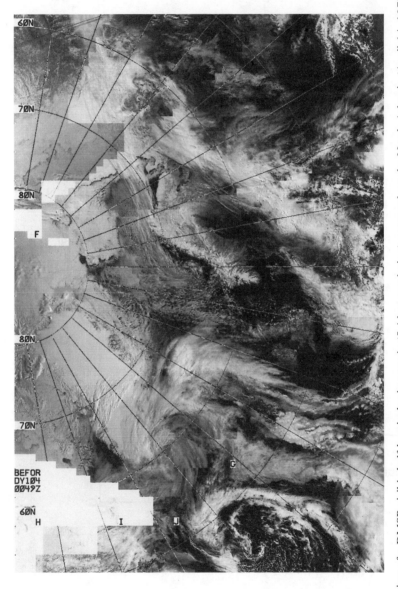

Fig. 3.3 Section of a DMSP visible half-hemispheric mosaic (5.4 km resolution) centered on the North Atlantic (April 14, 1977). Dominant features of the mosaic are the high reflectivities of clouds and cloud systems, and also the snow and ice cover of Greenland, pack ice, and mountains of Scandinavia. (From Carleton, 1985c.)

(4) *Cloud patterns* Much information on cloud type can be obtained from examining clouds in the context of the larger-scale atmospheric (synoptic) circulation. These include the vortical structures associated with cyclones and the extensive bands of cirrus commonly associated with fronts and jet streams (Fig. 3.3). The typical cloud patterns associated with different sectors of mature frontal cyclones are shown, with a Southern Hemisphere orientation, in Fig. 3.4. They include the cloud masses associated with subsynoptic-scale cyclones ('polar lows').

The earlier cloud classifications made much use of the technique of pattern recognition of cloud fields. They form the basis of the *nephanalysis* technique, or the subjective standardized classification of cloud fields appearing on satellite images in the context of the larger circulation features (see, e.g., Hopkins, 1967; Rowles, 1978; Barrett and Harris, 1977). For visible channel imagery, classification is best achieved using the morphological and textural characteristics since albedo differences between cloud types at different levels tend to be relatively small, except in the case of thin clouds. Thus, cirrus appears fibrous and is often arranged linearly (Fig. 3.3),

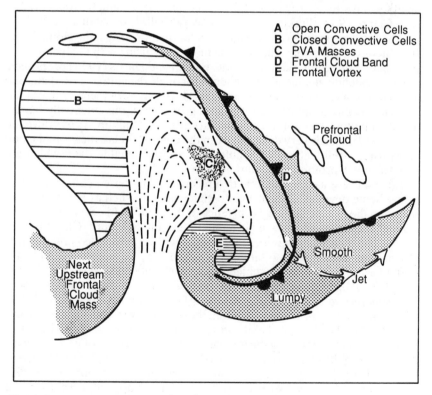

Fig. 3.4 Schematic of the typical satellite-observed cloud patterns associated with a mature oceanic extratropical cyclone (Southern Hemisphere orientation). (From Guymer, 1978.)

whereas cumulus has harder outlines and tends to be in the form of cells separated by clear areas. The latter results from convection of the Benard type, whereby clouds form in the regions of ascent and dissipate in regions of descent and adiabatic warming. The cumuliform clouds show a tendency to be arranged in roughly hexagonal patterns with diameters of about 10–100 km when viewed from high altitudes (Agee, 1987). This is known as MCC (meso-scale cellular convection), and is distinct in this discussion from MCS, which is addressed in Chapter 4. The amount of open space between cumuliform elements is related to the static stability. Fields of cumulus that are mostly open (MOP) on the nephanalysis tend to occur in unstable conditions (e.g., in the cold air behind a middle-latitude depression: Kruspe and Bakan, 1990). The generalized climatology of MCC is given in Fig. 3.5, where it is seen that areas of open MCC tend to be associated with warm ocean currents off the east coasts of continents, and closed MCC with cold currents of the west coasts.

Layered clouds at middle levels typically exhibit a striated form, often occurring in association with cirrus and cirrostratus in frontal cloud bands (Fig. 3.3). Although a secondary-level satellite product, operational neph-analyses have long proven useful for oceanic cloud climatological studies (e.g., Clapp, 1964; Bjerknes et al., 1969; Godshall, 1968; Sadler, 1969).

Where the nephanalyst has available a thermal IR image, the cloud classi-fication based on visible data is made more convincing with the inclusion of information on relative cloud altitude. Barrett and Harris (1977) proposed a detailed satellite nephanalysis scheme for high-resolution TIR data, and applied it to a case using DMSP imagery. It is only very recently that it has proven possible to automate, even in a gross sense, the essentially human ability of pattern recognition of cloud systems (cf. Hopkins, 1967; Burfeind et al., 1987). This is very different from the computer classification of cloud types using combinations of visible and infrared satellite data, which are showing much promise (e.g., Garand, 1988, 1989).

Table 4 presents one example of a recent satellite-based cloud classifica-tion scheme developed according to characteristics such as height, albedo, directionality, shape and co-occurrence of the cloud fields. Some types have distinct anomalies of surface air temperature and humidity (Garand et al., 1989). The classification is intended for use with an automated precipitation scheme for ocean areas (Garand, 1989). It may be compared with, for example, Table 5, which is a subjective cloud classification developed for use with DMSP 2.7 km visible and IR imagery in a study of land surface–atmosphere interaction for the Midwest US in summer (Brinegar, 1990). Cloud cover and type is classified from the once-daily (5 a.m. local time) pass for $1° \times 1°$ grid cells. The classification of rain-producing clouds (types 4, 5, 8) was tested by comparing the number of rain days in each of 25 US climate divisions with the number classified from the satellite data. A rain day occurs when at least 50% of the stations in a climate division report measurable precipitation. The results for the summer of 1987 appear in Table 6. The values in parentheses show the effect of including the following morning's image, which almost always results in an improved correspondence between the satellite classification and ground truth. This occurs because of the

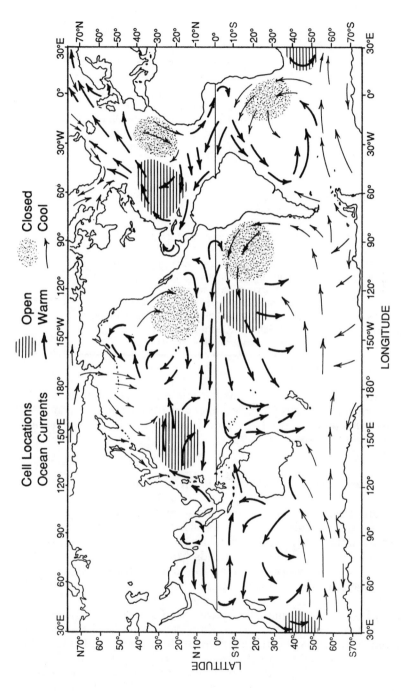

Fig. 3.5 Global climatology of meso-scale cellular convection (MCC), depicting the most favored regions of MCC over the ocean. (From Agee, 1987.)

Table 5 Cloud cover/cloud form categories: Midwest drought study using DMSP 2.7 km data

I. Cloud cover (amount of a 1° × 1° grid cell that is filled)

Code	Coverage
1	Clear (no clouds evident)
2	Scattered clouds (mostly clear; a few clouds)
3	Partly cloudy (about 40–70% filled)
4	Mainly cloudy (about 70–90% filled)
5	Overcast (totally filled)

II. Cloud form/type

Code	Coverage
0	Clear skies
1	Small cumuliform
2	Stratiform (low-level, mid-level)
3	Strato-cumuliform
4	Cumulonimbus/cu. congestus (hard outlines)
5	Anvil cumulonimbus (fuzzy dense cirriform tops)
6	Anvil cirrus (associated with 5)
7	Cirriform (not apparently associated with 5)
8	Thick middle and high clouds together (solid and bright in IR): nimbostratus
9	Cirriform overrunning stratus

Source: Brinegar, 1990.

dominantly nocturnal nature of summertime precipitation in much of this region. Verification results for the summer of 1988 are almost identical with those for 1987, even though that summer was considerably drier over much of the Midwest US (Brinegar, 1990).

(b)   Cloud fields, air mass modification and jet streams

Clouds are tracers of energy and moisture fluxes occurring in three dimensions, and are indicators of atmospheric static stability. The coupling between the Earth's surface, the boundary layer and the free atmosphere is most apparent for cumuliform cloud fields. Convective cloud fields developing over land surfaces tend to display a strong diurnal character that contrasts with the generally much reduced diurnal change for ocean surfaces in the same latitude (see, e.g., Short and Wallace, 1980; Minnis and Harrison, 1984a,b). Figure 3.6a,b,c shows a sequence of GOES-W visible images of the south-west US over a 6-hour period on a summer day in 1982. Note the development of deep convective clouds (cumulus congestus, cumulonimbus) during the day over the strongly heated desert and mountain surfaces of Arizona, Utah and Colorado. These give rise to showers and thundershowers. The fibrous tops to the cumulonimbi in Fig. 3.6c are oriented parallel to the upper-tropospheric flow. They denote a flow pattern that

Table 6 Precipitation data/cloud analysis comparison for summer 1987*

Site	Climate divs used for each state	Classified days[1]	Actual days[2]	When classified days = actual days[3]	Percentage
Missouri	#2	15	11	8 (11)	53.3 (73.3)
	#3	14	14	8 (11)	57.1 (78.5)
	#5	18	12	9 (11)	50.0 (61.1)
	#6	23	21	13 (19)	56.5 (82.6)
Ohio	#2	11	19	10	90.9
	#10	14	19	9 (12)	64.2 (85.7)
Kentucky	#2	25	17	12 (15)	48.0 (60.0)
West Va.	#5	17	20	11 (14)	64.7 (82.3)
North Carolina	#1	17	20	10 (13)	58.8 (76.4)
Virginia	#6	18	16	10 (14)	55.5 (77.7)
Illinois	#1	21	16	12 (15)	57.1 (71.4)
	#5	12	20	11	91.6
	#6	13	12	8 (11)	61.5 (84.6)
	#7	18	19	14 (16)	77.7 (88.8)
Arkansas	#3	20	18	14 (17)	70.0 (85.0)
Michigan	#9	10	14	7 (8)	70.0 (80.0)
Minnesota	#8	13	18	8 (10)	61.5 (76.9)
Indiana	#1	18	25	12 (14)	66.6 (77.7)
	#6	12	22	10 (11)	83.3 (91.6)
Tennessee	#3	17	14	10 (12)	58.8 (70.5)
Wisconsin	#5	15	23	12 (13)	80.0 (86.6)
Iowa	#3	14	15	10 (11)	71.4 (78.5)
	#5	17	17	9 (15)	52.9 (88.2)
	#9	13	11	7 (10)	53.8 (76.9)
Oklahoma	#6	26	16	11 (16)	42.3 (61.5)

Range of percentages 42.3–91.6 (61.1–91.6)
Average = 63.9% (79.1)

Notes
* The data only cover the months of June and July in this year owing to a lack of images for August.
1. The number of days classified as having rain-producing clouds.
2. The number of days on which more than 50 per cent of stations in a climate division reported more than a trace of precipitation.
3. The number of times that the classified rain day was the same day as the actual rain day.

The numbers in parentheses are those in which the following day is considered to be an actual rain day, owing to the local time of satellite overpass (close to 0500 h).

Source: Brinegar, 1990.

Fig. 3.6a, b, c GOES-W visible image sequence of the south-west US for August 11, 1982 (a) 1916Z; (b) 2116Z; (c) 2316Z. Compare the diurnal cloud development over land with that over the adjacent ocean.

moves up from the south over Arizona before curving eastwards over Colorado and Wyoming. This is in association with the unstable western edge of the Bermuda subtropical anticyclone that extends into the desert south-west US at this time of year. A climatology of this 'monsoonal' cloudiness is given by Carleton (1985a), and associations with the dominant atmospheric circulation controls are provided by Carleton (1986, 1987a) and Carleton *et al.* (1990). The much more conservative nature of the low-level strato-cumulus and stratus fields off southern California and the Baja coast contrast with the cloud fluctuations over the adjacent land surface, although these clouds undergo some diurnal variability (Duynkerke, 1989).

Cloud patterns typical of middle- and higher-latitude ocean areas in the cool season are shown in Fig. 3.7, which is a GOES-W visible image of the North-east Pacific. The bright cloud band running north–south demarcates a cold or occluded front associated with a strong upper-air trough. The front separates a cold air mass exhibiting deep convection to the west from warmer air having scattered stratiform clouds to the east. In a study of atmospheric stability differences between these convective and stratiform cloud fields associated with mature extratropical cyclones in the Southern Hemisphere, Zillman and Price (1972) showed that the open cellular fields in the cold sector of these systems have characteristically steeper lapse rates (unstable) and a lower and warmer tropopause (Fig. 3.8). This contrasts with the pre-frontal warm sector (stratiform clouds), which has typically more stable lapse rates, often with lower tropospheric inversions, and a higher and colder tropopause. In addition to these differences in static stability, there are major differences in precipitation rate and amount associated with the convective and upslide (frontal and pre-frontal cloud) areas. Figure 3.9 shows, for satellite-viewed winter depressions influencing the State of Victoria (Australia), that the pre-trough upslide cloud region has generally lighter precipitation than the post-trough sector in which heavier convective showers typically occur. The heaviest precipitation falls along the front. However, these associations may not be universally applicable. Satellite microwave analysis of fronts in the Australian region (K. Katsaros, personal communication, 1990) suggests that they may be drier than those in the North Pacific, probably owing to the passage of the pre-frontal air over the continent. A strong continental influence on SMMR-derived mean monthly precipitable water vapor fields has been identified by Stephens (1990).

Other features of interest in the satellite image in Fig. 3.7 are the generally clear skies over the extreme eastern Pacific and western US, associated with an upper-air high-pressure ridge. The cloud streamers in the lower right of the image are cirrus clouds associated with the high tropospheric subtropical jet stream. Most of the cirrus is located on the equatorward side of the jet, in accord with dynamical considerations.

Textural changes in satellite-viewed cloud fields tend to be indicators of air mass modification, associated with processes such as:

(1) changes in Earth surface type (e.g., from land to ocean; mountains to plains; irrigated to dry land);

Fig. 3.7 GOES-W visible image for the north-east Pacific Ocean (April 9, 1988). The image shows a well-developed frontal cloud band extending south from a cyclone in the Gulf of Alaska. Refer to the text for discussion.

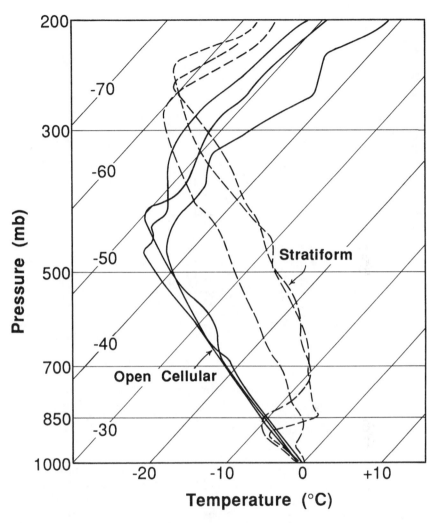

Fig. 3.8 Temperature profiles characteristic of the post-frontal (open cellular) and subsided stratiform cloud sectors of mature oceanic mid-latitude cyclones in the Southern Hemisphere. All soundings are referenced to a common surface air temperature and departures of SLP from 1000 mb are disregarded. (From Zillman and Price, 1972.)

(2) changes in the vertical gradients of temperature and moisture and, hence, in the vertical fluxes of sensible and latent heat;

(3) changes in the horizontal gradients of temperature and moisture and in the rate of cloud cell growth;

(4) changes in the depth of the boundary layer.

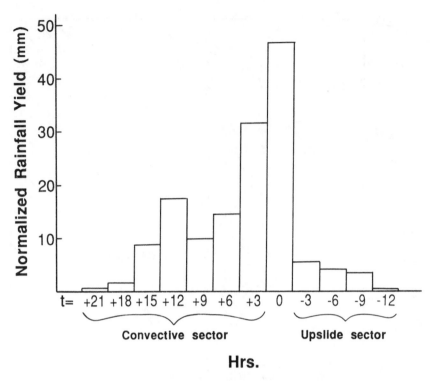

Fig. 3.9 Histogram of the combined normalized three-hourly rainfall totals for Victoria, Australia, associated with the pre-frontal upslide and post-frontal convective sectors of Southern Ocean cyclonic systems. The time t_0 refers to the frontal passage. (From DelBeato and Barrell, 1985.)

The meso-scale processes of air mass modification tend to be most evident in winter-time satellite imagery when cold air is transported rapidly equatorward and is destabilized (as behind a major synoptic-scale cyclone). These processes were the subject of an intensive investigation using satellite and conventional data in the East China Sea in 1974–5, known as the Air Mass Transformation Experiment (see, e.g., Lenschow and Agee, 1974). An example of such air mass modification appears in Fig. 3.10, which is an IR image of the western North Atlantic. A *cloud-free path* (Chou and Atlas, 1982) offshore gives way to cloud streets of small cumulus. Further downstream (south-eastward) these streets give way to larger and deeper cumuliform cells that may produce showers. These open cumuliform fields show a tendency to develop off the east coasts of continents in winter in contrast with the closed stratocumulus fields off the west coasts in summer (Agee, 1987). Note in Fig. 3.10 the occurrence of these convective cloud fields to the west and south-west of a cyclonic disturbance (comma-shaped cloud vortex of brighter, hence deeper, clouds). Chou and Atlas (1982)

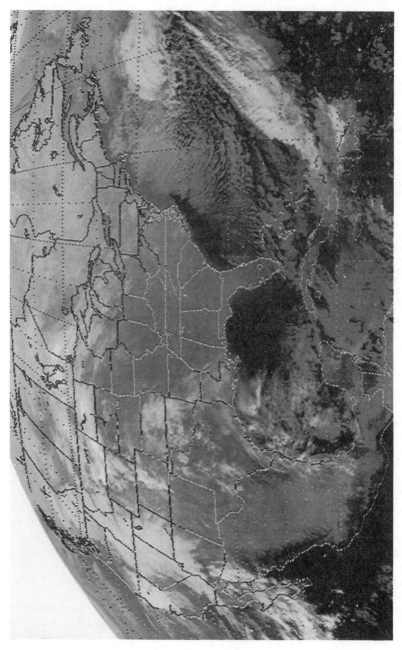

Fig. 3.10 GOES-E TIR image for January 29, 1988, showing a cold air outbreak from the eastern US moving over the warm waters of the Atlantic. Refer to the text for discussion.

found that the larger the land air temperature—SST difference, the greater the turbulent fluxes and the shorter the cloud-free path. Note the shorter cloud-free path off the north-east US compared with that off the south-east US. A discontinuity in the SSTs is also apparent in the image; the greyer tones corresponding to colder water than the black tones off the south-east US coast. Since the length of the CFP is sensitive to the initial (shore) lapse rate, Chou and Atlas propose future direct measurement by space-based lidar of the height of the boundary layer for refinement of the flux nomograms.

Similar cloud streets are observed when air moves rapidly equatorward from ice- and snow-covered surfaces (e.g., Fig. 3.7). A satellite study of cloud streets forming in the region of the Bering Sea by Streten (1975a) shows a progressive increase in cell size with increased distance from the shore or ice edge (Fig. 3.11). The rate of cloud cell growth slows with reduced sea–air temperature difference. Moreover, months with larger sea–air temperature contrast (winter) have cloud cells forming further away from the coast or ice edge compared with months having reduced sea–air temperature contrast (Fig. 3.11). The latter contrasts with the Chou and Atlas (1982) study. Walter (1980) used NOAA-4 imagery to study winter cold air outbreaks in the Bering Sea, and focused particularly on the processes involved in the transition from cloud rolls (streets) to cells. Large values of the ratio of roll wavelength to height of the boundary layer characterize these roll clouds. These are significantly different from the theoretically expected aspect ratios (about 2–3:1) for circular roll vortices, and may indicate the presence of a strong inversion at the top of the boundary layer.

The organization of the cumuliform cloud cells ('enhanced cumulus') into a cold air cyclone may occur in association with a jet or short amplitude trough (i.e. a PVA maximum). In Fig. 3.10 this is evident in the extreme right-hand side of the image by the bright (i.e., cold and high) clouds that comprise the comma shape. The cloud vortex and its associated cloud band are composed of thick high- and middle-level clouds. Note particularly the fibrous structure to the cloud band extending south-west out of the low. This marks the location of a jet stream. The jet axis is characteristically located on the poleward edge of a sheet of cirrostratus or cirriform streamers, and marks the discontinuity between ascent of air on the warm side and descent of air on the cold side. Figure 3.12 shows a similar cold air outbreak off the east coast of the US and a jet stream that accompanies the cold front cloud band. The jet emanates from the subtropics and transports moisture and energy into higher latitudes. The atmosphere off the east coast at this time is strongly *baroclinic*: that is, there are strong horizontal gradients of temperature and a marked vertical shear of the horizontal wind (i.e., a *thermal wind* (V_T) exists). This can be seen in the image (Fig. 3.12) from the very different orientations of the low-level cloud streets (north-west to south-east) and higher-level cirriform cloud of the jet (south-west to north-east). We can infer that *cold air advection* is taking place since the vertical wind shear from low to high tropospheric levels is anticlockwise. A clockwise turning of the wind with height in the Northern Hemisphere would denote *warm advection*, as occurs ahead of warm fronts, warm sector squall lines and meso-scale convective systems (MCSs). When upper- and lower-level winds are oriented in

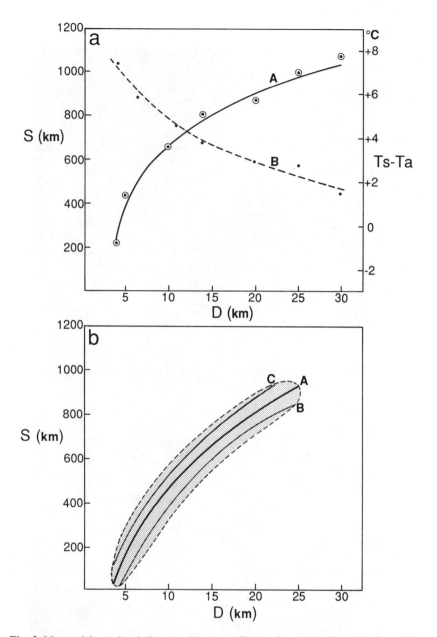

Fig. 3.11 (a): Mean cloud element diameter (D: km) for an Alaska cold air outflow situation in relation to (A) distance S(km) along the cloud streets as measured from the coast or pack ice edge, and the estimated sea–air temperature difference $T_s - T_a$ along the same path (B).

In (b) satellite-observed cloud element diameter (D: km) is plotted against distance (S) for a multi-month satellite study. The curve A is for all months of data; B is for winter months (JFM); C is for transition months (April, October). The shaded area encloses 75% of all observations. (From Steten, 1975a.)

Fig. 3.12 GOES-E TIR image for February 22, 1988, showing a jet stream off the US east coast, low-level cloud streets, and SST features.

approximately the same direction, there is no vertical wind shear (i.e., $V_T = 0$) and the atmosphere is said to be *barotropic* or *equivalent barotropic*. Such a condition tends to occur particularly in anticyclones and over low and subtropical latitudes.

Weston (1980) studied the attributes of satellite-observed cloud streets that develop over Britain during the period March and April 1977. He found the spring to have the highest frequencies of cloud streets, and these cover a wide range of topographies. Cloud streets tend to occur with anticyclonic curvature of isobars in cold advection, and to be oriented very closely with the mean direction of the wind in the boundary layer (Fig. 3.13). They exhibit aspect ratios of about 3:1. Convective cloud development over land in the absence of strong synoptic-scale forcing is discussed in the context of land surface heterogeneities in Chapter 6.

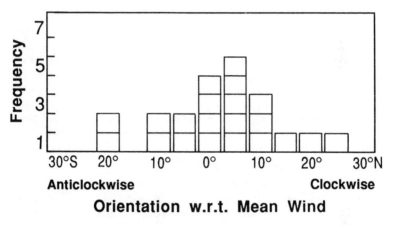

Fig. 3.13 Orientation of cloud streets over land (Britain) with respect to the mean wind direction in the convective layer, in degrees (21 cases) for March and April, 1977. (From Weston, 1980.)

2 Satellite cloud retrieval

The retrieval of cloud information from satellite data may be by manual or automated means. The former is implicit in the typical satellite nephanalysis methods noted in Section 1a, where cloud fields are classified according to certain broad-scale characteristics. For clouds occurring on the scale of a pixel or a small collection of pixels, manual methods may still be employed but they are time-consuming compared with the use of automated statistical classifiers (Parikh, 1977).

(a) Threshold methods

The commonly used methods of automated cloud extraction are the threshold and bi- (or multi-) spectral statistical techniques. The former is a cloud–no cloud decision made from an examination of individual pixels (Rossow, 1989). The cloud elements need to attain a certain concentration (percentage cover) within a given pixel in order to be detected by the sensor. This is a function of the sensor resolution as well as the location of the cloud elements within a particular pixel (Henderson-Sellers and Wilson, 1983). Similarly, when the sky is mostly cloud covered but with gaps present, the satellite may register a complete overcast for that pixel. The threshold technique can be used for estimates of cloud amount; however, it tends to exaggerate the tendency noted from surface observations of the bimodal structure of cloud cover. That is, clear or overcast skies tend to occur more often than partly cloudy skies. Changes in the system resolution and reflectance threshold values can change the computed mean cloud cover by up to a factor of 2 (Ruprecht, 1985). This is evident both globally and for latitude bands, and is especially noticeable at coarser resolutions (Shenk and Salomonson, 1985; Arking and Childs, 1985). Comparisons of cloud amounts derived simultaneously from meteorological satellites, such as GOES, with satellite data having much higher resolution (e.g., Landsat) confirm the impact of partially filled cells on the retrieved cloud amounts (Minnis and Wielicki, 1988). Analysis of cumulus cloud fields appearing on Landsat 80 m^2 resolution MSS data (Wielicki and Welch, 1986) reveals the following cloud characteritics relating to sensor resolution.

(1) The selection of the cloud brightness count (reflectance) threshold strongly determines the cloud fraction, especially at lower counts (Fig. 3.14) since reflectance variations are dominated by small cells comprising the clouds and these are not uniform reflectors (cf. 'torn-paper'-type clouds).
(2) The *number* of clouds is strongly determined by the reflectance threshold used. As the reflectance threshold is increased, the number of large clouds decreases but the number of small clouds remains the same (Fig. 3.15). The size distribution does not appear to be dependent on surface type (ocean, land).
(3) Cloud fractional coverage becomes increasingly sensitive to the chosen brightness threshold as sensor resolution is degraded (Fig. 3.16). However, this sensitivity turns out to be less than that suggested on theoretical grounds. Knowledge of the effect on cloud detection of changing sensor resolution is important for devising appropriate sampling strategies for different cloud types. In the near-IR, cloud types can be differentiated statistically according to their characteristic texture and brightness (Welch et al., 1988). Thus, cirrus clouds appear very different to cumulus and stratocumulus, although separation of the last two is complicated for stratocumulus that is breaking up.

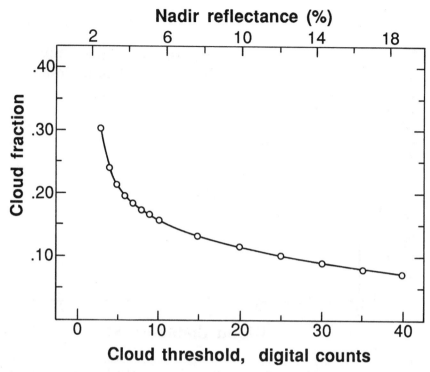

Fig. 3.14 Cloud fraction derived as a function of cloud reflectance threshold for a cloud field in the western Atlantic from Landsat data. (From Wielicki and Welch, 1986.)

In thermal IR wavelengths, where the threshold value is given by the change (usually downward) of the radiant temperatures as cloud cover increases, the larger the CTT decrease, then the higher and thicker the clouds (see, e.g., Rasool, 1964; Lo and Johnson, 1971). Figure 3.17 shows the distribution of IR counts for a 1° × 1° latitude/longitude cell in low latitudes, and their separation according to cloud level by comparison with NMC temperatures at constant pressure level. Problems with the IR threshold method again involve the assumptions that no pixels are partially filled and that the clouds are opaque. Thus, thin cirrus has a warm radiometric count relative to thick cirrus, and will appear to be located at a lower altitude. In addition, this method relies on fairly homogeneous surface conditions within a grid cell for determination of the 'background' (clear sky, highest temperature) surface and, accordingly, the low cloud amount (Lo and Johnson, 1971). For IR data in general, the presence of surface-based temperature inversions or near-isothermal conditions complicates the assignment of cloud altitude. This problem becomes most serious over snow- and ice-covered surfaces in winter. Clouds can then be warmer than the surface and, accordingly, appear grey against the white (cold) background.

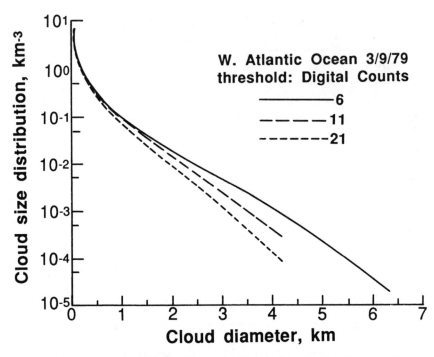

Fig. 3.15 Cloud size distribution for a western Atlantic cloud field with different cloud thresholds (6, 11, 21 digital counts). Values are given as the number of clouds per 1 km² surface area per 1 km cloud size class interval. (From Wielicki and Welch, 1986.)

These warm tones can be used as indicators of clouds and inversions or, on steeply sloping ice surfaces, turbulent mixing of air and 'warm' katabatic winds (Bromwich, 1989; Rasmussen, 1989).

(b) Statistical techniques

An improvement over simple thresholding methods are the statistical techniques of cloud classification that compare adjacent pixels in the field rather than on a pixel-by-pixel basis. Such spatial coherence methods attempt to deal with the occurrence of partially filled cells prior to the derivation of clusters (see, e.g., Coakley and Bretherton, 1982). They are made on the basis that the standard deviation of clear and cloudy regions is less than that of partially filled cells. This is the preferred algorithm for ISCCP (Rossow *et al.*, 1985). Classification is facilitated by comparing pixels in two (bispectral) or more (trispectral, multispectral) non-adjacent bands (see, e.g., Reynolds and Vonder Haar, 1977; Ebert, 1989). Ideally the pixel resolutions should be the same in both bands, but this is not always possible (e.g., with the

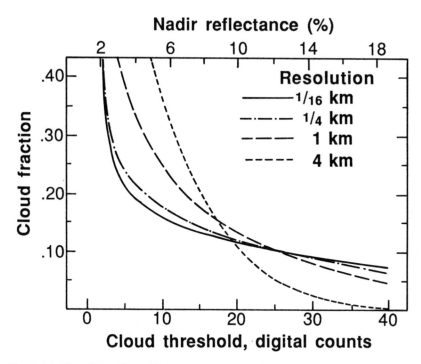

Fig. 3.16 The effect of satellite sensor resolution on the total cloud fraction for a western Atlantic cumulus field (1/16 km; 1/4 km; 1 km; 4 km). Cloud fraction is given for each spatial resolution as a function of a cloud reflectance threshold. (From Wielicki and Welch, 1986.)

GOES VISSR). In such cases it may be necessary to smooth spatially the radiometric counts of the higher resolution (usually the visible) sensor (see, e.g., Minnis and Harrison, 1984b). Cloud-group identification is made by deriving clusters on one-, two-, or three-dimensional histograms of the spectral responses of features in the scene (Fig. 3.18). Differentiation of discrete cloud layers (low, middle, high) according to their temperature of emission can ideally be made according to well-defined breaks of radiance counts in one-dimensional frequency histograms. This is maximized where a single cloud covers a uniform surface (Bolle, 1985). Confusion with surface characteristics increases as the resolution decreases and the surface becomes more mixed, especially over land—i.e., the histograms start to overlap.

Bispectral methods (see, e.g., Reynolds and Vonder Haar, 1977) identify clusters of two-dimensional histograms (e.g., Fig. 3.19). It can be seen that, even though cloud variations occur in a continuum, there are 'kernels' that have a high probability of corresponding with particular cloud or surface types in that scene. Thus, low albedo surfaces (cloud-free scenes over ocean and snowfree land) are also relatively warm, while increased visible and

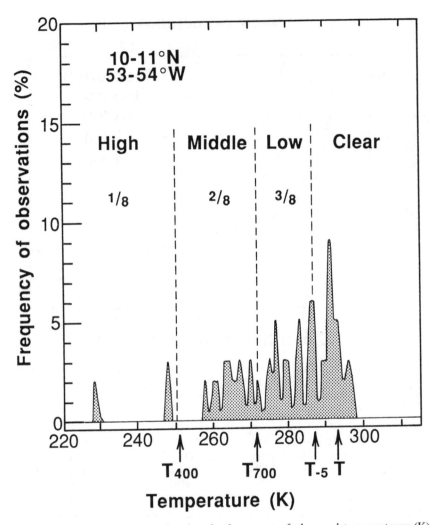

Fig. 3.17 Sample histogram showing the frequency of observed temperatures (K) from ITOS 1, and the classification of cloud amount types. (From Koffler *et al.*, 1973.)

decreased IR counts denote increasing cloud cover altitude. Note that, in accord with physical modeling considerations (Platt, 1983) separate layers of low and middle/high cloud behave differently in the plots. The steeper slope to the upper-level cloud group (Fig. 3.19) implies that the rate of decrease of IR counts with increased cloud altitude and thickness of an unbroken layer is generally greater than the albedo increase. The opposite is the case for the low cumuliform clouds, which may be broken but have a uniform optical depth. Note also that the coldest and brightest cluster represents convective (cumulonimbus) clouds, or HRCs (see Gadgil and Guruprasad, 1990). The

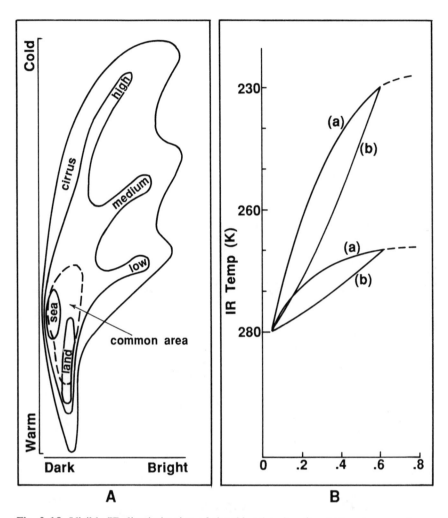

Fig. 3.18 Visible/IR discrimination of cloud level and surface (A) and comparisons with theoretical curves obtained by Platt (B). In (B), the envelopes enclose two layers of clouds with varying optical thickness (curves a) and partial coverage of the pixels (curves b). (From Sèze and Desbois, 1987.)

automated separation of classes using the bispectral technique still requires a threshold approach for separating clouds from the surface before running the classification over an entire area. Improvements in the use of this technique (see, e.g., Sèze and Desbois, 1987) take account of the slower change through time in background surface characteristics compared with the transient fluctuations arising from cloud effects; however, thin cirrus remains a problem with its wide range of possible emittances (Chapter 2).

Some bispectral cloud retrieval algorithms make use of bands other than

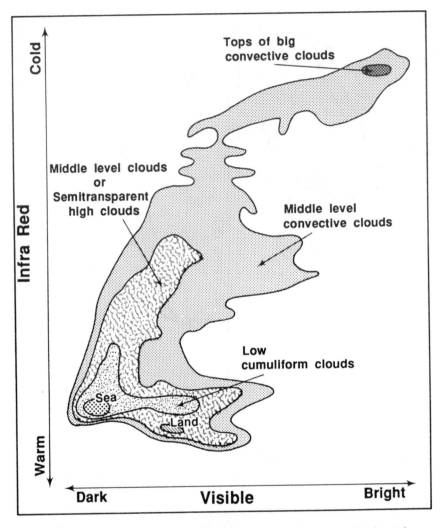

Fig. 3.19 Example of a visible-IR bidimensional histogram for clouds where classes can be identified visually. (From Desbois *et al.*, 1982.)

the visible-TIR couplet. The Nimbus-7 global cloud climatology algorithm compares the TIR with a UV band averaged at 0.37 μm (Hwang *et al.*, 1988). This technique is only possible for daylit scenes; also, the resolution of the TOMS UV band is coarser than the TIR. Both bands have difficulty detecting clouds over snow and ice. An advantage of this band is its ability to detect thin clouds, although it may underestimate the amount (Stowe *et al.*, 1988). At night, only the TIR band is used in a thresholding (cloud–no cloud) manner. As a principle of remote sensing, the amount of information about a target increases when the number of non-adjacent bands is increased. The addition of a third channel, such as the 3.7 μm (channel 3) of the NOAA AVHRR (Arking and Childs, 1985) or the water vapor absorption band (5.7–7.1 μm) of Meteosat (Desbois *et al.*, 1982; Saunders and Hunt, 1983), greatly improves cloud-type discrimination, particularly for semi-transparent cirrus clouds (Szejwach, 1982). With three channels, the data may be displayed as two two-dimensional histograms. Figure 3.20a,b shows the clusters obtained for an active portion of the ITCZ over Africa and the tropical Atlantic. Comparison of Fig. 3.20 with Fig. 3.19 reveals that classes 1, 4 and 5 have small variance and correspond, respectively, to the edges of convective masses and semi-transparent cirrus. They are classes difficult to separate visually. Note for the histogram of the IR and water vapor channels (Fig. 3.20b) that the classes are stretched and exhibit a more linear relationship. Moreover, the detectability of clouds from snow- or ice-covered surfaces (Kidder and Wu, 1984; Ebert, 1989) can also be improved if the third channel is in the reflective, or near-IR since the clouds do not exhibit strong wavelength dependence in this band. This may be particularly critical for thin cirrus clouds (McGuffie and Henderson-Sellers, 1986) and for which the relationship between emissivity and cloud thickness is poor. Arking and Childs (1985) describe an algorithm used to retrieve cloud parameters, notably cloud fraction within the satellite field of view (FOV), cloud optical thickness, CTT, and a cloud microphysical parameter, using three of the NOAA AVHRR channels (0.73, 3.7 and 1.1 μm). The cloud microphysical parameter includes such properties as particle size, shape and thermodynamic phase, and is determined from the 3.7 μm band. Information in this channel helps constrain the estimates of CTT and the cloud optical thickness derived from the other two channels.

Many modifications and combinations of the statistical cloud retrieval methods just described have been developed. Minnis and Harrison (1984a,b,c) developed a *hybrid bispectral threshold method (HBTM)* for use with geostationary visible and infrared data. This technique takes advantage of the high frequency (hourly) of the geostationary data and the ability to evaluate the diurnal cycle of brightness temperature and cloud cover. It has also been used to investigate the ability to detect thin cirrus clouds (Minnis *et al.*, 1987). Clear-sky brightness (visible) and temperature (IR) data are modeled for different surfaces and for different times of the day. Clouds at different levels are retrieved by thresholding the IR T_{BB}, and maps of mean cloud fraction for the total cloud and at different levels can be derived.

(c) Cloud detection and surface type

The ability to derive reliable cloud information from satellites is also dependent, to a large degree, on the nature of the underlying surface. The task is relatively easy over the ocean, but much more difficult over land with its characteristic large spatial variations in reflectance, emissivity and IR temperature. Even over oceans, thin cirrus may go undetected (Matthews and Rossow, 1987). The task becomes considerably more

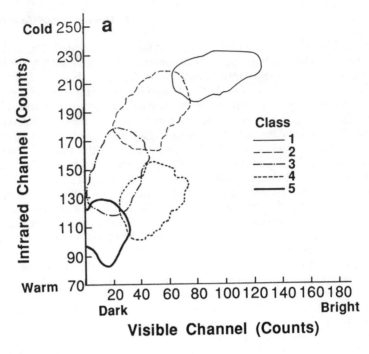

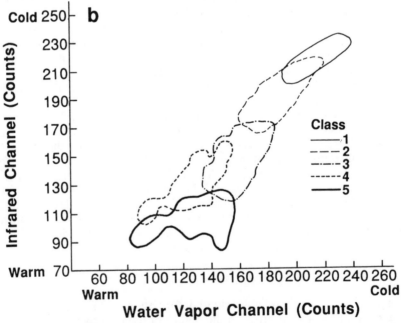

Fig. 3.20a, b Bidimensional clusters for clouds using (a) visible–IR combination, and (b) IR–water vapor channel combination. The classes are defined as (1) large opaque convective clouds; (2) edges of (1); (3) semi-transparent cirrus; (4) low clouds; (5) land and sea. (From Desbois *et al.*, 1982.)

difficult over high albedo land surfaces (e.g., deserts) in visible wavelengths and especially over snow and ice in both visible and IR (Tsonis, 1984). Thus, the early satellite cloud climatologies that utilized composite minimum brightess (CMB) techniques (see, e.g., Miller and Feddes, 1971) showed the subtropical deserts to be quite 'cloudy'; a result of their high surface albedo. The problem of cloud detection in the high latitudes using satellite information was apparent in the cloud climatology of Henderson-Sellers and Hughes (1985) and Hughes and Henderson-Sellers (1985), which showed chaotic variations over snow and ice surfaces. These studies utilized the US Air Force (USAF) three-dimensional (3-D) or real time (RT) nephanalyses. The 3-D nephanalyses integrate surface-observed, radiosonde, aircraft and other data with the satellite (DMSP) information. The cloud detection problem in the Arctic has also been examined by McGuffie *et al.* (1988), who point up the lack of comparability of different satellite cloud climatologies and the inability of visible and IR data for cloud mapping over some Arctic surfaces. The performance of the RT nephanalyses in the climatically sensitive marginal cryosphere of the East Canadian Arctic and Greenland was investigated explicitly by Barry *et al.* (1984). These researchers found substantial differences in cloud amount and variability between the oceans and ice-covered surfaces in this region, but were unable to confirm to what degree this is real or an artifact of the RT product. Certainly, the central Arctic is less cloudy than the ocean zones in spring, but probably not in summer when extensive stratus clouds occur (Kukla and Robinson, 1988). A more definitive appraisal of the performance of the RT nephanalysis in the marginal cryosphere of Alaska and northern Canada has recently been undertaken (McGuffie and Robinson, 1988). The RT product was compared with both the Air Force snow-cover model and independent analyses of the DMSP imagery during the early melt season (April, May) of 1984. The strong influence exerted by conventional surface observations in the RT product was noted, as was the likely effect of occasional discrepancies in the snow-cover extent. The need for further studies of surface–cloud interactions in the marginal cryosphere is clear.

Manual interpretation of cloud cover over high albedo surfaces in the Arctic forms the basis of studies by Barry *et al.* (1987) and Kukla and Robinson (1988), who used the DMSP visible and IR image pairs. In both studies, the former for spring (April–June) and the latter for late spring and summer (mid-May to mid-August), the surface was differentiated from the cloud largely on the basis of recognizable features of the Arctic pack ice, such as leads, and also by cloud textural attributes. The Barry *et al.* (1987) study identified associations between the distributions of middle-level clouds and synoptic circulation types, and also reported an overestimate of the low cloud fraction compared with previous studies for April. They suggested that this association arises from the effects of Arctic haze; however, this was disputed by Curry (1988), who favors an ice crystal cause for the discrepancies in optical depth occurring at that time. For both DMSP climatologies, the cloudy nature of the Arctic in spring appears to accompany, rather than to precede, the regional cloud maximum, thereby ruling out local sources of moisture as important in early cloud development.

Automated methods to retrieve cloud over high albedo surfaces have tended to make use of the NOAA AVHRR channel 3 (3.7 μm), either alone (Kidder and Wu, 1984) or in combination with one or more of the other visible and IR channels. These methods have met with considerable success when 'verified' against subjective interpretation of scenes and surface observations, and when used in a supervised classification. Even the detection of thin clouds may be possible when the radiance differences between channels 3 and 4 are used in conjunction with the difference values of channels 4 and 5 (thermal IR), except perhaps over the high, cold interior of Antarctica (Yamanouchi *et al.*, 1987). Key and Barry (1989) demonstrate the utility of combining the AVHRR (for cloud) and Nimbus-7 SMMR (for surface character) in cloud detection over polar surfaces.

The ability to model the surface to obtain a reliable clear-sky (minimum brightness, maximum temperature) non-cloud-contaminated value is a critical step in the ISCCP global cloud climatology (Rossow *et al.*, 1985; Rossow, 1989). Physical models of expected clear-sky brightness can be calculated given certain known parameters, particularly the sun–earth–satellite geometry and the type of underlying surface (see, e.g., Minnis and Harrison, 1984b). Ideally, diurnal variations of the clear-sky brightness may be determined since even land surface albedo is partly dependent on solar elevation as well as other surface-dependent factors (e.g., soil moisture for bare soils). In IR wavelengths, the determination of a clear-sky temperature is required, and this too can be model estimated (Minnis and Harrison, 1984b).

(d) Comparisons with surface observations of clouds

The range of satellite orbital configurations, sensor types, and cloud retrieval algorithms has forced the realization that a single homogeneous global cloud climatology may not be possible, nor particularly desirable; especially given the large number of potential applications involved (Hughes, 1984). The utility of having more than one global satellite cloud climatology is implicit in the two main programs that have developed to study clouds: the ISCCP and the Nimbus-7 Global Cloud Climatology. The ISCCP began in July 1983 for a projected five-year period, while the Nimbus-7 climatology is a six-year data set begun in 1979. A key element of ISCCP involves validating the satellite-derived cloud parameters with surface-observed cloud character-istics, and this has necessitated various field experiments for marine stratus (FIRE) and continental cirrus clouds (Cirrus IFO), both in the United States and elsewhere. The general problems involved in comparing satellite- with surface-based observations of a particular weather or climate variable have already been discussed (Chapter 1, Section 6). Satellite-derived and surface observations of cloud are most nearly comparable for circular areas having a diameter of about 50 km, or about $\frac{1}{2}°$ latitude (Barrett and Grant, 1979; Henderson-Sellers *et al.*, 1981). This is similar to the cell-size employed in the USAF RT nephanalysis. For clouds at a single level, the differences in cloud amount between the satellite and surface observer include observer bias, such as a preference for reporting 3, 5 or 7 oktas cover; the time of day of the observations (day versus night); and the fact that clouds towards the observer's horizon are viewed increasingly from the sides and give the impression of increased coverage of the sky dome (Malberg, 1973). The use of all-sky cameras has been advocated as a means to help reduce these observer biases. The result is that surface observers have tended to over-estimate cloud cover slightly compared with the satellite retrieval on a climatological (monthly, annual-averaged) basis (see, e.g., Barnes, 1966), especially over land (Minnis and Harrison, 1984a). The differences also tend to be greater at low than at high latitudes (Table 7; also Malberg, 1973; Minnis and Harrison, 1984a). This apparently results from the greater occurrence of cumuliform clouds and partly cloudy skies in lower latitudes. Thus, in addition, there may be satellite–surface observer differences related to cloud *type*. A surface-based observer may be able to detect very thin cirrus and cirrostratus that is missed by the satellite sensor (Henderson-Sellers *et al.*, 1987). There is considerable evidence for the underreporting of this cloud type by surface observers when compared with all-sky camera photographs, at least over the ocean (McGuffie and Henderson-Sellers, 1988). However, on climatic time scales, good agreement was found for cloud amounts using Landsat data for Britain compared with surface-observed and satellite analysis. This extended over all clouds combined and also when separated according to broad cloud type (Barrett and Grant, 1979). At the same time, cloud cover amounts that change on very short temporal scales may not be captured by 3- or 6-hourly standard synoptic observations.

Table 7 The mean annual differences in cloudiness between ground observations and satellite observations averaged over 10° latitude belts

60°–70° N	50°–60° N	40°–50° N	30°–40° N
0.94	1.11	1.30	1.48

Source: Malberg, 1973.

Comparisons between all-sky camera and SMS (Synchronous Meteorological Satellite) cloud amounts have been made for the GATE (GARP Atlantic Tropical Experiment) observing periods (Ackerman and Cox, 1981). Differences in total cloud varied in magnitude between observing periods. Estimates of the vertical structure of the cloud field by the two methods are only really comparable with the assumption of non-overlapping bases or tops, and for small cloud amounts. A study of the 3-D/RT nephanalysis data for the North Atlantic (40°–60° N) in January 1979 (Tian and Curry, 1989) shows that assumptions of random overlap of cloud layers may not be generally applicable. For adjacent layers that contain cloud, the assumption of maximum overlap may be the best in determining total cloud amount from satellite data (Tian and Curry, 1989).

Results from the ISCCP Special Study Area in Europe, conducted for the period July 22–August 10, 1983, show that 64% of the differences between surface-observed and satellite (Meteosat)-derived estimates of *total* cloud are within ±1 okta (Henderson-Sellers *et al.*, 1987). However, when considering individual *layers*, the accuracy drops to 50% for the high cloud fraction compared with 64% for the low clouds. The accuracy of retrievals is greatest under wholly clear or cloudy skies (Fig. 3.21), and this underlines the strongly bimodal structure of total cloud cover noted elsewhere. It has been suggested (Henderson-Sellers *et al.*, 1987; Sèze *et al.*, 1986) that the determination of lower-level cloud from satellite altitudes when obscured by higher cloud might be successfully made using surface-based cloud climatologies such as that of Warren *et al.* (1985). These authors defined the probabilities of the co-occurrence of different cloud types according to geographic location, from surface observer data. Data such as these could be used to obtain the probabilities of obtaining a particular cloud type for a

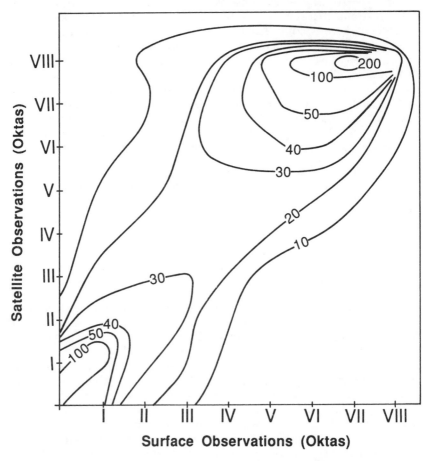

Fig. 3.21 Scattergraph showing the relationship between total cloud amount retrievals (oktas) for surface observer and satellite retrievals during the ISCCP regional validation study for Europe (1983). (From Henderson-Sellers *et al.*, 1987.)

given satellite retrieval of other type(s) (see Goodman *et al.*, 1990). For example, there is a tendency for cirrus and altostratus to occur together but not cumulus and altostratus. Also, there are some strong differences in cloud type co-occurrences over land compared with ocean and also with latitude.

Satellite detection of clouds and comparisons with surface observations are also complicated where surface characteristics change markedly over short distances, as along coasts and over surfaces of variable reflectance; especially since cloud amounts also tend to vary abruptly between land and ocean in the same latitude (Henderson-Sellers, 1978; Warren *et al.*, 1985; Minnis and Harrison, 1984a). The problem becomes more acute as the sensor spatial resolution increases, since there is increased probability of pixels having variable Earth surface characteristics. Combinations of two or more spectral bands show promise for distinguishing cloud-free from cloud-contaminated pixels in these high-resolution data (Sakellariou and Leighton, 1988).

3 Satellite cloud climatologies

Cloud climatologies incorporating a satellite component have been derived in any of four ways. These are:

(1) multiple exposure of visible and infrared images to produce mean, maximum and minimum cloud amount distributions;
(2) subjective manual interpretation of imagery, or nephanalysis;
(3) automated or semi-automated (supervised) cloud retrieval using threshold and statistical techniques applied to the satellite radiation data, most notably for ISCCP and the Nimbus-7 cloud climatologies, and also as part of other initiatives (e.g., ERBE);
(4) satellite and conventional observations blended into one product; most notably the USAF RT nephanalysis.

Climatologies derived in any of these four categories are likely to have problems that stem from, but are not limited to, the following.

(1) The presence or absence of surface observations, their density, and the way in which they are combined with the satellite fields.
(2) The nature of the underlying surface, particularly in marginal zones, and how this changes temporally.
(3) The sensor spectral characteristics. These include spectral sensitivities; vertical versus limb sensing; broad-band versus narrow-band; wide field of view (WFOV) versus narrow field of view (NFOV); the sensor spatial and temporal resolutions, and whether different sensors are combined to produce a single climatology. For example, the width of the DMSP IR channel changed after June 1979 from 8–13 μm to a narrower 10.4–12.5 μm.
(4) The cloud retrieval algorithms used, and whether changes in the algorithm have occurred through time, as with the USAF RT nephanalysis (McGuffie *et al.*, 1989).

(5)   The effects of atmospheric aerosol loading in certain regions, such as the Sahara or the Arctic (see, e.g., Shine *et al.*, 1984; Minnis and Harrison, 1984b; Barry *et al.*, 1987), which influence the atmospheric optical depth.

The following summarizes the development of satellite-based cloud climatologies. A comprehensive review of these and intercomparisons between satellite and conventional cloud analyses is given by Hughes (1984). The following overview shows that, despite the many potential sources of error and lack of homogeneity among analyses, most cloud climatologies agree on the broad-scale features of the atmospheric circulation and on representation of its seasonal and interannual variability (see, e.g., Koenig *et al.*, 1987). On the other hand, the magnitudes of cloud amount (total, by layer) tend to differ between cloud climatologies, and this reflects the different sensors and techniques used (Hughes, 1984). At the same time, these differences may have important consequences for estimates of the planetary albedo, the global radiation balance, and the cloud-climate sensitivity (cf. ISCCP and Nimbus-7 cloud climatologies: Hwang *et al.*, 1988).

(a)   Climatologies based on multiple image photographic techniques

Early satellite cloud climatologies made extensive use of compositing techniques to extract the broad-scale cloud cover features of Earth's atmosphere. These essentially extract that component of the planetary albedo that is transient (clouds) from the more conservative changes associated with the underlying surface. Multiple exposure of images (Booth and Taylor, 1969; Kornfield *et al.*, 1967; Bristor *et al.*, 1966) or computer processing of brightness counts in grid cells (Taylor and Winston, 1968) over a given time period (e.g., 3, 5, 15, 30 days) were made to extract either the 'average' or the 'lowest' reflectance. This is a threshold technique. The former method results in a composite average brightness (CAB) mosaic that displays the *persistence* of cloud cover for the period, as well as high albedo surfaces, such as snow/ice and also deserts (Kornfield *et al.*, 1967). These data are useful for identifying features such as monsoon cloudiness and semi-permanent cloud bands. The latter indicate extratropical storm tracks, the ITCZ, and zones of tropical–extratropical interaction in both hemispheres (Taylor and Winston, 1968; Kornfield and Hasler, 1969; Streten, 1973).

The composite minimum brightness (CMB) mosaic essentially displays the surface albedo, except in areas where persistent clouds have occurred over the averaging period. These proved useful for studying the snow and ice covers in both hemispheres (Streten, 1968), prior to the advent of passive and active microwave sensing that penetrates cloud cover and the polar darkness. A threshold brightness technique was applied to TIROS III vidicon photographs by Arking (1964). Despite the variations between orbits in the threshold value and other limitations of the TIROS instrument and its data coverage, the resulting zonal mean global cloud cover distributions for summer 1961 compared favorably with earlier conventional cloud climatologies over low and middle latitudes.

A three-year (1967–70) global meso-scale (40 km^2 resolution) atlas of relative cloud cover (brightness) was derived by Miller and Feddes (1971). Maps of mean relative cloud cover and classes of cover (in oktas) were presented for both hemispheres. The locations of the ITCZ, middle-latitude storm tracks and summer monsoon cloudiness are particularly well displayed in these analyses.

The multiple exposure technique has also been applied to TIR imagery. In this case, the product is a composite maximum temperature (CMT) display that is useful for identifying freezing compared with above-freezing surface temperatures. A contamination problem occurs when 'warm clouds' located in inversions overlie cold, often snow- and ice-covered surfaces.

(b) Climatologies based on nephanalysis

Cloud climatologies have also been derived based on subjective interpretation (nephanalysis) of the original image data. Thus, there is an additional step in the development of the cloud climatology involving someone other than the cloud climatologist. This additional step introduces a potential bias; in the estimation of cloud coverage by the nephanalyst and also when converting the neph codes into actual cloud amounts (Clapp, 1964). The former tends to overestimate large cloud amounts and underestimate small cloud amounts, while the latter may be reduced if consistency in the technique is assured (see, e.g., Bjerkness et al., 1969). Spatial variations in data coverage of the nephanalyses are an important limitation (Clapp, 1964). The nephanalysis has been applied particularly to data-remote lower-latitude oceans (see, e.g., Godshall, 1968; Bjerknes, 1969), culminating in the atlases of Sadler (1969), Sadler et al. (1976), and the HRC climatology of Garcia (1985). For areas where they overlap, these climatologies are broadly comparable. They can be used to identify relationships between observed cloud cover features and atmospheric and oceanic features (Saha, 1971) and their anomalies.

Figure 3.22 compares summer zonal cloud amounts in the Arking (1964) and Clapp (1964) satellite estimates with the cloud climatology derived using conventional data by Landsberg. Differences are large in the tropics and Southern Hemisphere and small in the northern subtropics and extratropics. Some of these differences may arise from anomalies in the general circulation, while others result from differences in data density and their analysis.

Much more recently, subjective interpretation of higher-resolution visible and IR imagery from the NOAA AVHRR and DMSP polar orbiters has resulted in climatologies of Arctic cloudiness, mainly for the spring and summer seasons (Barry et al., 1987; Kukla and Robinson, 1988).

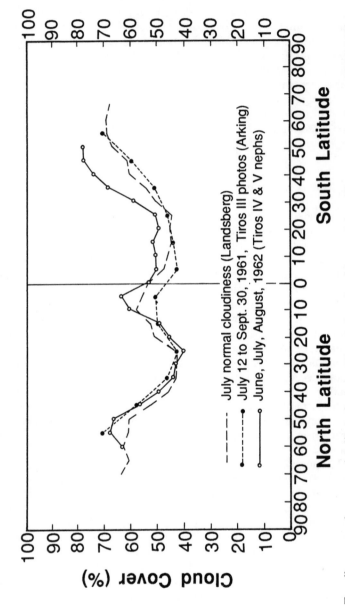

Fig. 3.22 Zonally averaged summer cloudiness from satellite- and surface-observed data. The curves shown are: TIROS III data for July 12–September 30, 1961 (short-dashed curve with heavy dots) after Arking, 1964; surface cloud climatology (dashed curve) after Landsberg; and TIROS V and VI data for June–August 1962 (solid curve). (From Clapp, 1964.)

(c)   Climatologies based on automated retrieval of cloud parameters

Much has already been said concerning the more recent cloud climatologies derived using automated retrieval techniques. Cloud climatologies of this type are mainly those developed in association with multi-year large-scale observing programs (Nimbus-7 Global Cloud Climatology, ISCCP). At the time of writing, the ISCCP climatologies are more regional in scope, partly because of the ongoing intensive field programs to validate oceanic stratus and continental cirrus clouds. However, the Nimbus-7 climatology is essentially complete (Stowe *et al.*, 1988, 1989). Note that these climatologies would not be expected to be identical owing to the differences in sensors and satellites used and time periods investigated, although there are periods of common overlap (Hwang *et al.*, 1988; Duvel, 1988). Here we will note some characteristics of these cloud climatological studies.

(1) *NOAA SR cloud climatology*   Radiance information in the narrow-band visible and IR channels of the NOAA SR is the basis of an ongoing global cloud climatology (Rossow *et al.*, 1989a,b). Surface and cloud effects on the daytime upwelling radiation in both channels are first separated by a minimum brightness/temperature threshold technique, and cloud optical properties are then determined by comparing satellite radiances with model simulations that account for the effects of satellite viewing geometry. Thus, twin products of surface climatology and cloud climatology are derived. The ultimate objective of this climatology is to help improve climate modeling and to facilitate an appraisal of the cloud-radiative forcing (Chapter 6). Comparisons of the cloud amount results for January and July 1977 with existing cloud climatologies (Fig. 3.23a,b) show some differences, especially over polar regions in July and the northern mid-latitudes and subtropics in January. Rossow *et al.* (1989a) ascribe these differences to the cloud retrieval algorithm used in their analysis, rather than to real changes in the atmospheric circulation. Further validation is required.

(2) *Nimbus-7 Global Cloud Climatology*   Unlike the NOAA and ISCCP climatologies, the Nimbus-7 Global Cloud Climatology is complete and covers the period April 1979–March 1985 (Hwang *et al.*, 1988; Stowe *et al.*, 1988, 1989). Cloud retrieval utilizes a bispectral algorithm performed on radiances in the 0.37 μm TOMS and 11.5 THIR wavelengths, and conventional data are also incorporated. Unlike the NOAA SR climatology, night-time (near midnight) as well as daytime (near noon) cloud cover is retrieved; the former tending to give greater cloudiness globally (about 56% compared with 49%) and indicating a substantial diurnal cycle. This data set shows the cloud cover maximum in the summer hemisphere and minimum in the winter. Comparisons of the July 1983 zonally averaged cloud amounts with those for the modified ISCCP (Stowe *et al.*, 1989) show broadly similar shapes to the curves, but greater (reduced) Nimbus-7 amounts for the tropics (higher middle latitude of the Southern Hemisphere) (Fig. 3.24). The differences are viewed as a function of the different cloud detection algorithms used, as well as the different spectral sensitivities of the satellite sensors.

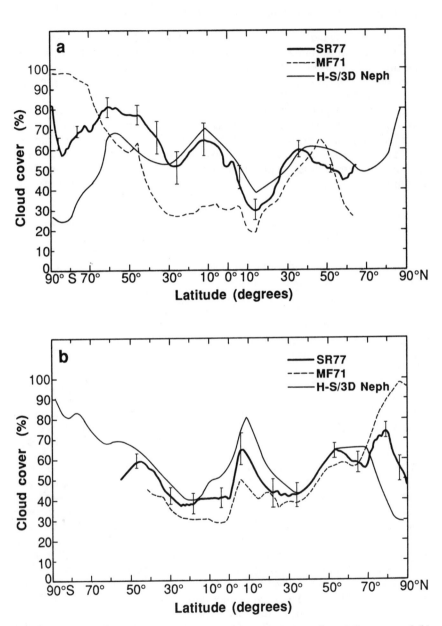

Fig. 3.23a, b Comparisons of zonally averaged cloud cover for (a) January and (b) July, from different satellite sources. These are NOAA-5 SR, 1977; Miller and Feddes (1971); Henderson-Sellers (1986). (From Rossow *et al.*, 1989.)

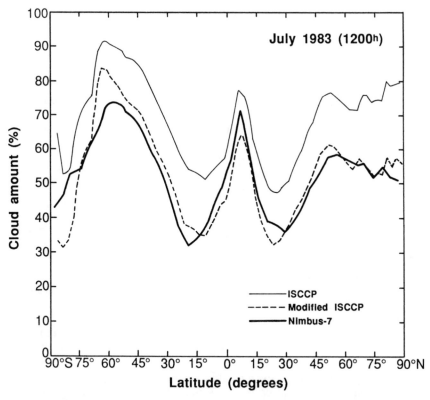

Fig. 3.24 Comparisons of zonally averaged daytime cloud amounts (local noon) during July 1983 for the Nimbus-7 preliminary ISCCP and modified ISCCP estimates. (From Stowe *et al.*, 1989.)

(3) *International Satellite Cloud Climatology Project (ISCCP)* In keeping with the international character of this programme, several sub-climatologies of satellite-observed cloud covers are becoming available that utilize one satellite platform. In particular, the Europeans have published ISCCP-related climatologies using Meteosat data (Raschke *et al.*, 1987; Duvel, 1988; Coulmann *et al.*, 1986). Since high-resolution sampling of the diurnal cycle is possible, Duvell (1988) showed that the time scale of cloud cover fluctuations for the METEOSAT viewing area between about ±50° latitude is shortest over convective areas (1–2.5 days) and greater than 9 days over subsidence areas. The ISCCP climatology will continue to evolve over the next few years. Although not explicitly connected with ISCCP, Kodama and Asai (1988) used three years' (1984–6) data from the Japanese GMS to describe the characteristics of clouds with tops above the 700 mb level and their association with dominant features of the atmospheric general circulation.

(d) Climatologies based on integrated satellite–conventional data

Complementary to satellite cloud climatologies derived exclusively from satellite information (ISCCP, Nimbus-7) are those where there is substantial inclusion of surface and other conventional observations of clouds. The best-known operational product of this kind is the USAF 3-D (RT) nephanalysis. Cloud retrieval is based essentially on a threshold technique that undergoes various statistical manipulations (Hughes and Henderson-Sellers, 1985; Bunting *et al.*, 1983). The final multi-level RT product for 46 km resolution boxes is the result of considerable 'massaging' of the satellite cloud and conventional meteorological fields, including manual adjustments ('boguss-ing'). Substantial use has been made of the RT neph data for comparison with other satellite cloud climatologies (Rossow, 1989a; Stowe *et al.*, 1989), as well as for more regional-scale investigations (Schulz and Samson, 1988; McGuffie *et al.*, 1989). The year 1979, which coincides with the FGGE intensive meteorological observing period, has been extensively studied (Henderson-Sellers and Hughes, 1985; Hughes and Henderson-Sellers, 1985; Henderson-Sellers, 1986; Koenig *et al.*, 1987; Tian and Curry, 1989). The climatology captures the broad-scale zonal cloud cover features associated with atmospheric circulations and their seasonal variations (Fig. 3.25). These include the cloud minima in the subtropics, cloud maxima in the mid-latitude westerlies, and seasonal migration of the cloudy ITCZ. However, problems were detected in the RT climatology. The temporal variability of cloud amount is influenced by the inclusion of surface observations of clouds in data-rich areas. Also, the assumption of an emissivity of 1.0 in the algorithm influences particularly the retrieval of cirrus clouds, especially in the tropics (Henderson-Sellers, 1986). These have apparently resulted in an overestimation of the summer monsoon cloudiness over South-east Asia, underestimation of the marine stratus off the west coasts of the Americas and Africa and, in particular, the chaotic variations in cloud cover over the Arctic. Similar seasonal and geographically dependent discrepancies to those found by Henderson-Sellers and Hughes are noted by Koenig *et al.* (1987) who, moreover, point out the biases in cloud amounts for different layers. In the tropics especially, the RT cloud retrieval algorithm underestimates the high cloud and overestimates the thickness of the middle-level clouds, at least for January and July of 1979.

(e) The cirrus cloud/high cloud fraction: climatologies

The importance to the atmospheric and surface radiation and energy budgets of cirrus clouds has been noted earlier. So too has the attendant difficulty in the accurate satellite retrieval of thin cirrus and the T_s. Using GMS data, Nitta (1986) showed the climatic associations of high-level (above 400 mb level) clouds in the western Pacific with tropical convection, extratropical cyclone activity, SSTs and dominant teleconnection (ENSO) patterns for the 1978–83 period. While some advances have been made with regard to

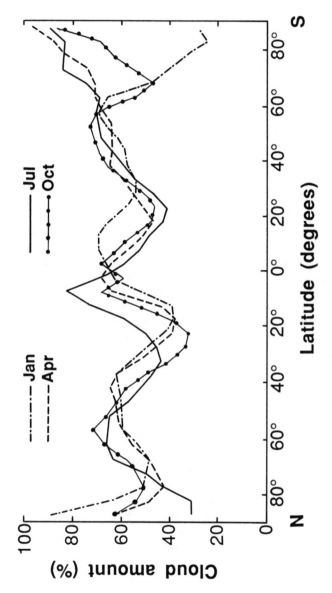

Fig. 3.25 Area-weighted zonally averaged 3-D nephanalysis cloud amounts for mid-season months of 1979. (From Hughes and Henderson-Sellers, 1985.)

detecting thin cirrus using water vapor absorption data in conjunction with TIR (Desbois *et al.*, 1982), other attempts at detecting these clouds and obtaining climatologies of cirrus clouds have adopted completely different approaches. These include the use of data from limb sounding sensors (see, e.g., Woodbury and McCormick, 1983, 1986) and vertical temperature and moisture sounders (Barton, 1983b; Susskind *et al.*, 1987; Prabhakara *et al.*, 1988; Wylie and Menzel, 1989; Menzel *et al.*, 1990). Barton (1983b) used data from two channels (2.63 μm and 2.7 μm) of the Selective Chopper Radiometer (SCR) of Nimbus-5 to detect 'upper-level' (cirrus) clouds independent of clouds at lower levels, for a period of $2\frac{1}{2}$ years and over latitudes between $+60°$ and $-60°$. The zonal variations are shown in Fig. 3.26 and reaffirm the shift to greater high-level cloudiness in the summer hemisphere. Maps showing the distribution of clouds obtained for the DJF and JJA periods showed maximum amounts over the maritime continent, and the strong seasonal variation corresponding with monsoon circulations and extensive cloud bands and jet streams. Interannual variations in cirrus cloud cover are shown to have a marked ENSO signal in the tropics and subtropics.

Data from the HIRS2/MSU sounders on the NOAA satellites were used by Susskind *et al.* (1987) to retrieve global-scale cloud parameters such as cloud fraction, effective cloud fraction (fraction × emissivity) and cloud-top pressure for June 1979. In this way, high-level clouds are apparent by their low cloud-top pressures; however, a distinction between high-level convective and cirrus clouds could not be made in this study. This is an important distinction since both the cloud emissivity and satellite estimates of precipitation are strongly influenced by the mode of cirrus formation. Cirrus clouds in a layer produced by advection appear broadly similar in visible and IR sensors to those derived from convection and the upper remnants of

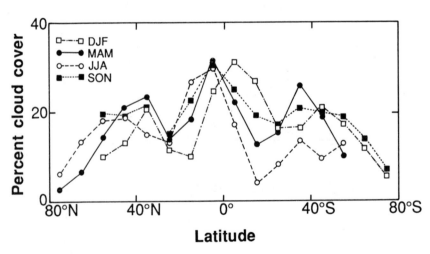

Fig. 3.26 Seasonal variations in the zonally averaged distributions of 'high cloud' amounts. (From Barton, 1983b.)

cumulonimbus clouds. Both day and night measurements can be made for estimates of diurnal variability (cf. Barton, 1983b).

Data on the extinctions of the solar disc for altitudes close to the tropopause at sunrise and sunset were measured by the limb-sounding SAGE (Stratospheric Aerosol and Gas Experiment) satellite. These have been used to derive a near-global climatology of high-level clouds by Woodbury and McCormick (1983, 1986). This technique has a coarse resolution (100 km²), a somewhat arbitrary definition of 'high cloud', and does not permit estimates

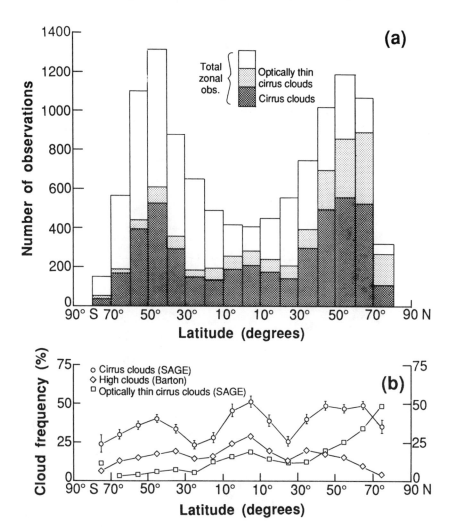

Fig. 3.27 Zonal statistics of high cloud/cirrus cloud for the period February 1979 to November 1981: (a) number of observations of SAGE cirrus and thin cirrus; and (b) frequency of occurrence of SAGE cirrus, SCR high clouds (from Barton, 1983b), and SAGE thin cirrus. (From Woodbury and McCormick, 1986.)

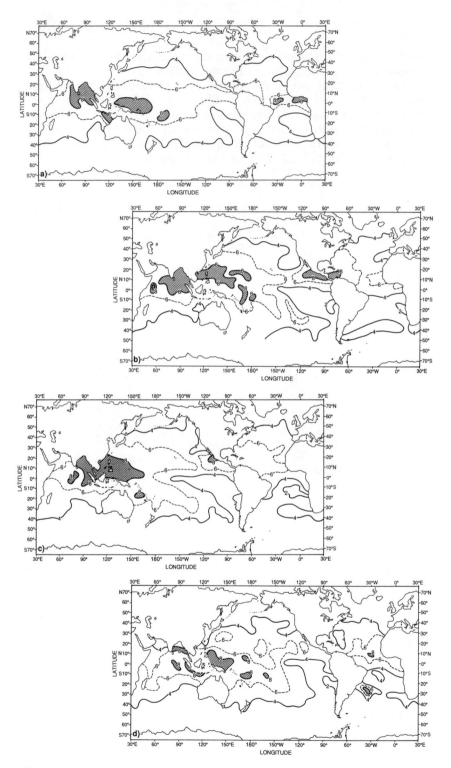

of cloud amounts. However, it does permit differentiation of thin as opposed to thick cirrus. Figure 3.27 shows the full 34 months (February 1979–November 1981) zonal cirrus cloud results, and compares them with the climatology of Barton (1983b). Relatively high frequencies of cirrus are found in the tropics, associated with active convection; and also in middle latitudes, associated with storm systems, fronts and jet streams. These associations were confirmed in the mapped (spatial) distributions of these clouds. The SAGE cirrus curves in Fig. 3.27 are broadly comparable with those of Barton (1983b) in shape, but are consistently greater in magnitude. These differences are probably related to the lower threshold required to register cirrus clouds in the limb-sounding method compared with the vertical-sounding technique. For the most part, the frequencies of optically thin cirrus are considerably less than for thick cirrus, but appear relatively high for the higher northern latitudes. However, the latter might also be explained by the presence of Arctic haze and volcanic dust (Woodbury and McCormick, 1986). Thin cirrus seems to exhibit little seasonal change compared with thick cirrus.

A climatology of thin cirrus clouds over the oceans between $+50°$ and $-50°$ latitude was developed by Prabhakara $et\ al.$ (1988) using two channels of the IRIS on Nimbus-4. The technique was validated by comparison with measurements taken using an aircraft radiometer and by radiative transfer calculations. The presence of thin cirrus can be determined from values of the difference in temperature between the 10.8 μm and 12.6 μm bands; the temperature difference being largest for small optical depths and when compared against relatively low brightness temperatures in the 10.8 μm channel. According to these criteria, thin cirrus is observed to occur about 100–200 km away from high-altitude cold clouds, apparently as a result of spreading. The seasonal distribution of these clouds is shown for April 1970–January 1971 in Fig. 3.28a–d. The results compare favorably with those of Barton (1983b) and Woodbury and McCormick (1986) and also with aircraft-level observations of 'high clouds' derived by Jasperson $et\ al.$ (1985).

Suggested further reading

Barrett, E. C., 1974: $Climatology\ from\ Satellites$. Methuen, London. 418 pp.
Henderson-Sellers, A. (ed.), 1984: $Satellite\ Sensing\ of\ a\ Cloudy\ Atmosphere:$ $Observing\ the\ Third\ Planet$. Taylor and Francis, London and Philadelphia. 340 pp.
Hobbs, P. V. and Deepak, A., 1981: $Clouds:\ Their\ Formation,\ Optical\ Properties\ and$ $Effects$. Academic Press, New York. 497 pp.

Fig. 3.28 (opposite) Distribution of optically thin cirrus by season for the year 1970–1. These are: (a) April and May 1970; (b) June, July, August, 1970; (c) September, October, November, 1970; (d) December 1970 and January 1971. (From Prabhakara $et\ al.$, 1988.)

4 Cloud systems

A key legacy of the application of satellite cloud imagery to synoptic weather analysis and forecasting in the 1960s and 1970s were the insights that were provided into the dynamics of synoptic-scale systems, particularly cyclones (see, e.g., Boucher and Newcomb, 1962; Widger, 1964; Oliver and Anderson, 1969; Burtt and Junker, 1976). With the improved spatial and temporal resolutions of subsequent satellite sensors, a similar impact is occurring in regard to meso-scale systems. Foremost among these are the meso-cyclones in cold air streams ('polar lows') (see, e.g., Forbes and Lottes, 1985; Carleton and Carpenter, 1989, 1990; Heinemann, 1990), and the large thunderstorm complexes known as meso-scale convective systems (MCSs) (Maddox, 1980; Velasco and Fritsch, 1987; Augustine and Howard, 1988). Both groups of meso-systems interact strongly with the synoptic-scale flows. However, at still smaller scales are cloud streets, lee wave clouds, tornadoes, squall lines, and the rain bands associated with tropical and extra-tropical synoptic cloud systems (MPAs: meso-scale precipitation areas). Since the meso-scale is characterized by relatively short time scales (minutes to hours, up to about a day) and large ratios of vertical to horizontal motion, high-frequency imaging is necessary to ascertain reliably the features of these systems for climatological research. In this chapter we survey the meteorological and climatological aspects of synoptic and large meso-scale cloud systems that have been, and continue to be, revealed by satellite data.

1 Cold air vortices

Characteristic satellite-viewed cloud signatures are associated with organized convection; that is, where the local-scale vertical motions giving rise to individual cumulus and cumulonimbus clouds are augmented by frictional convergence of air at lower levels and divergence at upper tropospheric levels (a process known as CISK: conditional instability of the second kind). The latter is often in association with the jet stream. Accordingly, these meso-scale systems tend to display some degree of cyclonic rotation (positive vorticity). They include, particularly, the vortices occurring in cold air streams over the ocean (variously known as the 'polar low', 'cold air instability low', 'Arctic/Antarctic instability low'). Smaller-scale systems of similar appearance are observed over large lakes and inland seas in winter. For example, they have been studied in connection with heavy snowfalls in northwesterly flow for southern Japan (Sakakibara et al., 1988) and the Baltic (Andersson and Nilsson, 1990)).

(a) Lake vortices

Satellite imagery first revealed the existence of winter-time meso-scale cold air vortices over the Great Lakes, although their existence was argued on theoretical grounds much earlier (see discussion in Pease *et al.*, 1988). Forbes and Merritt (1984) studied the satellite aspects of Great Lakes meso-scale vortices for winters of 1978–82, and found a total of fourteen vortices. Vortices are most common over Lake Michigan, and consist predominantly of the spiraliform (multiple-banded) signature type. Great Lakes meso-scale vortices may produce little significant snowfall when moving ashore, and tend to be accompanied, on average, by SLP falls of only about 1 mb. While there is apparently little synoptic-scale forcing in terms of vorticity or thickness advection, most meso-scale vortices occur under conditions of weak gradient flow and with high pressure over or west of the region and a weak thermal trough over the lakes. The last is associated with diabatic heating. Lake vortices may develop from shoreline parallel cloud bands that result from opposing land breeze circulations and convergence over the central part of the lake. Over southern Lake Michigan, vortex stretching may be enhanced by the shoreline curvature, perhaps accounting for the large numbers of meso-scale vortices seen there. A detailed case study utilizing Landsat imagery and a numerical meso-scale model confirmed the role of surface heating in the spin-up of these circulations, and showed that this can be largely independent of the larger-scale atmospheric dynamics (Pease *et al.*, 1988).

(b) 'Polar lows'

(1) *Satellite characteristics*   Meso-vortices in cold air that develop over open ocean are often termed 'polar lows'. They tend to be larger than their counterparts over the Great Lakes, and available synoptic information suggests that they are generally deeper. This may result from a combination of enhanced synoptic-scale forcing and a greater fetch of the cold air over open water. The term 'polar low' has been used differently by researchers, but the term is now becoming increasingly recognized as referring to a wide range of possible polar air vortex signatures; not just the subsynoptic-scale 'comma cloud'. 'Polar lows' have been studied extensively using satellite imagery in combination with conventional meteorological and aircraft data (see, e.g., Shapiro *et al.*, 1987) for the Norwegian and east Greenland seas. They may be accompanied by high winds and heavy precipitation and, accordingly, are also known as 'Arctic hurricanes' or 'Arctic instability lows' (Businger and Reed, 1989). Some, but by no means all, of these systems may represent the explosive cyclogenesis events discussed by Sanders and Gyakum (1980). A special program to study the dynamics of these systems for the Norwegian Sea area was established and known as the Polar Lows Project (Rasmussen and Lystad, 1987; Shapiro *et al.*, 1987).

Figures 4.1 and 4.2 are high-resolution (2.7 km) DMSP images of two

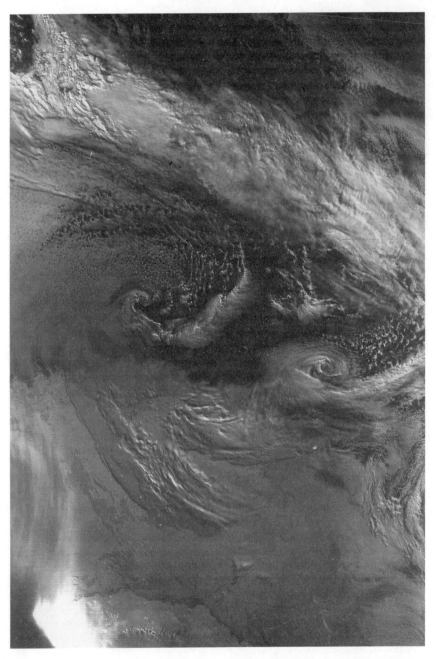

Fig. 4.1 Portion of a DMSP visible orbital swath (2.7 km resolution) for the southern Indian Ocean south of Africa. A line drawn horizontally across the image approximates 50° S; the lowermost border lies at about 70° S. Refer to the text for a discussion of cloud features.

Fig. 4.2 Portion of a DMSP visible orbital swath (2.7 km resolution) showing a shallow spiraliform polar air vortex in the southern Indian Ocean (center right).

types of polar air vortex forming over higher latitudes of the Southern Hemisphere in winter. They have a broadly similar appearance to those identified over the Northern Hemisphere oceans (see Businger and Reed, 1989). The two (inverted) comma cloud systems in Fig. 4.1 have formed just equatorward of the Antarctic sea ice margin, and poleward of the main polar front. The latter is the bright cloud band running from north-west to south-east in the upper part of the image. Note the change in cloud structure of the frontal band from predominantly convective elements in the north-west section to more cirriform overlying stratus in the south-east portion. The subsynoptic-scale comma clouds tend to be baroclinic features: they are initiated by PVA aloft, and development of the vortex and accompanying frontal band is enhanced by latent heat and sensible heat release from the ocean in the cold air (i.e., CISK). Polar air vortices of the type shown in Fig. 4.2, which have a dominantly spiral structure, have been shown in Northern Hemisphere studies to be shallow systems. Rasmussen (1981) first studied the spiraliform polar low type using satellite imagery. He suggested, from case studies, that these vortices may be predominantly warm-cored and form in an essentially barotropic atmosphere (i.e., be CISK-generated). The baroclinicity is confined more to the lower troposphere, at least in the early stages of development, and the role of diabatic heating at that time may be more pronounced compared with the comma clouds. A tendency for comma clouds to be more frequent than spiraliform polar lows over the North Pacific compared with the North Atlantic (see, e.g., Carleton, 1985b) may indicate real geographic differences in the relative importance of CISK and baroclinic instability (Sardie and Warner, 1983; Businger, 1985, 1987). In the North Atlantic, the elevated Greenland ice sheet and sea ice edge are juxtaposed with the North Atlantic Drift current, and this permits rapid modification of cold air masses and the spin-up of the spiral polar low. In the North Pacific, this situation does not exist (Businger, 1985, 1987). The geographical differences in the two meso-cyclone types now seem less distinct than previously thought; spiraliform polar lows occur over the North Pacific (Yarnal and Henderson, 1989a), and there is strong satellite evidence to support the non-exclusivity of the spiral and comma-cloud types. In a study of Antarctic meso-cyclones for the sector 100° E eastward to 80° W undertaken using DMSP imagery, it was found that about 15–25% of systems that were first identified as being either comma cloud or spiraliform changed signature to the other type within the first 12–24 hours of satellite-observed existence (Fitch and Carleton, 1990). These observations suggest a cooperation of both CISK and baroclinic instability in most polar low developments, but to slightly different degrees depending on location and signature type (cf. Sardie and Warner, 1985; Bonatti and Rao, 1987).

Other configurations of polar lows occur, some of which may not persist beyond a few hours. They are weaker systems than those depicted in Figs. 4.1 and 4.2 (Forbes and Lottes, 1985). However, examples of other, more persistent, polar low signature types include the multiple vortices that develop from an old cold-core low (e.g., Fig. 4.3) and which resemble mini-comma clouds in a 'merry-go-round' configuration (Zick, 1983). In addition there are spiraliform developments initiated on a 'boundary layer front',

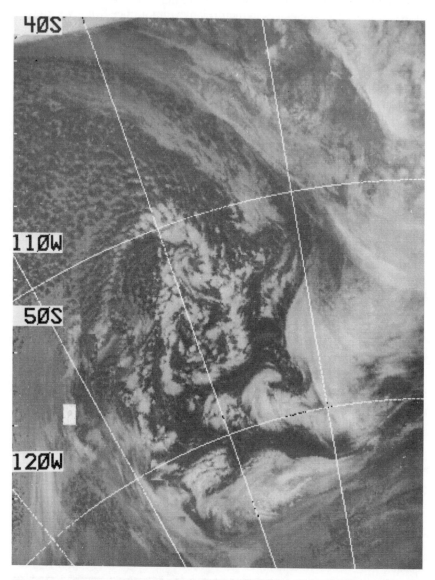

Fig. 4.3 Portion of a DMSP mosaic (5.4 km resolution) for the south-east South Pacific Ocean, showing multiple polar air vortices developing around an old cyclonic vortex (center right of image). A frontal wave cyclone is located to the south-west (lower center).

particularly near the sea ice margin (Fig. 4.4). Presumably these differences in signature type and mode of formation are indicative of different physical processes, or at least changes in their relative importance from type to type (Businger and Reed, 1989). Polar air vortices are currently the subject of much research, both in terms of the associated dynamics and their climatological regimes.

(2) *Classification and meteorological attributes* A satellite-based classification of meso-scale vortices was developed by Forbes and Lottes (1985) from analysis of satellite imagery for the north-east Atlantic in winter 1981–2. It was hoped that the identification of distinctive cloud configurations and the derivation of mean synoptic-scale parameters from conventional analyses could be used to improve the forecastability of polar lows. Table 8 presents the cloud configuration categories and also gives relative frequencies and SLP deficits (anomalies) for the major types. Some meso-cyclones, particularly the comma and spiral types (Category 1) continue to develop under favorable broader-scale conditions. These appear to include a moderate degree of baroclinicity, relatively weak winds, large lapse rates, and stronger PVA compared with non-developing systems. The results of this study were mixed from the standpoint of an ability to forecast which vortices would continue to develop.

Table 8 Observations by category of cloud pattern for North Atlantic polar lows

Category	Description	Total no. of observations	No. with ΔP	Mean ΔP
1	Comma, deep spiral	84	54	10.1 ± 6.1 mb
2	'Merry-go-round' or ring of vortices	3	3	$8.7\ (\pm 3.2)$
3	Crescent	17	13	3.5 ± 2.4
4	Oval, solid mass	10	3	$5.7\ (\pm 1.2)$
5	Multiple deep bands	108	34	4.0 ± 2.3
6	Multiple shallow bands	36	12	3.5 ± 2.2
7	Single deep band	2	2	$4.0\ (\pm 1.4)$
8	Single shallow band	1	1	0.0
9	Swirl in cumulus streets	72	26	2.9 ± 2.1
Strong comma	(diameter >3.5 deg. lat. or >0.5 revolution)	44	27	12.7 ± 6.6
Minor comma	(diameter ≤3.5 deg. lat. and ≤0.5 revolution)	40	26	7.0 ± 3.5

Source: Forbes and Lottes, 1985.

(3) *Climatologies* Analyses of meso-scale vortices for the North Pacific that have utilized satellite data have, until quite recently, tended to be case studies rather than climatologies. Exceptions were the studies by Reed (1979) and Mullen (1979). Some case studies have identified smaller-scale

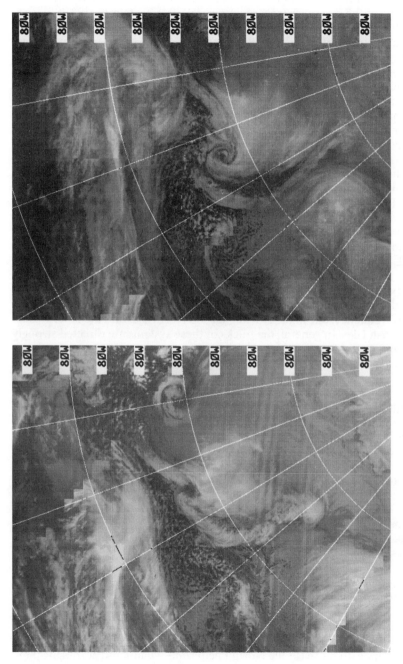

Fig. 4.4a ,b DMSP TIR sequence (5.4 km) for the south-east South Pacific showing (a) development of a polar air vortex on a 'boundary layer front' located just equatorward of the sea ice margin; (b) twelve hours later the spiral vortex of the polar low is clearly apparent in image center.

131

precipitation features within meso-cyclones; they help improve our understanding of the dynamics (see, e.g., Reed, 1979; Locatelli *et al.*, 1982; Reed and Blier, 1986a,b; Businger and Hobbs, 1987).

The North Pacific studies by Reed (1979), Mullen (1979) and Yarnal and Henderson (1989a,b) note the predominance of relatively large comma-cloud systems that are often deeply baroclinic. The climatology of Yarnal and Henderson is for seven cool seasons (1976/77 to 1982/83) using DMSP imagery. These authors found that about two-thirds of polar-air vortices continue to evolve past the incipient, cyclogenetic stage. However, there are differences in the degree of continued vortex development according to predominant type (about two-thirds of all comma clouds; two-fifths of spiral developments) and, accordingly, location. Broadly similar results are found for the Southern Hemisphere (Carleton and Carpenter, 1989, 1990). Although the satellite-observed spatial regimes of comma clouds and spiral polar lows overlap, and have maximum frequencies in the north-west Pacific basin, the spiral systems show a tendency to predominate slightly further east of the comma clouds.

Climatological studies of polar air vortices for the North Atlantic (Forbes and Lottes, 1985; Ese *et al.*, 1988) show a tendency for these systems to occur in deep cold air flows, and as part of a 'seesaw' in atmospheric circulation between Greenland and the Barents/Norwegian seas. For a 36-day period in the winter of 1981–2, Forbes and Lottes found two areas of high vortex frequency: one in the Norwegian and North seas and the other west of the British Isles. In general, the tracks of these systems are directed strongly equatorward, at least in the developing stages.

Over the Southern Hemisphere oceans, polar air vortices (mainly comma clouds) become more frequent with increasing latitude, and may account for up to 50% or more of cyclogenesis in the winter season when considered over all latitudes (Streten and Troup, 1973; Carleton, 1979). Recent studies of these systems using higher-resolution DMSP data (Carleton and Carpenter, 1989, 1990) show that comma clouds tend to predominate over a wide range of ocean latitudes, whereas the less commonly observed spiral type tends to occur in Antarctic latitudes near the sea ice edge (e.g., Fig. 4.5). Polar lows exhibit variations in intraseasonal (winter) and interannual regimes that are consistent with changes in Antarctic sea ice extents, SSTs and the phase of ENSO. Thus, high frequencies of polar lows over the Southern Hemisphere oceans in winter tend to occur in areas in which the sea ice margin is displaced anomalously equatorward for the season (many cold air outbreaks), the sea–air temperature difference is large (destabilization of cold air), and the phase of ENSO favors a trough over, and slightly to the east of, the region concerned. For the winters studied (1977–83), these active polar low areas oscillated between the southern Indian Ocean and the south-west Pacific, and this is consistent with observed changes in the long waves for those years (Carleton and Carpenter, 1990).

The climatologies of Carleton and Carpenter (1989, 1990) suggest broad similarities with those for the North Pacific and Atlantic. Accordingly, there is growing interest in investigating whether the dynamics of meso-cyclones (not just comma clouds) in the Antarctic region are similar to those in the

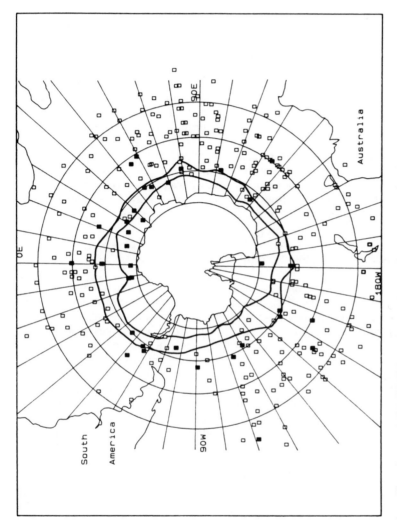

Fig. 4.5 Locations of satellite-observed (DMSP TIR 5.4 km resolution) polar air cloud vortices for the seven July months of 1977–83. Comma clouds are shown as open squares; spiraliform vortices as filled squares. The maximum and minimum extents of the sea ice for the seven July months are shown by heavy lines. (From Carleton and Carpenter, 1990.)

Northern Hemisphere. Fitch and Carleton (1990) have investigated meso-cyclone regimes for the half-hemisphere centered on the Ross Sea. Five winter and transition months in 1988 (March, April, June, August, October) have been studied from DMSP imagery. Figure 4.6 shows meso-cyclone tracks for the month of August. Large numbers of systems develop near the sea ice margin off East Antarctica, and these may be associated with strong katabatic cold air flows off the ice sheet. A similar mechanism may also operate for the Ross Ice Shelf frequency maximum. While the influence of the katabatic winds on meso-cyclone development in this sector is still under investigation, case studies suggest that katabatic surges may precede, by up to 72 hours, outbreaks of polar lows over ocean latitudes east and south-east of New Zealand. At that time also, most meso-cyclones occur on the cold side of a 1000–500 mb thickness anomaly minimum (cold pool) located north of the Ross Sea. The latter suggests a cooperation between the synoptic- and meso-scale circulations.

A similar type of climatological investigation has very recently been undertaken by Heinemann (1990) for the Weddell Sea region, only for the summer season. Most meso-cyclones tend to develop over the ice-free region, and are associated with offshore flow and a cold pool aloft. Extension of these Antarctic studies to include additional years is proceeding.

2 Meso-scale convective systems

Like the polar low, the MCS is a large meso-scale phenomenon that interacts with the synoptic-scale circulation. Also in common with polar lows, the MCS was first adequately described from meteorological satellite imagery, particularly in the IR (Maddox, 1980). MCSs are large regional-scale systems (length scales at least 250 km) and are composed of individual thunderstorm clusters. These cells of 1–2 hours' duration readily lend themselves to satellite analysis from high-frequency geostationary IR data (see, e.g., Adler and Fenn, 1979; Purdom, 1984; Velasco and Fritsch, 1987). Such data may be useful in identifying those storms which are likely to produce 'severe' (hail, tornadic) weather (Adler *et al.*, 1985). Increasingly, passive microwave data are providing information on deep convective activity, because of the association with heavy precipitation (Spencer and Santek, 1985). This is possible even over land where advantage is taken of the horizontal and vertical polarization of higher frequency microwave channels.

MCSs are 'right-moving' storms. Individual convective cells may be generated on the warm side of the system, and then propagate subsequently to the poleward side where they dissipate. In the US Great Plains and Midwest, this movement is often associated with a low-level (about 900–850 mb) wind speed maximum comprising a low-level jet (LLJ) that assists in the advection of moisture into the storm. However, the dominant direction of air flow at and above the mid-troposphere is usually from the southwest (i.e., a thermal wind exists) and is characterized by warm air advection, since the wind shear is clockwise with height. This gives a general storm direction in the Northern Hemisphere of west to east. These conditions are often met

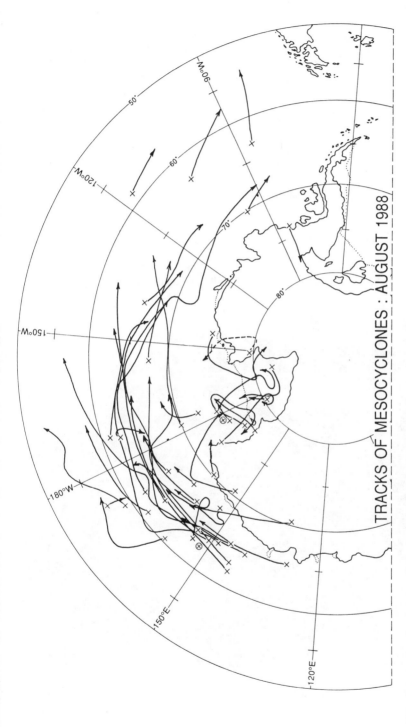

Fig. 4.6 Antarctic meso-cyclone tracks for the half-hemisphere from 100° E eastward to 50° W, as identified from twice-daily DMSP TIR mosaics (5.4 km resolution), August 1988. The beginning and end points of each system are shown, respectively, by the ×s and arrow-head symbols.

TRACKS OF MESOCYCLONES : AUGUST 1988

135

over middle-latitude land areas in the warm season and equatorward of a cold or quasi-stationary front with an associated jet maximum aloft. Enhanced upper-air divergence associated with the jet encourages thunderstorm development and lowered surface pressures. Persistence over time and the propagation of the MCS is encouraged by replenishment of warm moist air at lower levels via the LLJ. Accordingly, MCSs tend to be most intense during the night when the LLJ is characteristically strongest and the vertical temperature lapse has been steepened by nocturnal radiative cooling at cloud-top altitudes. A similar diurnal variation of CTTs and cloud development has been noted in satellite studies of tropical cyclones (Steranka *et al.*, 1984). Given the great vertical and substantial horizontal extent of MCSs, they are ideal subjects for satellite climatological investigation.

(a) Satellite characterization

A satellite example of an MCS is given in Fig. 4.7, which is a GOES-E IR image that has been enhanced to show the coldest and highest cloud tops.

Fig. 4.7 GOES-E enhanced TIR image (August 13, 1982). Prominent features of this image are the low CTTs associated with deep convective systems; particularly the MCS over the central US.

Grey-scale enhancement involves an expansion of a section of the blackbody temperature range to the full scale. This enables a certain temperature interval (e.g., that representative of only thick high and middle clouds, or SST gradients) to be examined at a finer grey-scale resolution. The large elliptical system in the central US is an MCS that has developed just south of the jet stream (note the linear cloud streaks to the north and north-west). This is the typical appearance of an MCS in its mature stages. The radiometrically cold feature along the west coast of Mexico may also qualify as an MCS, of the type identified in the climatological study of Velasco and Fritsch (1987). The high cold cloud tops of thunderstorm cells in the Intertropical Convergence Zone (ITCZ) are evident in the lower center and left of the image.

It is only quite recently that features resembling the MCSs of the US Great Plains have been identified and described for the desert south-west US during its summer 'monsoon' season of July and August. Figure 4.8 is an enhanced IR image of one such feature located over Arizona on August 12, 1982. These systems seem to form along a so-called 'monsoon boundary' (Adang and Gall, 1989) that separates drier air at mid-tropospheric levels over California and the eastern Pacific from the more humid air associated with the Bermuda High to the east. This feature can be clearly detected using

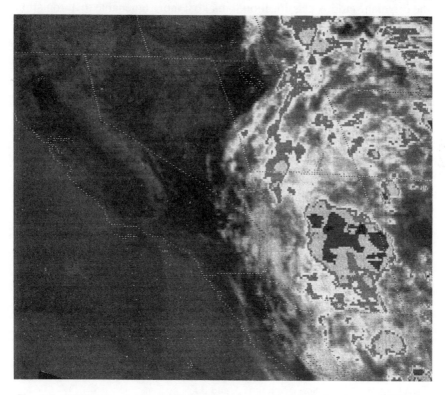

Fig. 4.8 GOES-W enhanced TIR image (August 12, 1982) showing an MCS-type feature located over eastern Arizona.

data in the water vapor absorption band. Owing to the generally drier air in the sub-cloud layer of this region, systems of this type tend not to produce the copious rainfalls typical of those in the more humid central US. A synoptic climatological study of these systems from enhanced GOES IR imagery has very recently been completed (Perry, 1990).

(b) Classification of MCSs

Maddox (1980) defined and used a classification of MCSs based on geostationary satellite-enhanced IR descriptors; in particular, size, shape (ellipticity) and CTT (refer to Figs. 4.7, 4.8, Table 9). Maddox's classification of MCSs into types A and B is made according to the size and areal extent of the lowest CTTs in the enhanced imagery: either $\leqslant -32$ °C, or $\leqslant -52$ °C. As a rule, MCSs appear characteristically egg-shaped at maturity with the major axis oriented in the direction of storm propagation (Fig. 4.7). Although Maddox specified a minimum duration time of 6 hours or longer for MCSs, some cases may persist for a couple of days before dissipating.

More recent studies (e.g., Augustine and Howard, 1988) have modified the original Maddox classification to make it more amenable to automation (Table 10). Thus, the warmer threshold temperature of $\leqslant -32$ °C has been dropped owing to the subjectivity involved in its determination, while the critical $\leqslant -52$ °C isotherm is retained. Using the satellite criteria, the time-dependent associations between enhanced IR signatures of 122 MCSs and precipitation have been documented for the central US by McAnelly and Cotton (1989). They find that, on average, MCSs produced a rainfall volume of 3.46 km^3 during their life cycle, over an area of 3.2×10^5 km^2 and at an average depth of 10.8 mm. In accord with previously observed relationships between cloud area and precipitation (see, e.g., Stout et al., 1979), the precipitation maximum occurs early on in the life cycle of MCSs. For the entire life cycle, the heaviest precipitation tends to be displaced some 50–100 km south of the centroid location of the IR cloud shield (i.e., where the mature thunderstorms predominate).

Table 9 Meso-scale convective system defining criteria

Size	A: Contiguous cold cloud shield (IR temperature $\leqslant -32$ °C) must have an area $\geqslant 100\ 000$ km^2
	B: Interior cold cloud region with temperature $\leqslant -52$ °C must have an area $\geqslant 50\ 000$ km^2
Initiation	Size definitions A and B are first satisfied
Duration	Size definitions A and B must be met for a period $\geqslant 6$ hours
Maximum extent	Contiguous cold cloud shield (IR temperature $\leqslant -32$ °C) reaches maximum size
Shape	Minor axis/major axis $\geqslant 0.7$ at time of maximum extent
Termination	Size definitions A and B are no longer satisfied

Source: Maddox, 1980.

Table 10 Modified meso-scale convective system (MCC) criteria

Size	Contiguous cold cloud shield (IR temperature $\leqslant -52\ ^\circ$C) must have an area $\geqslant 50\ 000$ km^2
Initiation	Size definition is first satisfied
Duration	Size definition must be met for a period $\geqslant 6$ hours
Maximum extent	Contiguous cold cloud shield (IR temperature $\leqslant -52\ ^\circ$C) reaches maximum size
Shape	Minor axis/major axis $\geqslant 0.7$ at time of maximum extent
Termination	Size definition is no longer satisfied

Source: Augustine and Howard, 1988.

(c) Satellite climatologies

Given the ease with which MCSs may be identified on enhanced IR imagery, summaries of the centroid locations and tracks of these features have been made on a more-or-less annual basis for the US (see, e.g., Rogers *et al.*, 1985). There is a seasonal variation in the locations of MCS development, being generally further south in the late spring and early summer, and further north in the mid- to late summer (Fig. 4.9). This movement is associated with the heating of the land mass and retreat of the upper westerlies during this time. Substantial interannual variations in MCS occurrence have been noted, apparently in response to the disposition of favorable larger-scale synoptic forcing patterns.

Extension of the satellite climatology of MCS into South America, where many of the same forcing conditions occur in summer, has been made by Velasco and Fritsch (1987) using GOES-E full disc imagery. These workers find that South American MCSs occur with about the same frequency as those in North America and they also have a nocturnal frequency peak (Table 11). However, the cloud shield is observed to be about 60% larger for these systems (Fig. 4.10). South American mid-latitude MCSs exhibit a similar seasonal variation to those over the US. Velasco and Fritsch (1987) also identify MCS-appearing features over the *tropical* land areas, and even over warm ocean waters (see also Fig. 4.7). The latter may develop into tropical storm vortices. There may also be an ENSO signal in MCS occurrence, associated with the development of the warm water anomaly off the coast of Peru. The 1982–3 'warm' event was associated with a doubling of MCSs over South America.

MCSs over the south-west US and north-west Mexico (refer to Fig. 4.8; also Velasco and Fritsch, 1987) are associated with the summer rainfall singularity over the deserts, and the occurrence of a weaker but still persistent version of the Great Plains LLJ (Tang and Reiter, 1984; Carleton, 1985a). The seven-summer (July 1–September 15) climatology of Perry (1990) shows that MCSs tend to form near the north-west Mexico highlands. They track generally from south-east to north—west around the Bermuda

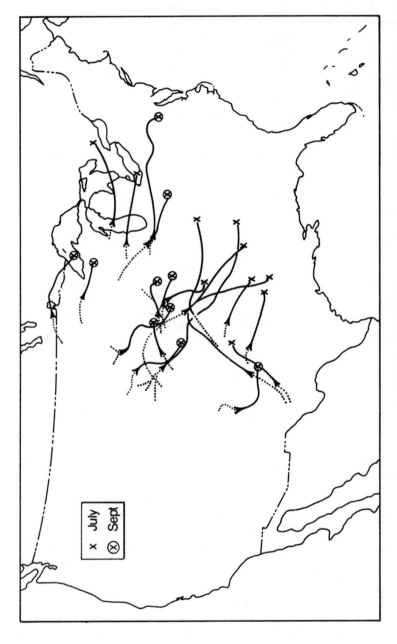

Fig. 4.9 Tracks of MCSs for the months of July and September 1985. Dotted portions of the tracks represent the MCSs' initial stages prior to initiation; circles indicate the storm positions at maximum extent; ×s mark the termination points. (From Augustine and Howard, 1988.)

Table 11 Comparison of satellite characteristics of US and South American mid-latitude MCSs

	Number of systems	Time, local					Area ($\times 10^3$ km^2)	
		First storms	Initiate	Maximum extent	Terminate	Duration (hours)	Shield	Active core
United States (1978)	43	1400	1930	0130	0630	11.0	308	139
United States (1981)	23	1640	2300	0300	0730	8.5	310	192
United States (1982)	37	1530	2100	0100	0530	8.5	281	181
Mid-latitude South America (1981–2)	22	1830	2030	0200	0800	11.5	397	189
Mid-latitude South America (1982–3)	56	1930	2200	0330	0930	11.5	519	187

Mid-latitude South America is south of 20 °S. Cold cloud shield area is for $T \leqslant -32$ °C for US systems. For all other systems it is $T \leqslant -40$ °C and 42 °C for 1981–2 and 1982–3, respectively. T core $\leqslant -52$ °C for US systems. For all other systems it is $T \leqslant -60$ °C and 64 °C for 1981–2 and 1982–3, respectively. Times are rounded to the half hour.
Source: Velasco and Fritsch, 1987.

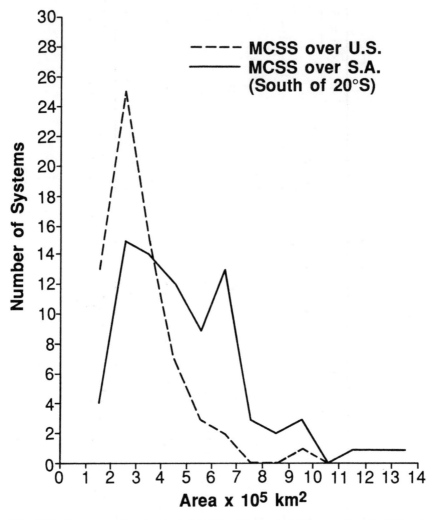

Fig. 4.10 Frequency distribution of MCS cold-cloud shield areas for the US and South American middle latitudes (south of 20S). These are derived from analysis of GOES imagery. (From Velasco and Fritsch, 1987.)

High when the ridge axis is displaced northward and moist subtropical air feeds in from the Gulfs of California and Mexico (e.g., Fig. 4.11). Examination of the rainfall characteristics of these storms is ongoing.

(d) Tropical superclusters

Satellite monitoring of OLR variations for the western tropical Pacific reveals that the dominant intraseasonal variations associated with the 30–60

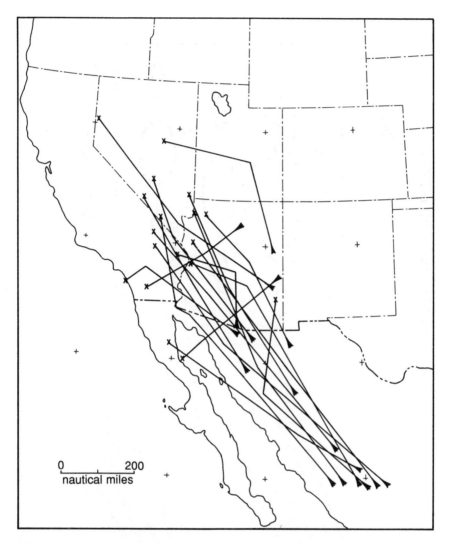

Fig. 4.11 Track map of MCSs over the south-west US in summer (July 1–September 15) 1984 (centroid locations). End points are marked by Xs. (From Perry, 1990.)

day oscillation are composed of cloud clusters and groups of cloud clusters ('superclusters') that have some resemblance to MCSs. Superclusters have been studied from GMS IR imagery by Nakazawa (1988), who finds that these systems propagate eastward into the central Pacific on time scales of less than about 10 days and have a horizontal scale of several thousand kilometers. New superclusters are generated to the east of a mature-stage cluster, and result from low-level convergence near the equator (Fig. 4.12).

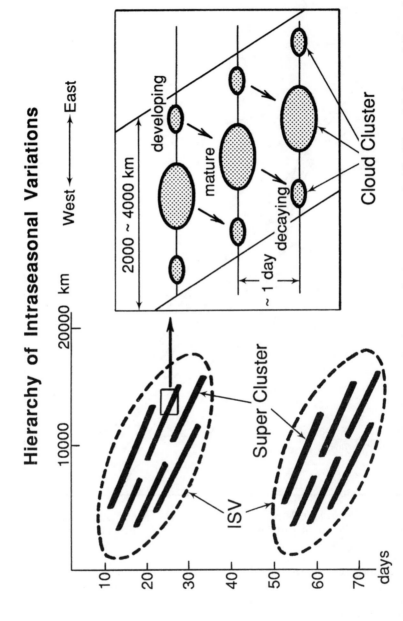

Fig. 4.12 Longitude–time schematic of satellite-observed cloud systems in the western tropical Pacific, and their associations with dominant intraseasonal variations (ISVs). Note that the superclusters move eastward while the individual clusters comprising them move westward. (From Nakazawa, 1988.)

Research is ongoing into the connection between the 30–60 day oscillation in the western Pacific and the lower-frequency variations associated with ENSO in the central and eastern Pacific (Lau and Chan, 1988). However, Nakazawa (1988) notes that marked supercluster activity occurred in the onset phase of the 1986–7 warm event. It remains to be seen whether such activity is a necessary precursor to ENSO events. Again, satellite data can probably best be used to examine this possibility.

3 Synoptic-scale cloud vortices

Probably the most dramatic features appearing on satellite cloud imagery are the organized vortices and cloud bands associated with cyclones and fronts. Both are usually characterized by cold signatures in the IR, resulting from vertical motions that are superimposed on broad-scale ascent over large horizontal distances. Even the earliest coarse resolution vidicon camera views from TIROS revealed different cloud vortex configurations corresponding more-or-less to the successive stages of the Norwegian frontal cyclone 'model'. Figure 4.13 is a DMSP IR view of a winter-time mature depression in the north-west Pacific Ocean. The tight spiral of middle and high cloud denotes the cyclone center, and the broad cloud band extending south of the system is the associated frontal cloud band. The latter is probably occluded along its northern section. Cold air flows south and east around the low, giving rise to the characteristic cumuliform cloud fields. The brightening of the southern portion of the cloud band and its anticyclonic curvature is an indicator of a new frontal wave cyclone (Burtt and Junker, 1976). This signature indicates upper-air divergence associated with a jet maximum or short-wavelength trough.

Non-frontal vortices forming in lower latitudes are typically associated with tropical depressions, tropical storms and hurricanes. They are often first evident in satellite imagery as cloud clusters that develop off the West African coast and west coast of Mexico, and which are indicative of easterly waves. Like the extratropical cyclones, tropical vortices show characteristic stages of development denoted by different configurations of the cloud vortex (see, e.g., Oliver, 1969; Dvorak, 1975; Lander, 1990). In this case, the main features of classification are the presence, absence and extent of a central cloud-free 'eye', the degree of circularity of the system, and the organization of the cloud shield. GOES satellite examples of the mature stages of Hurricane Gilbert, which was distinguished by having the lowest recorded sea level pressure for an Atlantic hurricane, are shown for September 12 (visible) and 15 (IR) in Fig. 4.14a,b. The eye is clearly evident, as are the spiral rain bands feeding the circular main cloud shield. Figure 4.14c shows a weakened Gilbert interacting with a jet stream and an extratropical cloud band associated with a depression over the Great Lakes. The tropical–extratropical cloud bands (TECBs: Kuhnel, 1989) that form from this process are a feature of the satellite climatology of late summer and early fall along the eastern coasts of Asia and North America (Erickson and Winston, 1972). A DMSP example of such a tropical storm–extratropical cloud band connection is shown for the western Pacific in Fig. 4.15.

Fig. 4.13 Portion of a DMSP TIR mosaic (5.4 km resolution) centered on the north-west Pacific. The spiral cloud vortex in upper center marks a mature frontal cyclone; the frontal cloud band extends south and east of the vortex. Refer to the text for discussion. (From Carleton, 1987b).

(a) Classification and vortex characteristics

The stages of development of extratropical cyclones, from development through maturity and dissipation, are evident by changes in the degree of organization of the central cloud vortex (Troup and Streten, 1972; Streten and Kellas, 1973), and classification schemes have been developed to describe this process and to relate the satellite signatures to conventional meteorological fields (see, e.g., Carleton, 1987b). Genetically based classi-fication schemes of cyclone development stages were developed early on in satellite meteorology, and have been confirmed and expanded upon in

Fig. 4.14a, b, c GOES-E TIR and visible sequence of Hurricane Gilbert. In (a) September 12, 1988, and (b) September 15, the eye and spiral rainbands are clearly seen. In (c) September 16, the hurricane has weakened by passage over land (Mexico) and interaction with an extended frontal cloud band associated with a cyclone north of the Great Lakes.

Fig. 4.15 Portion of a DMSP TIR mosaic (5.4 km) for the western Pacific on a day in October 1977. The dominant feature of the image is the well-developed TECB that extends from the tropics (extreme left) where it connects with a decaying tropical vortex; to the Bering Sea, where it ends in a cyclonic system at 60° N, 167° W. Other cyclones in various stages of development are located along the band over middle latitudes. (From Carleton, 1985c).

subsequent studies using high-resolution data. Although these classifications were originally developed to assist weather analysis and forecasting by numerical methods in data-void regions, particularly for the Southern Hemisphere (Kelly, 1978), the extraction of the recurring features across many cases led inevitably to their study synoptic climatologically (Streten and Troup, 1973; Carleton, 1979). The typical extratropical cyclone evolution stages are summarized below and in Fig. 4.16.

Frontal cyclogenesis (Stage 1 in Fig. 4.16) is usually first evident as a thickening or bulge along a pre-existing middle-latitude cloud band that exhibits some degree of anticyclonic curvature. This is associated with

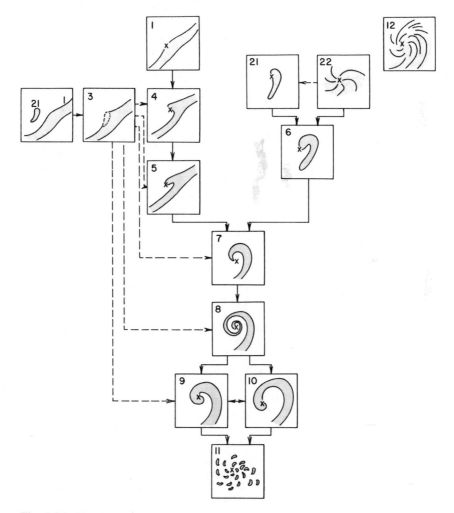

Fig. 4.16 Cloud vortex classification system suitable for use with higher-resolution image data (Northern Hemisphere perspective). The signature types are discussed in the text. (From Carleton and Carpenter, 1989.)

divergence in the upper troposphere on the eastern side of a trough and, consequently, enhanced convergence and ascent in the lower troposphere. TIR data also reveal this process to be associated with a brightening of the clouds (lower CTT) as they increase in thickness (e.g., Fig. 4.13). Most cloud vortex classifications have, as their second stage of development, the appearance of a wedge or slot of clear air on the poleward flank of the cloud development (stages 4 and 5 in Fig. 4.16). The 'dry slot' represents a zone of subsidence. When the cloud vortex configuration in this stage takes on a leaf-like appearance, it is often a satellite indicator of explosive development (Böttger *et al.*, 1975; Barrett *et al.*, 1990). Eventually, the dry slot penetrates and wraps around the center of the cyclone, giving rise to the familiar large 'comma' and ultimately spiral shape as the cyclone approaches maximum intensity (Carlson, 1980). The classification by Carleton contained two sub-stages in this stage of development (numbered 7 and 8 in Fig. 4.16), and these were defined according to the extent of penetration of the dry slot. These were found to be associated with pronounced differences in the composite vertical structure of systems in the Northern Hemisphere, with the deepest cyclones (lowest SLP and 500- and 300-mb heights) occurring in stage 7 rather than stage 8.

The disappearance or diffusion of the tightly wound spiral characteristic of 'maturity' has been generally taken to indicate the onset of dissipation or cyclolysis (stages 9 and 10 in Fig. 4.16). Unlike any of the previous stages, dissipation is usually protracted with little or no appreciable change in cloud vortex signature from image to image. In the Northern Hemisphere, about 25–35% of dissipating vortices are in the same stage on the succeeding 12-hourly image (Carleton, 1987b). This compares with only about 20% for the frontal wave (stage 1); 12–15% for the developing phase (stages 4 and 5); and only 7–10% for the mature cyclones (stages 7 and 8). Troup and Streten (1972) characterized dissipation as the appearance of a cloud-filled or essentially cloud-free vortex. Subsequently it was found that strong differences in vertical structure and deepening occurred if an asymmetric cloud band was evident (Streten and Kellas, 1973). Accordingly, the classification represented in Fig. 4.16 incorporates this regeneration category. Composite analysis of the associated SLP and 500- and 300-mb height anomalies confirms that substantial deepening takes place when a stage 10 development occurs. The most protracted stage of cloud vortex development follows the loss of the frontal cloud band. In this stage, the vortex is cold-cored throughout and may be characterized by multiple lines of cumuliform or stratiform clouds converging on the center. This decay stage is very different from a vortex that may also have a broadly similar appearance but is often located over lower-middle latitudes. In this type, the cloud is generally thicker and comprised of stratiform high- and middle-level cloud (low radiometric temperatures), particularly on its eastern side where the absolute vorticity is most positive (Fig. 4.17). Far from being a cyclolytic system, this cloud vortex type is cyclogenetic, persistent and associated with upper-air cyclones that have been cut-off from the main belt of westerlies ('cut-off lows').

Changes in the vortical cloud amount associated with successive stages of a

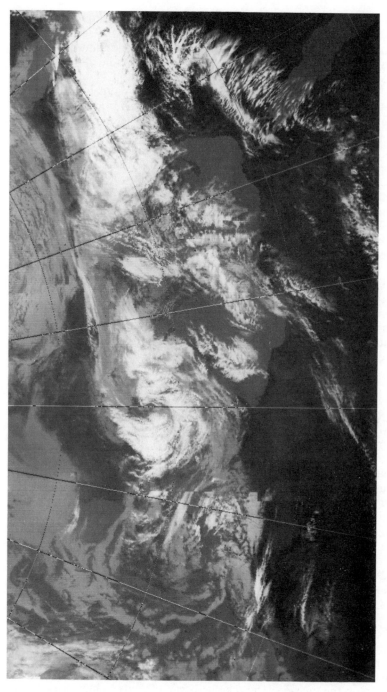

Fig. 4.17 Portion of a DMSP TIR mosaic (5.4 km) centered on the Mediterranean Sea/North Africa for a day in July 1978. The main feature of interest is the bright (cold) cloud vortex in the image center associated with an upper-level 'cut-off low'. Jet stream cirrus is evident as the linear streaks to the east and also south. (From Carleton, 1985c).

frontal cyclone in the Gulf of Alaska, which are considered to be typical of such systems, are shown in Fig. 4.18 (Streten and Kellas, 1975). The figure shows the increase in the total cloud amount as the cyclone matures (stage C), and also the out-of-phase behavior of the low and middle cloud amounts noted in Chapter 1, Section 6.

While satellite detection of frontal wave cyclogenesis largely confirms the Norwegian cyclone model developed much earlier from conventional observation data, the satellite has also revealed modes of extratropical cyclogenesis not explained by this 'model'. The polar air vortex is one such mode (Chapter 4, Section 1b). Another is the so-called 'instant occlusion', which represents a combination of the PVA max and frontal wave models of synoptic development (Anderson et al., 1969; Locatelli et al., 1982; Browning and Hill, 1985; McGinnigle, 1988). Efforts are being made to include prediction of this phenomenon in numerical models (McGinnigle, 1990). A sequence of images illustrating this cyclogenesis over the southern Indian Ocean, south-west of Australia, appears in Fig. 4.19a,b,c. As the enhanced cumulus feature (PVA max) approaches the frontal cloud band located to the east, it induces a wave on the band by vorticity advection (Fig. 4.19a,b). The final image of the sequence (Fig. 4.19c) shows the new vortex that results from this interaction. Because the development has a configuration resembling an occluded (dissipating) frontal cyclone, and appears to take place rapidly (say, between two images that are 12 hours apart), it was originally named the instant occlusion. However, analysis of the pressure/height fields for cases of this satellite-observed system (Carleton, 1985b; 1987b) seem to confirm its cyclogenetic, rather than cyclolytic, nature. Figure 4.20 shows typical composite SLP and upper-air height anomaly patterns for satellite-observed cyclogenesis over the Northern Hemisphere oceans. There are marked differences between the frontal wave (A1) cyclogenesis and the polar low 'comma cloud' (A2) type, particularly the strong westward tilt with height of the former (baroclinic) and the deep cold-core structure of the latter. Note the very deep pressure/height anomalies of the instant occlusion (A3) compared with those cloud vortices in the dissipation stage (Da) that have previously undergone the progression through the developing and mature stages (Fig. 4.20) of cyclone evolution.

Studies have sought to relate conventional observations over data-rich areas to cloud vortex types by deriving mean vertical profiles of temperature and humidity, SLP and tropospheric height anomalies (from the climatological mean values), typical patterns of relative vorticity, wind shear and thermal advection in the lower troposphere (see, e.g., Barr et al., 1966; Boucher and Newcomb, 1962; Widger, 1964; Kondrat'ev et al., 1970; Leese, 1962; Sovetova and Grigorov, 1978; Troup and Streten, 1972). It has been found that, in the majority of the cases, the cloud vortex center corresponds more closely to the upper-air height minimum than the surface cyclone center does. The distance between the two centers decreases with system maturity, both because of the decreased baroclinity (given by the vertical tilt of the system) and also the improved ability to detect the center of circulation on the imagery. The relative vorticity tends to increase into the spiral ('mature') stage and to decrease slightly thereafter, and the vorticity

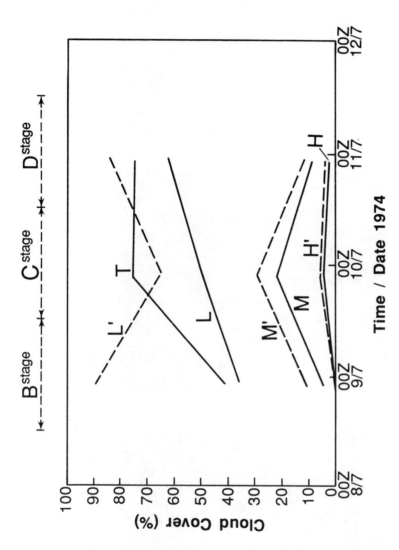

Fig. 4.18 Plot of the variations in satellite-viewed vortical cloud amounts at different levels (L, low; M, middle; H, high; T, total) for an extratropical cyclone. L´, M´, and H´ show, respectively, the corresponding variation in relative percentage distribution for each observation. (From Streten and Kellas, 1975.)

Fig. 4.19a, b, c DMSP TIR sequence (5.4 km) of an instant occlusion development over the southern Indian Ocean, south-west of Australia, for (a) 1326 GMT, June 20, 1989; (b) 0055 GMT, June 21, 1989; (c) 1016 GMT, June 21, 1989. Refer to the text for discussion.

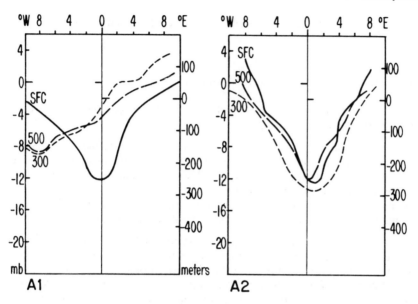

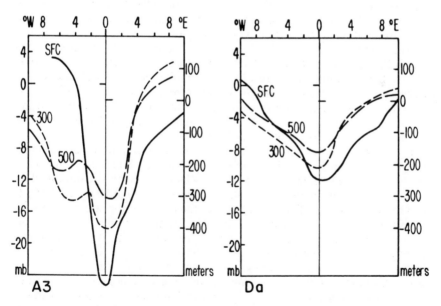

Fig. 4.20 West–east transects of composite SLP (mb) and upper-air (500, 300 mb) height (m) departures for satellite-viewed Northern Hemisphere cloud vortices in October 1978. Shown are the curves for the incipient frontal wave (A1); the comma cloud polar low (A2), the instant occlusion (A3) and the dissipating frontal vortex (Da). (From Carleton, 1987b.)

maximum moves closer to the cloud vortex center with increasing development.

(b) Climatologies and climatic associations

Satellite-based climatologies of cyclonic cloud vortices were, until quite recently, most concerned with:

(1) the development of statistical 'models' of pressure/height, vorticity, which provide clues on the associated dynamics (in addition, these data could be 'bogussed' into numerical weather analysis and forecasting models for data-void regions);
(2) mapping the distributions of the different types and stages of development for examination in a synoptic climatological framework.

To a certain extent, the contemporary studies of meso-cyclones (polar air vortices) using satellite data are being investigated in the above-mentioned contexts. Until quite recently, there was little attempt to take advantage of the meteorological information implicit in satellite data on synoptic-scale cloud vortices and apply it to the question of climate variations. An exception was the study of Streten (1975b), who investigated changes in cloud vortex occurrence and characteristic system intensities that accompanied the development of the 1972–3 ENSO over the South Pacific sector. He showed the movement eastwards of the major areas of cyclonic activity and an increase of about 20% in cloud vortex frequencies when comparing similar seasons of the two years 1971–2 and 1972–3, and a similar movement of the South Pacific Cloud Band (SPCB) during that time. Carleton and Whalley (1988) have extended these results by examining the statistical associations between Southern Hemisphere cloud vortex frequencies and the efficiency of the eddy sensible heat transport for the winters of 1973–7. They find that about 36% of the variance in the meridional heat transport by the vortices is explained by the phase of the ENSO in those winters.

Figure 4.21 shows the normalized mean distributions of cyclogenesis (stages 1 through 6, 12, 21, 22, in Fig. 4.16) for the Southern Hemisphere winter. It may be compared with Fig. 4.22 which shows the distribution of the combined mature and dissipating stages (stages 7 through 10 in Fig. 4.16). In common with the earlier cloud vortex climatologies for summer and the transition seasons by Streten and Troup (1973), the cyclones tend to migrate to the south-east as they evolve and to dissipate in the Antarctic coastal embayments in a basic three-wave pattern. A notable exception is the frequency maximum extending into East Antarctica in longitudes of Australia. However, unlike the summer and transition months, there is much more active cyclogenesis over high ocean latitudes in winter and this is predominantly of the 'polar air' vortex type (Section 1b, 3). This occurs particularly north of the Ross Sea and just west of Tierra del Fuego.

Carleton (1981a) presented a climatology of the 'instant occlusion' phenomenon for the Southern Hemisphere winter. The maximum frequencies of this vortex type overlap the preferred latitudes of the frontal wave (30–50° S)

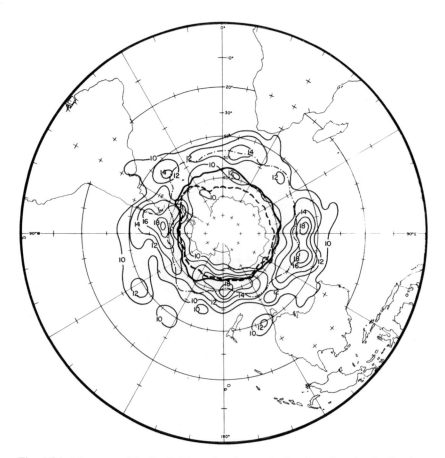

Fig. 4.21 Mean monthly distribution of cyclogenetic cloud vortices for the Southern Hemisphere winters (June through September) of 1973–7. Isopleths refer to area-normalized frequencies in each 5° latitude by 10° longitude unit. The OPF is shown by the dot-dashed line; the mean sea ice margin for the June and September months of those years are shown, respectively, by the heavy dashed and heavy solid lines. (From Carleton, 1981b, c.)

and polar air (40–65° S) cyclogenesis types. Note also in Fig. 4.21 the close association between cyclogenesis and the location of strong thermal gradients in the ocean (the ocean polar front: OPF) but not with the sea ice extent. The latter seems important only in regions where it is located adjacent to the OPF.

The recognition that regional-scale anomalies in sea ice and snow-cover extent are associated with climatic anomalies has led to the application of satellite information to studies of the accompanying variations in cyclone types, frequencies and statistical intensities. An early study by Schwerdtfeger and Kachelhoffer (1973) suggested a link between the latitude zone of

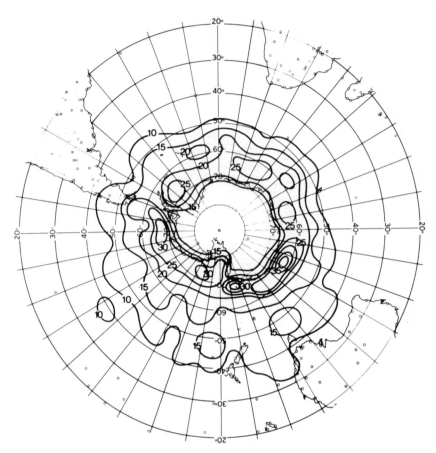

Fig. 4.22 Similar to Fig. 4.21, except that this shows the mature and dissipating vortices. (From Carleton, 1981b.)

maximum cloud vortex occurrence in the Southern Hemisphere and the seasonal movement of the Antarctic sea ice. Subsequent studies (e.g., Carleton, 1981b,c, 1983; Budd, 1982) revealed that this relationship is not a simple one. Although the latitude of maximum *cyclogenesis* shows some equatorward movement as the sea ice expands during the winter, the total frequencies of *all* cyclones actually shift polewards in association with the semi-annual oscillation of SLP, temperature gradients and winds. It was found that, on relatively short time periods and on regional scales, the Antarctic sea ice may interact strongly with the cyclonic cloud vortex regimes. Favored areas for this interaction appear to be the Ross and Weddell Sea sectors (Carleton, 1983), which are also the regions of maximum seasonal and interannual variability of the Antarctic sea ice extent (Cavalieri and Parkinson, 1981). The longitudes of maximum cloud vortex frequencies in the Antarctic tend to be located just to the east of maximum seasonal change of the sea ice extent, since cold air advection is favored on the western side of cyclonic systems (Fig. 4.23). A similar relationship has been noted for the North Pacific seasonal sea ice zone (see, e.g., Carleton, 1984b, 1987b). In addition, the large open-water areas (polynyas) well within the pack ice, as determined using all-weather microwave sensing, show some associations with satellite-observed regional cyclogenesis patterns (Carleton, 1981b,c).

The climatic role played by the large seasonal change in the Antarctic sea ice for the Southern Hemisphere is filled, in the Northern Hemisphere, by continental snow cover. Carleton (1984b, 1985b,c, 1987b) investigated the associations between the Northern Hemisphere snow cover and sea ice extents and the frequencies and characteristic intensities of cloud vortex systems for two years' mid-season months. He found that more extensive snow cover in April or January was associated with increased cyclogenesis, and that these increases were displaced to the south of the snowline. For the sea ice zone, the synoptic feedback appears to be highly regional: enhanced cyclonic activity may be associated with less ice in one region (e.g., the Okhotsk Sea) but more ice in another (e.g., the Bering Sea) (Carleton, 1987b). Clearly, longer-term studies are required.

Geosynchronous satellite data are ideal for high-frequency and high-resolution monitoring of tropical cloud vortices. Infrared data from such platforms have shown that the cloud cover undergoes strong diurnal and also semi-diurnal patterns that reflect, particularly, the patterns of convection associated with cloud-top radiational processes (Muramatsu, 1983; Steranka et al., 1984). Increasingly, also, passive microwave data on polar orbiters are yielding information on tropical cyclone evolution and dynamics; notably the vertical temperature structure and strength of the upper tropospheric warm core (Velden and Smith, 1983). A satellite climatological study by Velden (1989) used microwave sounding data for 18 North Atlantic tropical cyclones to identify a relationship between the warm core anomaly at 250 mb and SLP intensity. Controlling for the influences of storm latitude, surface intensity tendency and eye size reduces further the errors in the surface intensity estimation. Application of the Nimbus-7 TOMS to a census of western tropical Atlantic hurricanes is yielding important information on the dynamics of the exchange of stratospheric ozone across the tropopause (Rodgers et al., 1990); as it has for extratropical cyclones (Uccellini et al., 1985). The anticipated future deployment of a high-resolution microwave sounder in geostationary orbit (e.g., on GOES-NEXT) is expected to lead to refinement of the remote sensing of tropical cyclone parameters.

4 Cloud bands

(a) Fronts

Fronts are distinguished on satellite imagery as bands of middle- and high-level cloudiness associated with extratropical cyclones (e.g., Figs 4.13, 3.7).

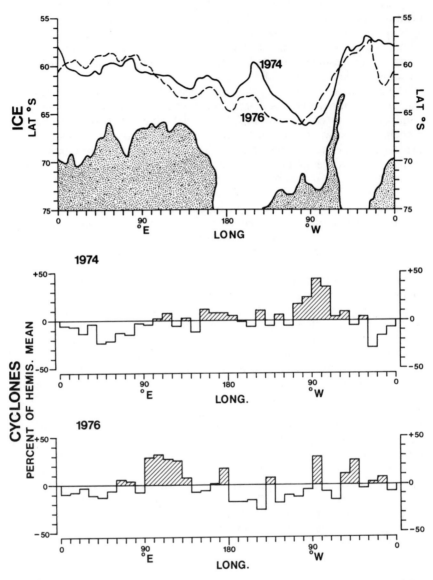

Fig. 4.23 Associations between the winter-mean locations of the Antarctic sea ice margin (upper panel) and the longitude variations of satellite-observed cloud vortices for the winters (JJAS) of 1974 and 1976. Cyclone frequencies are given as the percentage in a 10° longitude sector from the hemispheric mean. (From Carleton, 1981b.)

Along at least part of their length there is usually evidence of the importance of the jet stream in their development, from the presence of jet stream cirrus. While there are many studies of cloud vortices, little attention has been paid to developing satellite climatologies of fronts, although case studies are available (e.g., Egorova, 1970; Streten, 1977). Part of the reason for the lack of satellite frontal climatologies undoubtedly lies in the difficulties involved in treating quantitatively line features. These are distinct from discrete point phenomena, such as a cloud vortex center. Those 'frontal climatologies' that are available are largely a by-product of the hemispheric and global-scale cloud climatologies developed from multi-day compositing of visible (minimum brightness) and infrared (maximum temperature) mosaics (Booth and Taylor, 1969; Kornfield et al., 1967; Streten, 1968, 1970, 1973; Yasunari, 1977: see Chapter 3, Section 3a). However, satellite case studies of fronts (Streten, 1977; Streten and Downey, 1977) reveal that quite marked changes in cloud cover, cloud type and CTT may occur as fronts move and evolve (e.g., Fig. 2.15a,b). These are not well captured in composite (multi-day) analyses. Changes in system cloud cover and CTT are important dynamically since they imply changes in the available eddy potential energy of the storm. Further, the climatic fronts so defined from CMB and CMT analyses are also storm tracks along which frontal waves develop, mature and dissipate. Moreover, they are not necessarily frontal (i.e., deeply baroclinic) along their entire length. Those bands that develop in tropical latitudes and connect with frontal systems over middle latitudes (TECBs) tend to be upper-level features over lower and subtropical latitudes (Bell, 1986). The separate contribution to frontal activity of boundary-layer convergence that does not develop in association with frontal waves, but which appears to exhibit frontal characteristics (e.g., in polar air vortices) is also unknown. Because of the association of fronts with steep tropospheric gradients of humidity and the occurrence of precipitation, all of which are readily detected over the oceans by satellite passive microwave data (Katsaros et al., 1989; also Fig. 1.1 and Chapter 5), it seems likely that a large-scale climatology of fronts may be forthcoming from this source.

(b) Tropical–extratropical cloud bands (TECBs)

In strong cyclonic situations over middle latitudes, the polar front jet (PFJ) and the subtropical jet (STJ) may merge (Reiter and Whitney, 1969) to produce a band of middle- and high-level cloudiness that has marked latitudinal and longitudinal extent but limited width (Oliver and Anderson, 1969; Gray and Clapp, 1978; McGuirk et al., 1987). Where these bands connect with active tropical convection they are known as tropical–extratropical cloud bands (TECBs) (Kuhnel, 1989; also Fig. 4.15). TECBs facilitate the rapid transport of energy and moisture between low and higher latitudes and, in the fall season, are apparently involved in the build-up of the middle-latitude westerlies (Erickson and Winston, 1972). Their importance as precipitation producers in the subtropics and middle latitudes has been noted from case studies (Dickson, 1973; Downey et al., 1981; Harrison,

1984) and implied from studies of climate teleconnections (Douglas and Englehart, 1981). An analysis of synoptic precipitation mechanisms for the Australian state of Victoria (Wright, 1988) shows that between 25% and 50% of winter precipitation may be derived from the interaction of low-latitude cloud bands with frontal systems in that region.

The development of a tropical–extratropical interaction is largely dependent on the basic state of the tropical atmosphere (Lim and Chang, 1983), which has a strong longitude dependence. Convection in the tropics during northern winter tends to be favored under conditions of easterly, as opposed to westerly, upper flow (Liebmann, 1987). However, the propagation of energy and moisture out of the tropics is generally favored by upper-tropospheric westerlies (Arkin and Webster, 1985), since easterlies may lead to equatorially trapped waves (Lau and Lim, 1984). Incursions by middle-latitude troughs provide such a mechanism, and may result in the formation of a 'cloud surge' or 'moisture burst' (McGuirk et al., 1987; McGuirk and Ulsh, 1990). This may become a TECB and lead to downstream intensification of the STJ. Tropical–extratropical interactions are apparently favored for weak to moderate cold surge events since subsidence in the subtropics is enhanced in the case of strong surges (Lau et al., 1983).

For the globe, Kuhnel (1989) identifies 14 TECBs; 7 in each hemisphere, and with different seasonal preferences. This work follows on from the more general satellite image analysis of Kletter (1972). Most of these bands show some association with large-scale teleconnection patterns, most notably the ENSO. Climatologies of TECBs exist for the Southern Hemisphere (Streten, 1968, 1970, 1973; Harrison, 1984; Tapp and Barrell, 1984), owing to a heavy reliance on satellite information for the largely data-void oceans. The South Pacific Cloud Band (SPCB) has been a particular focus of study (see, e.g., Huang and Vincent, 1985). It is actively involved in extratropical cyclogenesis and exhibits an ENSO signature (see, e.g., Kuhnel, 1989) that is evidently the result of strong air–sea interaction. Evidence for a quasi-periodicity in longitudinal movements of the SPCB averaging about 20–25 days, and similar to the time scales of zonal indices and kinetic energy, has been identified from satellite analysis by Streten (1978b).

In the Northern Hemisphere, TECBs are often observed to link the tropical and eastern Pacific with North America and higher latitudes of the North Atlantic (Thepenier and Cruette, 1981). McGuirk et al. (1987) find that cloud surge events over the Pacific are followed by tropospheric-wide increases in the extratropical westerlies and fluxes of poleward-directed momentum. At that time, a TECB may fill the role of a transient Hadley circulation that becomes important when the standing Hadley cell is less intense (McGuirk et al., 1987). The latter would be expected to occur, for example, during positive (cold) phases of ENSO, and there is satellite evidence for this association. Comparisons of TECB occurrence for the North Pacific/Atlantic and the Australian region (Fig. 4.24) reveal a similar seasonal distribution, with maximum frequencies in the cool season (see also Kuhnel, 1989).

Figure 4.25a,b shows the occurrence of Northern Hemisphere TECBs for one month (January 1980), as analyzed using NOAA twice-daily images.

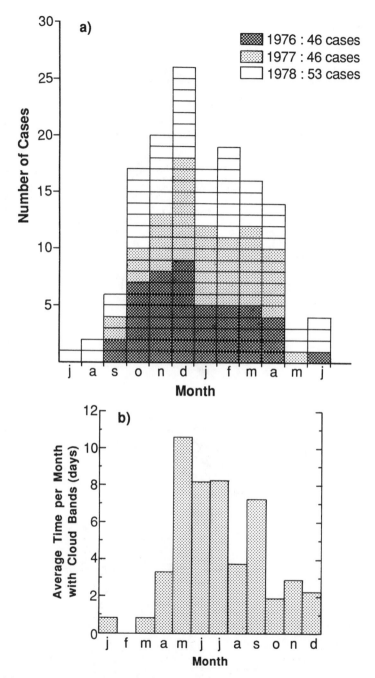

Fig. 4.24a, b Comparisons of the seasonal variation of TECBs for (a) the North Pacific/North Atlantic sector (three years: 1976, 1977, 1978), and (b) the Australian sector (four years: 1978–82). The data in (a) are plotted as frequencies of occurrence (from Thepenier and Cruette, 1981); in (b) as the average time per month (days) with cloud bands. (from Tapp and Barrell, 1984.)

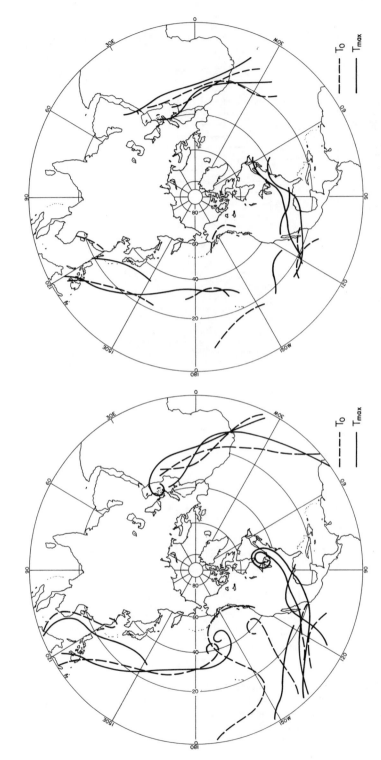

Fig. 4.25 Summary maps of Northern Hemisphere TECB activity for January 1980, derived from analysis of twice-daily NOAA IR mosaics. TECB characteristics presented are (a) median band axes at T_0 and T_{max}; and (b) jet axes at T_0 and T_{max} (refer to the text for explanation).

The median locations of a given cloud band and jet axes are shown for two times: T_0, which refers to the time the band moves out of the tropics by first crossing latitude 30° N; and T_{max}, or the 12-hour image preceding the first break in the TECB connection. In January 1980, 35 (60%) of all twice-daily images contained at least one TECB. Of those, 14% contained two or more bands on the one image (i.e., they exhibited simultaneity). There were eight total TECB events. While difficult to assess from the coarse resolution NOAA data, just under half of these TECBs (three) appeared to form without extratropical initiation. The mean band duration (T_0 through T_{max}) was about 2.6 days, and this compares well with the McGuirk *et al.* (1987) study. One TECB lasted for 4 days. The dominance of the Pacific Basin in TECB development is evident, at least for this month (Fig. 4.25a,b). Undoubtedly the use of large-scale and higher-resolution imagery (e.g., DMSP) would help resolve more details of TECBs in the Northern Hemisphere. Important questions that still need to be addressed include the relative frequencies of instant occlusion development related to TECBs (Thepenier and Cruette, 1981), and the ratio of tropical to extratropical initiation of TECBs, as well as their seasonal dependence. In addition, the association of TECBs with teleconnections other than ENSO, particularly the North Atlantic Oscillation, the North Pacific Oscillation, the Quasi-Biennial Oscillation, and the tropical intraseasonal variation associated with the 30–60 day oscillation in OLR, should probably be investigated.

Suggested further reading

Cracknell, A. P. (ed.), 1981: *Remote Sensing in Meteorology, Oceanography and Hydrology*. Wiley and Sons, UK. 542 pp.

Henderson-Sellers, A. (ed.), 1984: *Satellite Sensing of a Cloudy Atmosphere: Observing the Third Planet*. Taylor and Francis, London and Philadelphia. 340 pp.

Rao, P. K., Holmes, S. J., Anderson, R. K., Winston, J. S., and Lehr, P. E., 1990: *Weather Satellites: Systems, Data and Environmental Applications*. American Meteorological Society, Boston, Mass. 503 pp.

Twitchell, P. F., Rasmussen, E. A., and Davidson, K. L. (eds.), 1989: *Polar and Arctic Lows*. A. Deepak Publ., Hampton, VA. 421 pp.

5 Atmospheric moisture: water vapor and precipitation estimation

Water resides temporarily in the atmosphere in all three states. It is present as a gas (vapor) that is the result of evaporation and evapotranspiration of liquid water from the Earth's surface, and also from clouds and precipitation. Atmospheric water in solid form occurs as ice crystals in middle- and high-level clouds, and as precipitation in the form of snow and hail. The phase changes of water also represent energy conversions of substantial magnitude. They involve either the uptake (e.g., in evaporation, melting) or the release (in condensation, freezing, sublimation) of latent heat. For the Earth as a whole, and averaged over the entire year, the net radiative imbalance between the surface and atmosphere is redressed dominantly by the latent heat flux (LH). Thus, the energy budget of the Earth–atmosphere system is linked with the global water budget.

The quasi-horizontal transport of water vapor and liquid water (clouds) across latitudes is more a consequence of the atmospheric general circulation than a cause. As we have seen (Chapters 2 and 3), these quantities are tracers of air motions that can be sensed by satellites observing in the appropriate spectral bands: the IR window region for most optically thick clouds; the absorption bands for water vapor. In addition, use of the microwave region permits determination of the cloud liquid water, the water vapor and 'instantaneous' precipitation rate. Variations in the general circulation, such as ENSO, have a signature in the water vapor and cloud regimes, particularly over the global oceans. Here we focus on the satellite evaluation of atmospheric water vapor, and the estimation of precipitation. The latter has been the subject of a volume in its own right (Barrett and Martin, 1981). We will examine climatic scales beyond about 5 days for areas exceeding 10^4 km^2 (Arkin and Ardanuy, 1989), and with particular emphasis on intraseasonal and interannual variations. The last two involve, respectively, the monsoons and the ENSO and its teleconnections.

1 Atmospheric water vapor and precipitable water

Attempts to use satellite remote sensing of the relative humidity of the atmosphere have a long history. Timchalk and Hubert (1961) showed the value of TIROS-1 cloud cover imagery for deducing the general patterns of atmospheric drying (due to subsidence and advection), and moistening (from ascent) in a mid-latitude cyclone. In particular, they showed a strong positive association between the occurrence of low clouds (below 1500 metres) and the surface relative humidity. The relationship is evident for both frontal (upslide) and convective situations, although it is somewhat stronger in the latter owing to greater mixing of the air in the sub-cloud layer. A study by McClain (1966) showed the value of the interpretation of cloud cover categories appearing in TIROS visible imagery for determining the saturation deficit, or mean relative humidity of the 1000–500 mb layer on synoptic scales in middle latitudes. The method relies on the fact that the smallest saturation deficits (highest humidities) tend to occur with thick solid and widespread clouds (Timchalk and Hubert, 1961). These are distinct from the large saturation deficits that may occur with clear or cloudy skies, but where the latter may contain low and shallow stratiform clouds. The determination of cloud cover–humidity associations using satellite data is continuing (Garand *et al.*, 1989; Saito and Baba, 1988). Around the same time as the McClain study, Huang *et al.* (1967) noted consistent associations between TIROS IR temperatures and surface relative humidity, whereby low IR temperatures accompany extensive cloud cover and high relative humidity, compared with high IR temperatures; at least over the oceans. These authors also noted the potential value of a water vapor sensor at 6.5 μm, as was deployed on TIROS II, III and IV. Such a sensor proved useful for inferring the atmospheric mean relative humidity and precipitable water of the mid- to upper-troposphere (Raschke and Bandeen, 1967). The large-scale nature of these studies permitted the disclosure of the major features of the atmospheric circulation. Moist air and ascent occurs in association with the ITCZ and mid-latitude frontal zones; dry air in association with the subtropical highs.

Atmospheric water vapor retrieved from satellite sounders is either in the form of relative humidity, since it is also dependent on the air temperature in that portion of the vertical profile; or the column precipitable water. The latter is not as sensitive to an exact height assignment. Two main spectral regions of the EMS have been exploited in this retrieval. These are:

(1) the strong atmospheric water vapor absorption region between about 5.7 and 7.1 μm (refer to Chapter 2); and
(2) passive microwave frequencies around the 37 to 21 GHz region.

The water vapor absorption band has been used extensively to obtain operational temperature and moisture profiles (soundings) from geostationary and polar orbiter platforms. Microwave retrievals are the more recent to be developed and have utilized sensors primarily on the R&D polar orbiters Nimbus-5, -6 and -7, and also Seasat. It is anticipated that the new generation of GOES satellites will capitalize on developments in the microwave region with deployment of a passive microwave sensor for atmospheric water vapor and the determination of precipitable water.

The different spectral properties of the 5.7–7.1 μm (IR) and passive microwave methods for detecting atmospheric moisture have lent themselves to different and, in many cases, non-comparable climatologies. The infrared methods are most useful for detecting mid- to upper-tropospheric water vapor in cloud-free areas over both land and sea. Microwave methods are suitable for deriving column precipitable water and also cloud liquid water over the oceans at coarser spatial resolutions.

(a) Studies utilizing the water vapor absorption band

The spectral region between about 6 and 9 μm can be used for the determination of the layer relative humidity and the column precipitable water. The former represents the contribution from the upper troposphere, or the 600–300 mb layer (Eyre, 1981; Schmetz and Turpeinen, 1988), at around 6.3–6.7 μm (e.g., Fig. 5.1). The European METEOSAT was the first geostationary platform to provide half-hourly retrievals of upper-tropospheric moisture in clear-sky conditions for the 5.7–7.1 μm region (Houghton and Suomi, 1978). This built upon the experience gained from the early Nimbus satellites (Morel et al., 1978), particularly the Nimbus-4 THIR (Temperature and Humidity Infrared Radiometer). The VAS (VISSR Atmospheric Sounder) on the GOES platform (Allison et al., 1972; Steranka et al., 1973; Rodgers et al., 1976) performs a similar function to the water vapor sensor on METEOSAT, although it contains additional channels that facilitate the determination of the total precipitable water at subsynoptic scales (Chesters et al., 1983, 1987). Precipitable water is less influenced by the altitude of peak radiance than is the upper tropospheric humidity and, so, represents the moisture content of the lower layers once the surface contributions of temperature and moisture have been removed (Hayden et al., 1981). The surface moisture contribution can be accounted for by the 8.3 μm channel, in contrast with the 7.3 μm band where the altitude of the peak contribution to the radiance is located in the lower- to mid-troposphere. The wavelength dependence of the level of peak contribution to the water vapor radiance forms the basis of both the VISSR and TOVS retrieval methods.

(1) *Water vapor* Comparisons of a water vapor image, which is a map of individual soundings, with the corresponding IR image, reveals that high clouds in the latter are also evident in the water vapor. Because they are close to the level of peak emission (600–300 mb) they reduce the water vapor emission (Fritz and Rao, 1967; Davis, 1982). On the other hand, low clouds have a negligible impact since they generally lie below the level of peak contribution to the emission. High cloud effects in the 6.3 μm band are a fundamental influence on the water vapor signatures (Fritz and Rao, 1967), and the interpretation of water vapor *above* these clouds can be misleading. Thus, the determination of relative humidity in the water vapor channel is best obtained (1) for clear-sky conditions only, and (2) where utilizing both water vapor and IR window channels for detecting cirrus clouds (higher transmissivity in IR; opaque in water vapor). The IR image can be used to determine cloud-free scenes and, for low-level moisture, the surface temperature (see, e.g., Prabhakara et al., 1979; Chesters et al., 1983, 1987).

The altitude of the peak radiance in the water vapor channel, or contribution function $M(\log p)$, can be determined for cloud-free (i.e., uncontaminated) conditions by the following equation (after Eigenwillig and Fischer, 1982):

$$M(\log p) = \int_{\lambda_1}^{\lambda_2} A(\lambda) \cdot B(\lambda_1 \ T) \cdot \frac{\partial \tau(\lambda_1 \log p)}{\partial \log p} \ d\lambda \qquad (19)$$

where $A(\lambda)$ is the spectral response function for the water vapor channel;

p is the pressure;

T is the temperature;

$B(\lambda, T)$ is the Planck function (refer to Chapter 2);

λ_1 is the bottom of the atmospheric layer;

λ_2 is the top of the atmospheric layer;

$\tau(\lambda_1 \log p)$ is the transmission between atmospheric level p and the satellite sensor.

In middle latitudes, the drier the atmosphere the lower the level of peak contribution to the emission and the higher the radiative temperature, and vice versa (Poc and Roulleau, 1983). There is a strong dependence on the temperature profile (ΔT). If the profile is known from collocated radiosonde data, or from physical modeling, then a much more reliable estimation of the relative humidity is possible. Temperature profiles from forecast first-guess fields may also be used (Schmetz and Turpeinen, 1988). The twice-daily METEOSAT UTH (upper tropospheric humidity) product shows a satellite underestimation averaging about 4% (Turpeinen and Schmetz, 1989). This is largest at higher humidities and especially in middle latitudes (i.e., there is strongest agreement in the tropics). Figure 5.1 shows, for the same model temperature profile (ΔT), the strong dependence of the level of peak contribution on the humidity. It also emphasizes the importance of the contribution from the 550–400 mb layer at 6.7 μm. There are seasonal variations in the contribution function. For middle latitudes, the peak altitude is lower in winter (about 500 mb) and higher in summer (about 400 mb). This is shown by Fig. 5.2. A dependence on latitude in the same season (climate regime) is also evident (Poc and Roulleau, 1983). In summer, this involves higher altitudes in the tropics but a lower contribution and higher radiative temperatures in middle latitudes. Of particular importance for geostationary platforms, the spatial variation of the contribution function is also related to the satellite zenith angle. The peak contribution level rises as the zenith angle increases (Schmetz and Turpeinen, 1988). This may amount to a change of about 1.0 km for the same climate region. However, the difference is considerably larger (about 2.7 km) for a tropical and a mid-latitude winter situation if the satellite zenith angle is held constant at 62.7° (Poc and Roulleau, 1983).

A knowledge of the tropospheric level, or at least the dominant layer of contribution to the water vapor radiance, is crucial for tracking water vapor 'features' (Allison et al., 1972) on successive high-frequency images. These provide information on upper-tropospheric wind speeds (see, e.g., Eigenwillig and Fischer, 1982). These fuzzy-looking features (e.g., Fig. 2.4) occur in either clear-sky conditions or in the absence of opaque high-level clouds, and may last as long as 10 hours. They represent horizontal gradients in the humidity field (Eyre, 1981).

Given the generally lower temporal resolution of a polar orbiter over most latitudes compared with a geostationary platform, water vapor features have been used with some success to infer instantaneous wind flows (i.e., stream-

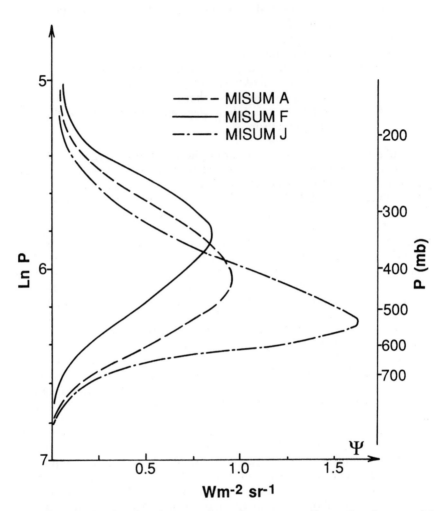

Fig. 5.1 Contribution functions to the water vapor radiance for three model atmospheres: the basic (MISUM A); the driest (MISUM J); the wettest (MISUM F). These assume the same temperature profile and show the importance of the contribution from the 550–400 mb layer. The Greek symbol ψ, denotes m(logp); Ln is the natural logarithm. (From Poc *et al.*, 1980.)

lines), rather than time-averaged trajectories (see, e.g., Steranka *et al.*, 1973). The latter requires a geostationary satellite. At these higher latitudes, advantage may be taken of the reduced Earth surface area, which gives substantial overlap of adjacent swaths from a polar orbiter (Kastner *et al.*, 1980; Turner and Warren, 1989). Eigenwillig and Fischer (1982) report reasonable success in determining wind speeds from tracking water vapor features in METEOSAT data, but their results for the wind directions are

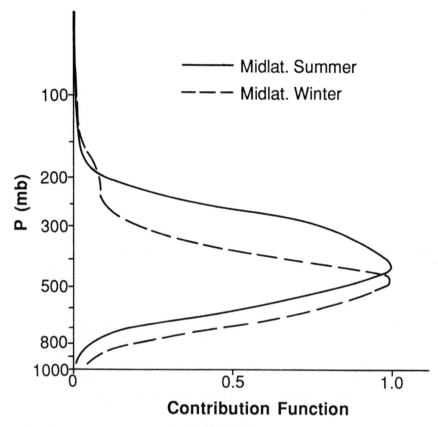

Fig. 5.2 Normalized contribution functions for model atmospheres; 'mid-latitude summer' and 'mid-latitude winter'. The nadir angle of 6.78° corresponds to 45° latitude at the 0° meridian. (From Eigenwillig and Fischer, 1982.)

generally less encouraging. The association of warm (dark) radiances with subsidence on the poleward side of jet streams, and cold (light) radiances on the equatorward side (see, e.g., Martin and Salomonson, 1970) has been examined for METEOSAT data by Ramond *et al.* (1981). They find differences in the interpretation of the 'dark band' between the polar front jet (PFJ) and subtropical jet (STJ) that can be explained by the different dynamics. For the PFJ, Ramond *et al.*, suggest that the dark band is a tracer of the tropopause break. The absence of a low-level baroclinic zone with the STJ means that what is probably being observed in that case is the zone of maximum subsidence on the poleward side. Satellite T_{BB} observations of warm dark bands investigated with the aid of a meso-scale model, reveal an association with time-averaged (rather than instantaneous) subsidence that is associated with short waves and polar jet streaks (Muller and Fuelberg,

1990). Timely identification of these features may be important for predictions of cyclogenesis and severe weather outbreaks.

(2) *Precipitable water* Moisture at lower levels, given by the total precipitable water, utilizes slightly longer wavelengths of the water vapor absorption band; 8–9 μm. Using data from the Nimbus-4 IRIS (Infrared Interferometer Spectrometer) for the global oceans between about 50° N–40° S, Prabhakara *et al.* (1979) identified a dependence of the precipitable water (w) on the strength of the trade inversion associated with subsidence in the subtropical high. This arises from the association between w and T_s (or SST). The latter can be determined remotely using the IR window data. Prabhakara *et al.* define the following index:

$$\frac{(\bar{w} - w)}{\bar{w}} \times 10, \tag{20}$$

or the departure of a measured value of the total water content (w) from the mean ($\bar{w}$). The latter is derived from climatology. In this index, positive values indicate a strong trade inversion (enhanced subsidence) and lower precipitable water; negative values indicate a weak inversion and higher precipitable water associated with convection. These authors mapped index values for three-monthly seasons of 1970.

Both the TOVS and VAS operational retrievals have proven invaluable for identifying meso-scale moisture gradients over land in clear-sky conditions preceding severe weather outbreaks (Hayden *et al.*, 1981; Chesters *et al.*, 1983, 1987). For the central and southern US, at least, there is a tendency for thunderstorms to develop along the edges of bands of vertical moisture gradient and differential moisture advection (Petersen *et al.*, 1984). Dry midtropospheric air in association with a jet maximum typically overtakes a moist layer at low levels. The greatest impact on the VAS retrievals is achieved when conventional surface data are included (Lee *et al.*, 1983). Additional improvement also occurs with the inclusion of radiosonde information (W. Robinson *et al.*, 1986). The radiosondes show the closest association with the satellite data when the VAS are smoothed by averaging over 30 km rather than the noisier 15 km nominal resolution of the instrument.

(b) Passive microwave studies

Satellite sensors to retrieve the atmospheric water vapor, total (column) precipitable water in the range 1–6 g cm^{-2} and the cloud liquid water using passive microwave radiometry include the Nimbus-E (-5) Microwave Spectrometer (NEMS) (Staelin *et al.*, 1976), the Nimbus-6 Scanning Microwave Spectrometer (SCAMS) (Grody *et al.*, 1980), and the Nimbus-7 and Seasat SMMRs (Wilheit and Chang, 1980; Prabhakara *et al.*, 1982; Alishouse, 1983; Njoku and Swanson, 1983). These sensors have the following characteristics pertinent to the retrieval of atmospheric moisture.

(1) Water vapor and precipitable water are retrieved over *ocean* areas (cf. Curry *et al.*, 1990; Jones and Vonder Haar, 1990), which provide a relatively constant emissivity and T_B background compared with a land surface of highly variable emissivity. Even so, the oceanic background T_B is influenced by factors such as T_s and the surface windspeed. Figure 5.3 shows the dependence of emissivity on varying SST for the five channels of the SMMR. The plots apply to a *smooth* sea surface. They show that emissivity changes most at the higher (37, 21 GHz) frequencies. Note that the remote sensing of SSTs needs to control for the atmospheric water vapor. By the same token, knowledge of the SSTs can be used to predict SMMR-derived precipitable water (Stephens, 1990). With surface wind speeds in excess of about 7 m s^{-1}, the sea becomes roughened and foamed and goes from being a specular reflector with low emissivity and T_B, to a Lambertian reflector with higher emissivity (Prabhakara *et al.*, 1982). Underestimation of the precipitable water by as much as 10% occurs for a surface wind of 30 m s^{-1}. Thus, low-level winds can be directly remotely sensed using the SMMR and other non-nadir viewing passive microwave sensors (Wilheit and Chang, 1980).

(2) Unlike the IR methods, microwave retrievals of atmospheric moisture can be made through extensive clouds that are non-raining (Chang and Wilheit, 1979; Prabhakara *et al.*, 1982). The value of combining the IR water vapor and microwave T_B for cloud identification (liquid water) has been suggested by Prabhakara *et al.* (1982), and applied to situations over land using the SSM/I by Jones and Vonder Haar (1990). Where rain situations occur, there may be significant underestimation of the column precipitable water due to strongly increased T_B from absorption and scattering (Staelin *et al.*, 1976; Grody *et al.*, 1980).

(3) The NEMS and SCAMS used two channels, both centered on the weak water vapor absorption band at 22.235 GHz and a contrasting window band between about 31.4 and 31.65 GHz. For the NEMS, water vapor and liquid water (cloud) retrievals utilized a linear combination of the microwave emission in the two channels for a representative set of oceanic and atmospheric conditions (Staelin *et al.*, 1976; Chang and Wilheit, 1979). In the case of the side-scanning SCAMS, a non-linear combination gave improved accuracy when compared with colocated radiosonde data, particularly for the large values of precipitable water typically found in the tropics (Grody *et al.*, 1980). The slightly coarser resolution of the SCAMS (150–300 km) contrasts with the NEMS (about 200 km).

(4) The Nimbus-7 and Seasat SMMRs are five-channel sensors. The precipitable water is obtained as a physically based retrieval that utilizes characteristics of radiative transfer in the microwave region (Prabhakara *et al.*, 1982). In this method, two channels (21 GHz, 18 GHz) are located on one side of the weak water vapor absorption band at 22.235 GHz. The difference in T_B between these two channels ($T_{21} - T_{18}$) is proportional to the amount of the absorbing gas (i.e., the precipitable water) (Fig. 5.4). Note from Fig. 5.4 that the increase in T_B spectra due to an increase in the cloud liquid water is generally small at the 18 and 21 GHz frequencies when compared with the effect of the water vapor. The T_B difference ($T_{21} - T_{18}$) is reduced as a result of increasing surface windspeed and also is dependent on

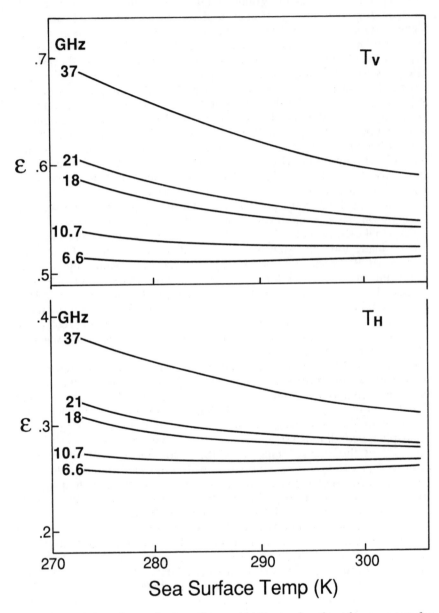

Fig. 5.3 Variation of smooth sea surface emissivity as a function of temperature for SMMR frequencies. (From Prabhakara *et al.*, 1982.)

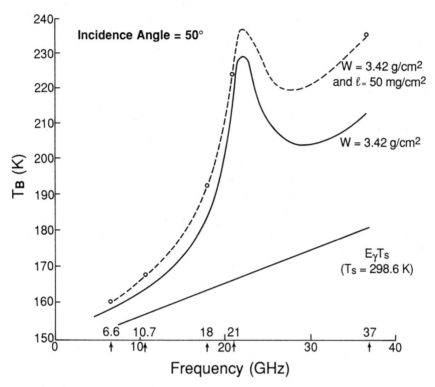

Fig. 5.4 Brightness temperature (T_B) spectra between 6 and 37 GHz for (1) no atmosphere (solid straight line); (2) a tropical model atmosphere with 3.42 g cm^{-2} of precipitable water (solid curve); and (3) addition of 50 mg cm^{-2} of cloud liquid water to case 2 (dashed curve). (From Prabhakara *et al.*, 1982.)

the cloud liquid water content for a given precipitable water (Prabhakara *et al.*, 1982). Below about 2.7 g cm^{-2}, precipitable water is *overestimated* in the presence of clouds, and *underestimated* above that value. However, Prabhakara *et al.* (1982) consider that the errors in retrieving precipitable water under average surface winds and cloud liquid water content are small, and this assertion seems justified by comparisons with colocated radiosonde observations. Similar good agreement has been reported for radiosonde–Seasat SMMR total precipitable water for the tropical oceans and Gulf of Alaska by Alishouse (1983). Wang *et al.* (1989) propose the use of higher microwave frequencies (183 GHz for water vapor absorption and the 90 GHz window) for more accurate determination of precipitable water in dry air (<0.5 g cm^{-2}), for improved spatial resolution and, possibly also, for retrievals over land (Kakar, 1983). Such a system is planned with the AMSU (Advanced Microwave Sounding Unit) on board the new NOAA polar orbiters as part of the Earth Observing System (Eos).

The value of SMMR-derived 'atmospheric water parameters' (water vapor, cloud liquid water, rain rates) for improving synoptic analysis and forecasting over the open ocean has been demonstrated for cases having high-quality *in situ* (ship) data by Katsaros and Lewis (1986) and McMurdie and Katsaros (1985). They find cold fronts to be located at the leading edge of the strongest gradient in the integrated water vapor field (refer to Fig. 1.1). Large values of cloud liquid water content are associated with the widespread cirrus shield associated with frontal cyclones, and meso-scale precipitation areas (MPAs) are also identifiable (Alishouse, 1983; McMurdie and Katsaros, 1985). The last seem to maintain their identity for some time and have the potential to be forecast using simple advection from real time satellite microwave data. Katsaros *et al.* (1989) examined a large sample of oceanic frontal systems, and found that a 'flag' to identify a water vapor gradient of 3.5 kg m^{-2}/60 km captured 86% of the fronts, while a precipitation flag identified 91%. These flags are complementary in that the gradient flag pertains mostly to the apex of the cold and warm fronts, and the weather flag defines the frontal zones further away from the cyclone center. The use of these techniques is suggested for operational use with the DMSP SSM/I (Special Sensor Microwave Imager).

(c) Climatologies

Atmospheric moisture (non-precipitation) climatologies from the water vapor absorption and passive microwave methods fall into the following three groups:

(1) global or near-global scale inventories of mid- and upper-tropospheric water vapor from the TIROS and Nimbus polar orbiters;
(2) analyses of the mean precipitable water and cloud liquid water over the oceans from the NEMS, SCAMS and SMMR (Seasat, Nimbus-7) sensors;
(3) analyses of the variability of atmospheric moisture associated with the ENSO phenomenon in the tropical and subtropical Pacific.

(1) Large-scale studies of mid- to upper-tropospheric water vapor Raschke and Bandeen (1967) compiled a monthly climatology of upper tropospheric relative humidity for the period February–June 1962 for latitudes between about 55° N and 55° S. Their study indicated the main features of the atmospheric circulation; notably the dry air associated with the oceanic subtropical highs and the higher humidities of the tropical land areas (ITCZ) and middle latitudes (polar front). In addition, they showed the migration of the ITCZ from low latitudes of the Southern Hemisphere in February/March into the Northern Hemisphere during May and June (Fig. 5.5). However, as acknowledged by these authors, and subsequently emphasized by Fritz and Rao (1967), the Raschke and Bandeen water vapor climatology is not strictly for clear air; there is considerable contamination by high-level clouds, especially in the ITCZ. In those areas, the climatology is probably capturing

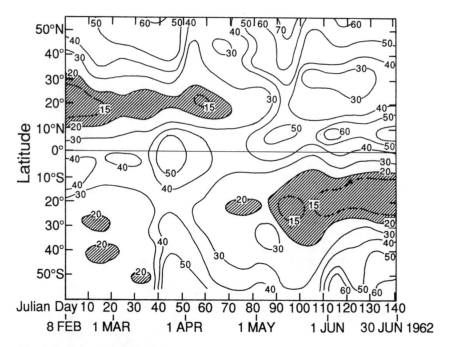

Fig. 5.5 Seasonal march of the zonal averages of mean relative humidity (%) of the upper-troposphere inferred from TIROS IV radiometric measurements. The averages were determined for 10-day periods within zones of approximately 5° of latitude. (From Raschke and Bandeen, 1967.)

the high humidities within the clouds so that the large spatial range of relative humidities shown in their maps should be reduced for clear-air-only conditions. Picon and Desbois (1990) investigated the statistical associations between METEOSAT mean monthly water vapor radiance fields and the atmospheric circulation features of the tropics derived from operational meteorological analyses. High radiances are generally associated with midtropospheric subsidence in the subtropical highs; low radiances with ascent (e.g., in the ITCZ). However, the advection of water vapor may significantly reduce the correlations between radiance and vertical motion in certain regions.

Lower-level moisture (precipitable water) over the oceans was determined from the 8–9 μm channel of the Nimbus-4 IRIS by Prabhakara *et al.* (1979). The parameter

$$\frac{(\bar{w} - w)}{\bar{w}} \times 10,$$

(refer to Section 1a, above), gives a measure of the strength of the trade inversion. For the period April, May and June 1970 (Fig. 5.6a,b) high values

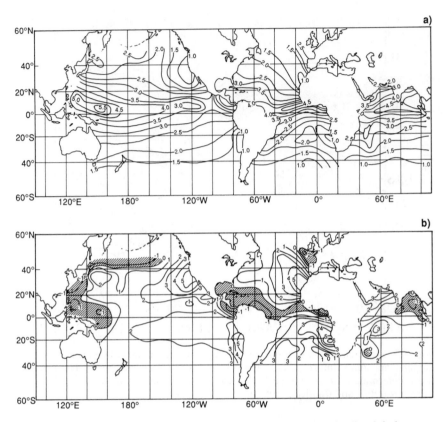

Fig. 5.6a, b Maps showing Nimbus-4 IRIS moisture parameters for the global oceans (50° N–40° S) in the period April, May and June 1970. These are (a) distribution of the total water vapor content (g cm^{-2}); and (b) distribution of the trade inversion index. Positive values of the index indicate inversion and negative values convection (stippled). (From Prabhakara *et al.*, 1979.)

of the precipitable water occur over the tropical oceans and in northern mid-latitudes over portions of the Kuroshio and Gulf Stream. These are associated with negative values of the index and weak boundary layer inversions. Conversely, lower values of precipitable water indicate strong inversion conditions associated with subsidence (positive values of the index). These are dominantly associated with cold ocean currents and upwelling in the subtropics. Prabhakara *et al.* (1979) also present similar maps for the 3-month seasons July–September and October–December 1970, and they determined the seasonal changes in the index

$$\frac{(\bar{w} - w)}{\bar{w}} \times 10,$$

(2) *Atmospheric moisture over oceans from microwave studies*   Using data from the NEMS sensor, Staelin *et al.* (1976) derived maps of precipitable water over the tropical and subtropical oceans for selected 4-day periods of 1973. For January they note a particularly sharp gradient in humidity between about 5° and 15° N and the influence of the trade winds. Latitude plots of NEMS water vapor and cloud liquid water for that month reveal a steady decrease with latitude of the former away from the peak in the ITCZ. The liquid water tends to be relatively constant with latitude and to have mean values ranging between about 1 and 20 mg cm^{-2}

For the two-week period August 18–September 4, 1975, Grody *et al.* (1980) present maps of mean precipitable water and cloud liquid water for the tropical and subtropical Pacific derived from Nimbus-6 SCAMS. For that period, the low values of precipitable water (25–45 mm) and cloud liquid water ($<2 \times 10^{-1}$ mm) in the dry zones of the central and eastern Pacific contrast with the considerably higher values (50–60 mm and 3–5×10^{-1} mm, respectively) over the western Pacific in a narrow band between about 5° and 12° N. The latter are associated with the tropical branch of the South Pacific Convergence Zone. The above patterns are considered representative of the 'normal' or even enhanced normal pattern for this time of the year, given a positive Southern Oscillation Index for much of the time period considered.

SCAMS data on 20 North Pacific extratropical cyclones were used by Viezee *et al.* (1979) to assess the feasibility of predicting coastal rainfall for the north-west US from the retrieved antecedent cloud liquid water values over the ocean. The experiment showed encouraging results for most cases, with the 72-hourly cumulative rainfall correlating most frequently with the SCAMS data available 60–72 hours previously.

Several authors (Prabhakara *et al.*, 1982; Njoku and Swanson, 1983; Chang *et al.*, 1984) have used the Nimbus-7 and Seasat SMMRs to derive mean patterns of precipitable water over the global oceans. The studies by Prabhakara *et al.* (1982) and Chang *et al.* (1984) are for the FGGE period (1979). The maps for January and July 1979 (Fig. 5.7a,b) show the expected associations between cold (warm) ocean currents and low (high) values of precipitable water. The high values over the maritime continent in both months are associated with active convection and low values of OLR. The zonally averaged water vapor for mid-season months of 1979 (Fig. 5.8) shows that the seasonal change is most marked for the Northern Hemisphere owing to its greater land area. Higher values of the water vapor occur further poleward in summer and more equatorward in winter. Chang *et al.* (1984) compare their Nimbus-7 SMMR climatology with climatologies derived using conventional data and find that their values are substantially higher over the Northern Hemisphere. They attribute this to the lack of oceanic data for the conventional climatologies and the ocean-only retrievals of water vapor for the SMMR. Accordingly, closer agreement between the satellite and conventional analyses is found for the Southern Hemisphere. Stephens (1990) demonstrates the close association between monthly mean bulk SSTs and SMMR-derived precipitable water for the global oceans, and the role of large-scale processes such as subsidence and moisture convergence for those water vapor patterns.

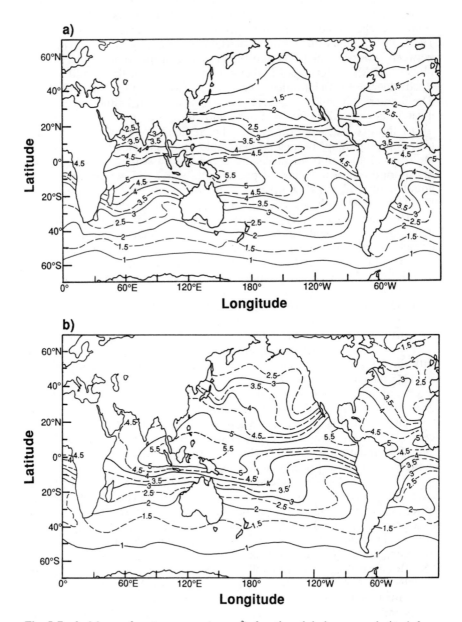

Fig. 5.7a, b Maps of water vapor (g cm^{-2}) for the global oceans derived from Nimbus-7 SMMR. These are (a) January 1979; (b) July 1979. (From Chang *et al.*, 1984.)

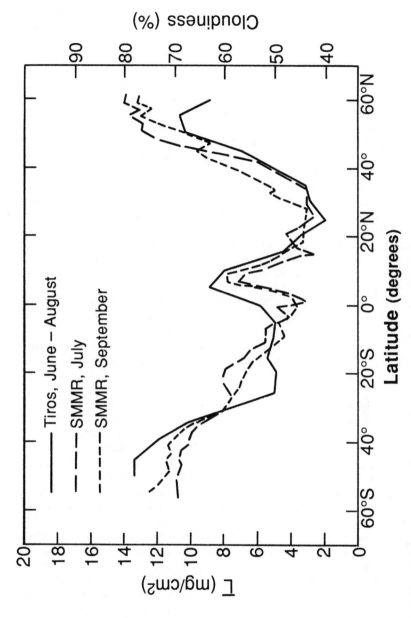

Fig. 5.8 SMMR-derived zonally averaged cloud liquid water (L⁻) for the periods July 11–August 10, 1978; and September 11–October 10, 1978. Also shown is fractional cloud cover deduced from the study of TIROS observations by Clapp (1964). (From Njoku and Swanson, 1983.)

Although of short duration (July 7 to October 10, 1978), the Seasat mission has provided a wealth of information for oceanographers, climatologists and glaciologists. Njoku and Swanson (1983) present a climatology of SST, wind speed, precipitable water and cloud liquid water for the Seasat mission and for sub-periods. Seasat analyses of precipitable water and cloud liquid water are comparable with other satellite analyses in terms of the spatial patterns disclosed, although there are differences in absolute values. These may be functions of real changes in the atmospheric moisture; the choice of averaging period; or differences in the spatial resolution of the sensors. Njoku and Swanson's analysis of cloud liquid water shows a generally close association with previous satellite observations of cloud cover (Fig. 5.8). Differences result from the oceanic nature of the SMMR climatology, and are particularly marked for the winter (Southern) hemisphere where the lower microwave values result from the reduced liquid water and enhanced ice content of clouds. The latter is essentially transparent to microwave radiation at SMMR wavelengths. The associations between Seasat SMMR-derived parameters is given for the full mission in Fig. 5.9. Both the zonally averaged SST and atmospheric water vapor decrease away from the tropics, and have a correlation of 0.87 for 90-day averages. The cloud liquid water content is high in the tropics, low in the subtropics and increases through middle to higher latitudes. On the other hand, wind speeds at the sea surface are lowest in the tropics and northern subtropics and increase with latitude, especially in the Southern Hemisphere (winter). The correlation between the wind speed and water vapor for the 90-day period July 11–October 10 is −0.79, while that for wind speed and SST is −0.88. Njoku and Swanson point out potential intercorrelations among satellite-retrieved parameters in these results, such as that between SST and atmospheric moisture.

(3) *Interannual variability and ENSO*   The relatively long-term (1978–87) and stable Nimbus-7 SMMR facilitates analysis of interannual variations of ocean climate in the context of a reliable background climatology (Chang *et al.*, 1984). These variations involve particularly the ENSO phenomenon of the tropical Pacific; notably the major 'warm' (El Niño) event of 1982–3. Prabhakara *et al.* (1985) examined SMMR-derived atmospheric water vapor for the 1982–3 ENSO event with respect to the seasonal means for the three years 1979–81. An analysis of the interannual variability for that period, determined as the monthly deviation of water vapor from the three-year monthly mean (Fig. 5.10), reveals that the largest non-seasonal variations occur over the Arabian Sea, the Bay of Bengal and in the tropical and subtropical regions of the Indian and west Pacific oceans. Large variations are also observed off the east coast of Japan, associated with cold air outbreaks. An EOF (empirical orthogonal function) analysis of these water vapor deviations reveals the presence of a signal reminiscent of ENSO, even in the non-extreme (ENSO) part of the record. Accordingly, Prabhakara *et al.* (1985) mapped monthly anomalies of water vapor for the onset, peak and withdrawal phases of the ENSO between January 1982 and July 1983. At the peak of the event in January 1983 (Fig. 5.11) positive anomalies occurred in

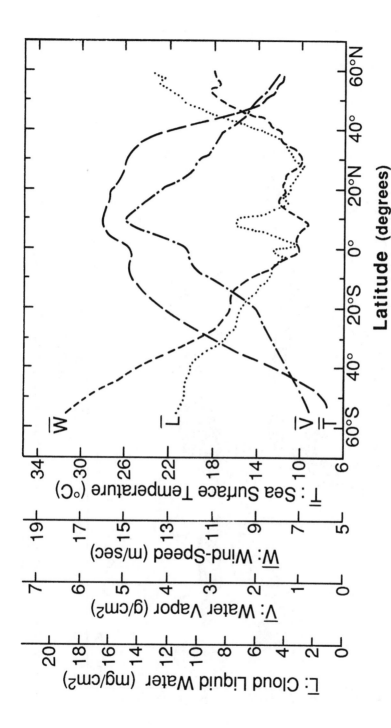

Fig. 5.9 Seasat-derived zonally averaged SST, wind speed, water vapor, and cloud liquid water for the period July 11–October 10, 1978 (full mission). The data are plotted on similar scales for comparison. (From Njoku and Swanson, 1983.)

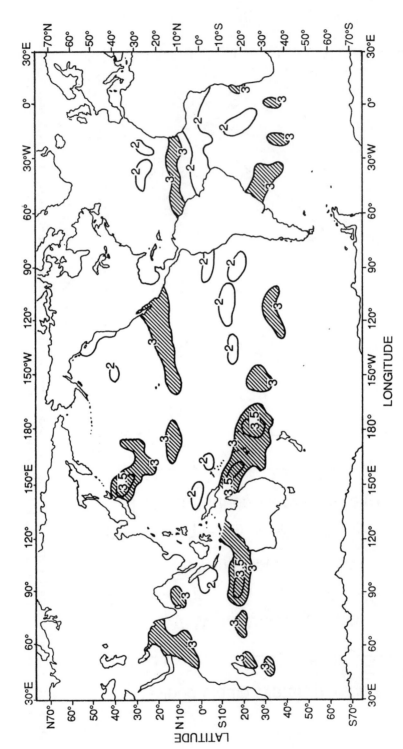

Fig. 5.10 Root-mean-square (RMS) deviation of water vapor (g cm^{-2}) for the global oceans derived from SMMR measurements for the period 1979–81. (From Prabhakara *et al.*, 1985.)

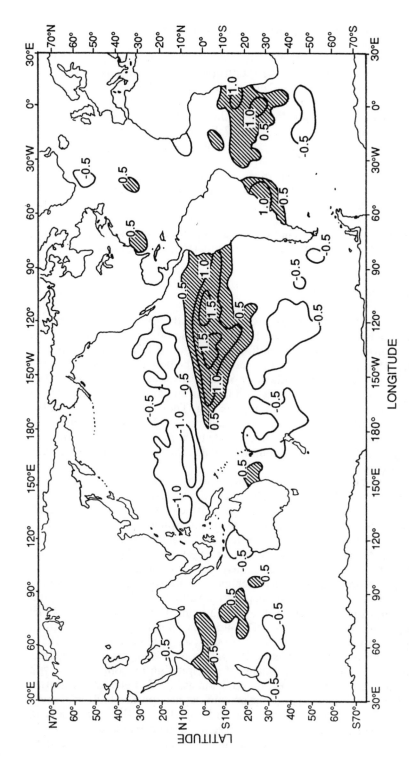

Fig. 5.11 SMMR-derived water vapor for the global oceans for January 1983, expressed as a departure from the 1979–81 means (g cm⁻²). Positive values are hatched. Note particularly the strong increases in the central and eastern tropical Pacific associated with ENSO. (From Prabhakara *et al.*, 1985.)

185

the central and eastern tropical Pacific and South Atlantic, and were flanked by negative anomalies in the subtropics. These results reveal the strengthened subsidence associated with subtropical highs and enhanced convergence over the central and eastern tropical Pacific. The latter was associated with strongly positive anomalies of SST. At the same time, they imply a reversal of the tropical zonal Walker Circulation, whereby descent (ascent) is enhanced over the western (central and eastern) Pacific, and its associated tropical teleconnections.

Liu (1987) presents an atlas of the monthly fields of water vapor for the 1982–3 ENSO, and also shows the interactions with SST and low-level winds. These vary regionally and according to stage of development of the ENSO. However, the close association between positive SST and positive water vapor anomalies is generally apparent, and there is some association with low wind speed anomalies (convergence) in the onset phase of 1982.

2 Precipitation estimation

Attempts to estimate precipitation using satellite data have a long history (see Barrett and Martin, 1981). Methods of precipitation estimation can be classified into two groups: the indirect and the direct methods (Arkin and Ardanuy, 1989). In the indirect method, cloud characteristics in the visible and IR channels are used as indicators of the occurrence of precipitation. The precipitation rate and duration is estimated from parameters such as cloud type, thickness, CTT, and cloud areal extent. They comprise the so-called indexing, life history and bispectral techniques. The direct method of precipitation estimation utilizes microwave techniques to obtain the 'instantaneous' rain rate (about a 20-minute average: Barrett et al., 1990); usually over the ocean.

Validation of the satellite precipitation retrievals is a critical component, as it is with all satellite-based techniques. Such 'ground truth' comprises surface-based (land, ship) observations of precipitation and weather conditions (e.g., cloud) (Garand, 1988, 1989); atmospheric soundings (Gruber, 1973; Wylie, 1979), and remote sensing, most notably surface-based radar. The radar–satellite mix dates back to the earliest studies of remote precipitation estimation in the synoptic context (e.g., Bristor and Ruzecki, 1960; Nagle and Serebreny, 1962; Blackmer and Serebreny, 1968). More recently, they comprised a critical component of the validation of satellite precipitation estimates for FACE (Florida Area Cumulus Experiment) and GATE; particularly for the association of cloud area, cloud thickness, fractional coverage, and CTT with precipitation rates (see, e.g., Griffith et al., 1978; Arkin, 1979; Lovejoy and Austin, 1979; Woodley et al., 1980). Of course, validation of surface radar estimates of precipitation by comparisons with gauges is also required (Barnston and Thomas, 1983; Petty and Katsaros, 1990). Bellon et al. (1980) describe an automated approach to precipitation estimation for eastern Canada using both visible and IR GOES data in conjunction with surface-based radar. Precipitation probabilities for daylight hours are derived and presented as maps. These can be sequenced to facilitate the tracking of major precipitation-producing features and for nowcasting. An evaluation of this so-called RAINSAT technique, which is similar to the UK Meteorological Office FRONTIERS, has very recently been performed for 1985 (King *et al.*, 1989). The greatest skill comes about when separating cloudy–no rain from cloudy–rain situations and, to a lesser extent, in assigning different levels of probability to the latter. Temporally, skill is highest between about 1 and 3 hours, but becomes zero after about 6 hours. Seasonal variations in precipitation probability were identified, and these are to be expected when considering changes in the dominant precipitation mechanisms. However, within-season variability is small, especially in summer and winter. The ultimate objective of such techniques is to be able to determine precipitation from the satellite data alone by extending the satellite–radar–precipitation relationships into areas lacking surface-based radar (Doneaud *et al.*, 1987).

To be reliable, rainfall estimation techniques clearly must be able to capture the dominant patterns and processes of precipitation. This involves particularly the strong diurnal variation of convective precipitation over the tropics and the humid mid-latitudes in summer, even where climatic time and space scales are considered (Arkin and Ardanuy, 1989). Moreover, a satellite technique of precipitation estimation for Western Europe is likely to emphasize *advective* processes compared with one for the United States Great Plains. In the latter, *convective* processes would more likely be emphasized, particularly in the spring, summer and fall (cf. Browning, 1979; Scofield and Oliver, 1977). At the same time, these schemes must be able to account for localized influences on precipitation. Regional and seasonal differences in the dominant mechanism of precipitation pose problems for the development of a single satellite rain estimation scheme. Wylie (1979) showed that a satellite scheme of rain rate estimation developed for the tropics does not transfer well to middle latitudes, especially for cool-season advective situations (Fig. 5.12): rain rates tend to be overestimated, and these are a function of differences in precipitable water, stability and cloud-top altitude. Similarly, satellite precipitation estimation schemes developed in the tropics and applied to convective situations in semi-arid regions such as the US High Plains, tend to overestimate actual amounts by a factor of 3–5, depending on area size (Griffith *et al.*, 1981). However, such techniques can still be used provided the different meteorological situations in drier regions are accommodated. Adjustments to the satellite estimates using gauge data or a cloud physical model are appropriate in these cases. Wylie and Laitsch (1983) confirm the improvement to precipitation estimation over the Great Plains when gauge data are included with the satellite analysis. Differences in rain rates between tropical and mid-latitudes can be adjusted either with the use of an appropriate cloud model, which may incorporate conventional sounding data (Wylie, 1979); or by the inclusion of a direct satellite microwave form of precipitation estimation (see, e.g., Barrett *et al.*, 1988).

Precipitation estimation through the use of a physical model involves parameterization of cloud growth processes. That is, the method is physically based rather than statistical. In Gruber's (1973) technique, convective precipitation per unit time (P) is established from the following equation:

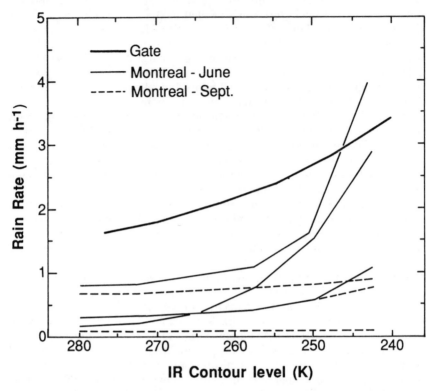

Fig. 5.12 Satellite-retrieved (IR) rain rates averaged over cloud areas for various cloud threshold temperatures (horizontal axis). Rain was measured by gauge-calibrated radars. Note the strong difference in the rain rates between tropical and mid-latitude situations for similar IR temperatures. (From Wylie, 1979.)

$$P = lQ_1/\Delta t \qquad (21)$$

where l is the fraction of a synoptic area covered by deep active convection; Q_1 is the amount of condensation heating according to moist adiabatic ascent; and Δt is a time parameter optimized for the precipitating lifetime of the convective elements. The parameter Δt is determined to be about 30 minutes and the precipitation is strongly influenced by the l parameter. Determination of the latter involves the use of satellite data: cloud top temperatures (IR) are thresholded in accord with surface-based radar observations. The application of this essentially tropical technique to a mid-latitude summertime convective situation gave encouraging results.

(a) Cloud indexing, life history and bispectral methods

Barrett (1970) pioneered the cloud indexing method of precipitation estimation over the western tropical and subtropical Pacific. In the original method, cloud cover characteristics on nephanalyses were related to the statistical probability and intensity of rain (e.g., Table 12). The technique was applied subsequently to the estimation of convective precipitation in arid Oman using DMSP imagery. Daily estimates were integrated over time scales of a month, once regression relationships had been established for different morphoclimatic regions of the Sultanate. The study revealed that use of the sparse gauge density alone has led to serious overestimation (by about 30%) of rainfall amounts in the north-east region. More recently, cloud indexing methods have been extended to include geostationary satellite data and to incorporate information from additional spectral bands. The ESOC (European Space Operations Centre) precipitation index (EPI) is derived using METEOSAT IR and 6.3 μm water vapor absorption band data (for upper-tropospheric humidity). The EPI is evaluated every three hours and summed over five-day periods. Its ability to estimate rainfall in different climatic regions of Africa was examined recently by Turpeinen et al. (1987), who compared the EPI with observed precipitation. For the short time period they studied (October 3–December 26, 1985), Turpeinen et al. (1987) found good agreement in the tropics but a poor performance in the subtropics. They attribute the latter to the occurrence of frontal precipitation and to the relatively short verification period. Automation of the process of cloud field extraction is also proving to give reliable estimates of oceanic precipitation (Garand, 1989).

Table 12 Rainfall probabilities and intensities as related to states of the sky

States of the sky (nephanalysis cloud categories)	Assigned probabilities of rainfall (relative scale range 0–1.00)	Assigned intensities of rainfall (relative scale range 0–1.00)
Cumulonimbus	0.90	0.80
Stratiform	0.50	0.50
Cumuliform	0.10	0.20
Stratocumuliform	0.10	0.01
Cirriform	0.10	0.01
Clear skies	—	—

Source: Barrett, 1970.

The utility of satellite nephanalyses for the *prediction* of rainfall at a single point in the mid-latitudes (Valentia, Ireland), was explored by Barrett (1973). The technique involves the determination of cloud indices associated with synoptic weather conditions upstream of the site, which are advected to the site over a given time period (e.g., 24 hours). Validation of the technique for January through June 1967 gave encouraging results, especially considering the complexity of middle-latitude weather.

Cloud spectral characteristics important in the indirect techniques of satellite precipitation estimation include, for the visible channel, the occurrence of highly reflective clouds (HRCs). In the IR, these correspond with low values of OLR. The probability of precipitation is increased with increased cloud brightness (albedo) for a given sun–cloud–satellite geometry because this is associated with greater cloud depth, or thickness (Griffith and Woodley, 1973).

In the IR, high cold clouds tend to be thick and rain producing. However, CTT alone is not a sufficiently reliable index of precipitation: thick cirrostratus clouds are radiatively cold but do not precipitate, and dissipating cumulonimbus cells in drier climates may give little precipitation, although they exhibit low CTTs. For GATE Arkin (1979) shows a strong relationship between the fractional coverage of high cold cloud (>10 km and 235 K) and 6-hourly accumulated rainfall ($r = 0.84$–0.88). Moreover, the relationship between OLR and tropical rainfall varies spatially. For example, OLR-derived rainfall tends to be overestimated in the drier eastern and central tropical Atlantic, and underestimated in the wetter western Atlantic (Yoo and Carton, 1988). These are associated with zonal differences in lower-level humidity and the height of the trade wind inversion. A similar result was found for the tropical Pacific (Motell and Weare, 1987). Accordingly, Yoo and Carton incorporate a measure of the trade-wind inversion height into their regression equation for precipitation estimation. The results are improved when verified against island rainfall observations.

As a general rule, the accuracy obtained by use of either visible-band or IR data for precipitation estimation varies with latitude. Visible data work best in tropical situations (dominated by convection), while IR data give better results in middle latitudes dominated by advective situations. Thus, HRCs in the tropics are strongly associated with surface-observed precipitation. For the period May 1971–April 1973 a correlation coefficient of $r = 0.75$ was determined between the occurrence of HRCs on polar orbiter imagery and island station rainfall in the tropical Pacific (Kilonsky and Ramage, 1976). Griffith et al. (1978) used geostationary satellite data to determine tropical convective precipitation on the basis of high cold clouds, and the technique was applied to precipitation estimation for the GATE experiment (e.g., Fig. 5.13a,b) (Woodley et al., 1980). The Griffith–Woodley technique is an example of a so-called 'life history' technique of precipitation estimation: one that relies on the time-dependent evolution of rain-producing cloud systems to infer precipitation amount, which is best monitored from geostationary platforms. Figure 1.14 confirms the ability of either visible or IR data to describe the life history of convective rain clouds, and also shows the decrease in precipitation according to stage of cloud development (a maximum in the early development stages prior to cirrus spreading). Development of the anvil top by spreading in the latter stages of growth gives an unrealistically large 'rain area' on the basis of the cold and bright spectral signatures. The cloud area may be of the order of 4 times the size of the precipitation area (Lovejoy and Austin, 1979). Accordingly, the rain area

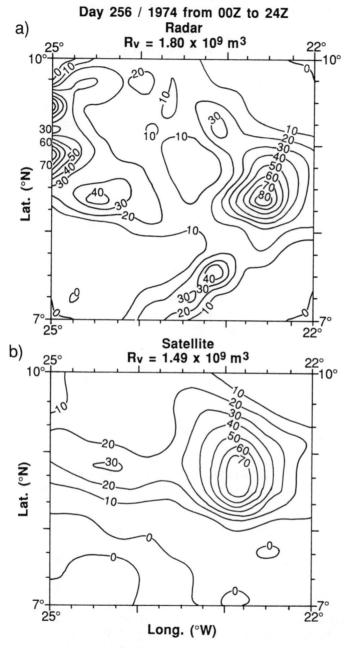

Fig. 5.13 Comparisons between ship-board radar (top) and satellite-estimated daily rainfall (mm) for GATE on September 13, 1974. The satellite estimates have been made with digital IR data of $1/3° \times 1/3°$ latitude/longitude using a computer method. Relationships derived over south Florida have been used in the satellite computation. Area- and time-integrated volumes are indicated for each product. (From Griffith *et al.*, 1978.)

can be effectively constrained with the use of ground-based radar and the rain rate tends to be increased compared with that derived from the satellite data alone.

Stout *et al.* (1979) show the value of satellite cloud area determined from high frequency (hourly) GOES data for estimating volumetric rainfall ($m^3 s^{-1}$) for GATE. Regression relationships were established between threshold cloud area and shipboard radar echoes for convective precipitation. Stout *et al.* (1979) also examined the problem of 'background rainfall' in the tropics, or precipitation that is not associated with deep convective clouds. They suggest that, although it may be important, it can be ignored when estimates include cumulonimbus-derived precipitation. For convective precipitation during GATE, satellite estimates improve as the space and time scales increase (e.g., Table 13). For space scales on the order of 2–3° of latitude, and in which the cloud life cycle may be ignored, a linear relationship of cloud area and rain rate is appropriate.

Table 13 Correlation between radar area within the 20 dBZ contour and the rainfall within that area for phases of GATE

Scale (deg.)	Phase		
	I	II	III
0.5	0.86	0.86	0.86
1.5	0.91	0.85	0.94
2.5	0.91	0.89	0.95

Source: Richards and Arkin, 1981.

Despite the different temporal resolutions of the satellite data used in the Kilonsky–Ramage and Griffith–Woodley techniques, comparisons for GATE (Garcia, 1981) reveal comparable performance over oceanic areas at the large scale. The inclusion of continental areas into the estimates reduces the ability of the Kilonsky–Ramage technique to capture the strong diurnal variations in west African rainfall which involve a nocturnal maximum. On smaller scales (GATE-B) both techniques are comparable and give rainfall volume estimates to within about 15% of radar estimates. The precipitation estimation results from GATE have been incorporated into a climatological analysis of precipitation from GOES data, the GOES precipitation index (GPI) (Arkin and Meisner, 1987). Much of the research deriving from the GATE studies forms the basis of the US NESDIS operational scheme for estimating heavy convective precipitation (Scofield, 1987).

The premise that precipitation-bearing features in middle latitudes tend to maintain their identity as they are advected forms the basis of Muench's (1981) nowcasting technique. Figure 5.14 shows the probability of receiving precipitation the following hour as a function of visible reflectivities and IR radiances, and confirms the bright–cold cloud relationship noted in other

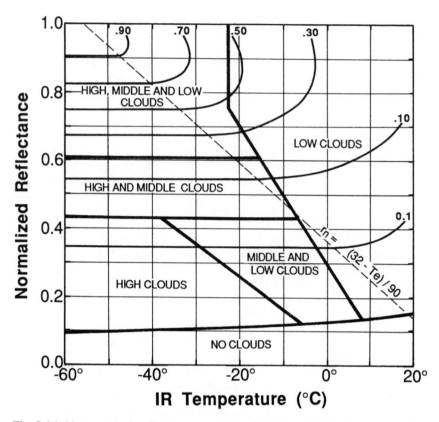

Fig. 5.14 Nomogram showing the probability of receiving 0.01 inches of precipitation the following hour, as a function of satellite-retrieved reflectivity and IR temperature ($Te = T_{BB}$, °C) using the Muench technique. (After Muench, 1981.)

studies. Initial cloud conditions are determined from a combination of geostationary satellite visible and IR data at high temporal resolution, and cloud features are extrapolated forward in time. The technique was tested on 12 spring and fall cases, with forecasts made out to 7 hours in $\frac{1}{2}$-hour increments. Precipitation forecasts were considerably better than persistence for periods exceeding about $1\frac{1}{2}$ to 2 hours, but worse on time scales shorter than this. Muench attributes the latter to problems in the initial specification of cloud conditions.

Three-hourly infrared imagery from the Japanese GMS system forms the basis of an estimation scheme for rainfall associated with extratropical cyclones moving over southern Australia (DelBeato and Barrell, 1985). The method capitalizes on the very different cloud and precipitation characteristics of maritime cyclones, according to sector (and cloud field) considered (refer to Fig. 3.9). Correlations between observed gauge precipitation over the State of Victoria and CTT coincident with each station revealed the need for imagery at a frequency greater than 3 hours. Accordingly, linear interpolation of cloud features to an interval of 90 minutes was necessitated to produce sufficiently high correlations between CTT and precipitation in the convective sector. These could then be used, in conjunction with other published data, to produce a curve of the relationship between CTT and 90 minute precipitation totals, which has the following form:

$$R_{90} = 0.00025(5 - CTT)^{2.61} \tag{22}$$

where CTT is in °C; R_{90} is in millimeters. Verification was performed for five cases by summing 90 minute increments over a 24-hour period for each case. The contribution of upslide precipitation ahead of the trough was assumed to be one-sixth of the value given in equation (22) (refer to Fig. 3.9). In general, mean district rainfall was overestimated by 22% over flat inland surfaces. Several reasons were suggested for these results, ranging from physical processes such as those associated with sub-cloud evaporation in the convective sector, to difficulties in estimating the upslide contribution and interpolation of cloud features. Martin *et al.* (1990) tested several GOES techniques for estimating rainfall amounts over Amazonia on daily to weekly time scales. They confirmed the ability of IR data to augment significantly the sparse gauge network in that region.

It was noted in Chapter 3 that cloud detection schemes tend to work best when using a combination of visible and IR channel data, and this is also the case for precipitation estimation (Lethbridge, 1967). As Fig. 5.15 shows for a day during GATE (Lovejoy and Austin, 1979), rain areas tend to be cold and bright; non-raining areas either warm and dull, cool and moderately bright, or cold and bright. A recent study by O'Sullivan *et al.* (1990) uses GOES visible and IR image pairs statistically to assign 10×10 pixel arrays to one of three classes of *instantaneous* surface rainfall rates. However, there are problems with this so-called bispectral method of precipitation estimation (Barrett and Martin, 1981). As with the assignment of clouds into the low, middle and high categories (Chapter 3), the main problem involves the areas of overlap in the temperature/brightness kernels. Thus, cold bright clouds may not give precipitation, but warmer less bright clouds may. This difficulty is associated with the generally smaller area of precipitation compared with the satellite-viewed cloud canopy which, in turn, varies according to whether the cloud is dominantly convective or advective (see, e.g., Barrett, 1973). In middle latitudes, the prevailing synoptic situation may be decisive for precipitation estimation by its influence on the cloud microphysics (condensation nuclei, drop-size distribution and concentration, liquid water content of the cloud) and, accordingly, the sublimation and coalescence processes. Aircraft studies conducted near Adelaide, Australia (Spillane and Yamaguchi, 1962) revealed strong differences in the precipitation efficiency of maritime compared with continental-derived clouds in that coastal region. Continental clouds tend to be deeper and to have lower CTTs than marine clouds for the occurrence of comparable precipitation intensities.

The satellite indirect techniques tend to underestimate daily and sub-daily precipitation amounts relative to verifying surface-based observations of

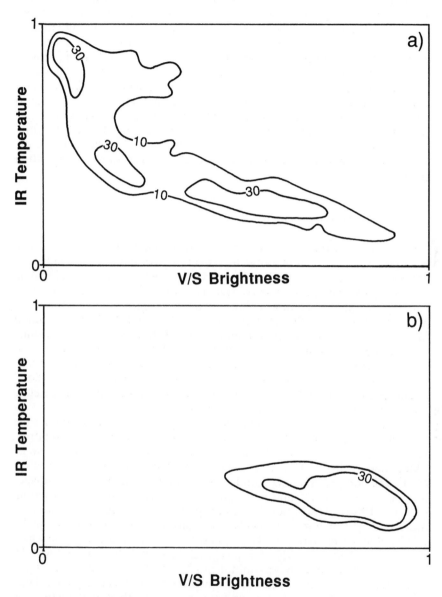

Fig. 5.15 Joint frequency distribution of SMS-1 visible (V/S) and IR data for a 400 × 400 km box centered at 09°00′ N, 22°40′ W in the eastern tropical Atlantic, 1300 GMT September 5, 1974; (a) no rain case; (b) rain case. Data are normalized to a scale 0–1. Note the overlap in the higher brightness, lower IR portion of the plots. (From Lovejoy and Austin, 1979.)

precipitation (see, e.g., Griffith, 1987). This is related to the differences inherent in a pixel (area-averaged) estimate contrasted with a point amount for the satellite and gauge situations. They are analogous to the validation of satellite observations of cloud cover from surface observations, and the determination of appropriate sampling strategies for clouds using remote sensing (refer to Chapter 1, Sections 5, 6). On these short time scales also, satellite navigation errors may have an adverse impact on the precipitation estimates. Increasing the time scale to the climatological (>5–10 days) reduces substantially the error in satellite precipitation estimates (see, e.g., Barrett, 1979), although actual amounts are still underestimated (Griffith, 1987; Flitcroft et al., 1989). The smaller differences are the result of the temporal and spatial smoothing of local effects and satellite navigation errors.

(b) Passive microwave methods

To date, space-borne microwave systems for precipitation estimation have been passive. That is, they sense the effect on the relatively low-intensity upwelling radiation from the Earth's surface and atmosphere of precipitating clouds, rather than utilizing active (radar) sensing. This is likely to change, particularly with the planned deployment of the Tropical Rainfall Measuring Mission (TRMM) satellite (Simpson et al., 1988) and the High-resolution Multi-frequency Microwave Radiometer (HMMR) of the polar-orbiting Eos (Earth Observing System) program (Wilheit, 1986; Arkin and Ardanuy, 1989). Thus, this method does not rely on establishing relationships between cloud parameters, such as cloud top height, CTT, or cloud brightness, but on relating the upwelling radiation received at the satellite sensor to a host of physical factors (Chapter 2).

The basis of the satellite direct technique of precipitation estimation is the attenuation by absorption and scattering of microwave radiation (approx. 0.1–10 cm, or 300–3 GHz: Chapter 2) in the presence of hydrometeors (Arkin and Ardanuy, 1989). At the lower frequencies sensed by ESMR-5 (19 GHz), non-precipitating clouds (reduced liquid water content, smaller drop sizes), ice clouds, and atmospheric gases are essentially transparent (Wilheit, 1986). Absorption by raindrops tends to increase the satellite-measured T_B. Using model calculations, Allison et al. (1974) developed a relationship between the ESMR-5 T_B and rainfall rate (Fig. 5.16), and which they then applied to a study of tropical cyclone rainfall. Note from Fig. 5.16 that the T_B change becomes asymptotic to 280 K, meaning that only rainfall rates less than 10 mm h^{-1} can be distinguished. At higher rates, the large extinction of radiation by the precipitation means that the top of the freezing level is observed. A similar effect has been noted for Nimbus-7 SMMR data at 37.0 GHz (Spencer et al., 1983a). Wilheit et al. (1977) quote a useful ESMR rain retrieval of about 1–20 mm hr^{-1} to within a factor of 2.

The intensity of the emitted microwave radiation is a function of the wavelength (λ), surface physical temperature (T) and surface emissivity (ε). The last is itself dependent on several physical factors, particularly the dielectric

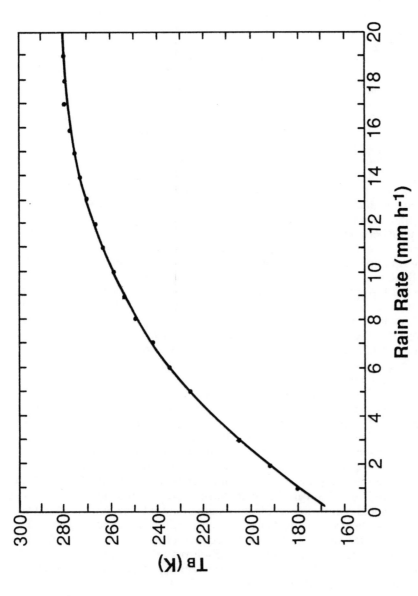

Fig. 5.16 Relationship of Nimbus-5 T_B (K) and rainfall rate in mm h^{-1} over ocean surfaces. (From Allison *et al.*, 1974.)

constant (Chapter 2). Heterogeneous surfaces, particularly over land, have variable emissivities. Moreover, in microwave wavelengths around 1.55 cm (19.35 GHz), the single channel ESMR−5, surface emissivities are relatively high over dry land (approx. 0.85–0.95) compared with moist land surfaces (<0.8) and water surfaces (about 0.4) (Schmugge *et al.*, 1977). Accordingly, the detection of precipitating clouds in the ESMR-5 was limited to oceanic regions with their low background emissivity (Wilheit *et al.*, 1977). Over land, precipitation could not be differentiated from changes in soil moisture. Subsequent microwave sensors utilizing dual polarization (Nimbus-6, Nimbus-7, DMSP SSM/I) have continued the monitoring of precipitation over oceans and, in addition, have improved the ability to detect heavy convective precipitation over land. These sensors utilize higher frequencies (e.g., 37 GHz of Nimbus-6 EMSR), dual polarization techniques and scattering-based algorithms for satellite rain determination (Chapter 2).

Spencer *et al.* (1983b) examined the utility of Nimbus-7 SMMR dual-polarized data for estimates of warm season rain rates over the central US by comparisons with ground-based radar. They found an approximately linear relation between T_B and rain rate up to at least 40 mm h^{-1}, which is such that lowest T_B corresponds with heaviest precipitation. Development of a stepwise multiple-linear regression model to specify the degree to which the SMMR data supply independent rain rate information revealed a multiple $r = 0.79$. Application of the statistical algorithm to an independent data set confirmed its utility, although there was a tendency to underestimate slightly the rain intensity at higher rain rates. Recent work utilizing the dual polarized 85.5 GHz channel of the SSM/I (Spencer *et al.*, 1989) holds promise for the ability to differentiate precipitation above and below the freezing level by comparison with the radiances at 19.35 GHz, and also to gain insight into the precipitation dynamics in different atmospheric environments.

In the microwave region, where radiation intensity is relatively low, the ground resolution for space-borne sensors is relatively coarse. For the Nimbus-5 ESMR, the nadir resolution was of the order of 25 km^2. The five channels of the SMMR (37, 21, 18, 10.7, and 6.6 GHz) had ground resolutions decreasing, respectively, from about 20 km at 37 GHz to about 70 km at 11 GHz (Spencer *et al.*, 1983c). For the 85.5 GHz channel on the SSM/I, ground resolution is about 15 km^2 (Spencer *et al.*, 1989). Improvements in the ground resolution in the microwave region from space-borne radiometers would require increasing the length of the antenna. At the same time, shortening the wavelength changes the absorption and scattering characteristics by water vapor, clouds and precipitation. Thus, precipitating areas detected in satellite passive microwave sensors exhibit the problem of *non-beam filling*: the pixel brightness temperature (T_B) may contain contributions from precipitating and non-precipitating cloud cells, particularly as spatial resolutions degrade. Comparisons of precipitation rates derived from ESMR-5 and -6 with surface observed (radar, gauge) precipitation, show an *underestimation* of the satellite estimates (Wilheit *et al.*, 1977; Weinman and Davies, 1978; Spencer *et al.*, 1983b; Beer, 1980). A recent study by Alishouse *et al.* (1990) uses the SMMR T_B to estimate the surface-based radar reflectivity and, accordingly, the fraction of the field of view of the radiometer that is filled with rain. Thus, the problem of estimating rain rates from measured radar reflectivities is avoided.

(c) Satellite precipitation climatologies

The application of the various techniques of satellite precipitation estimation to climatic scales has been mainly carried out for the data-sparse oceanic areas, and also the land areas of the tropics and subtropics. The latter includes particularly the monsoon regions of south and south-east Asia. These areas are dominated by convective precipitation, which lends itself to the application of the life-history techniques. In the case of the middle- and higher-latitude oceans, where widespread gradual ascent dominates, estimation of the 'instantaneous' precipitation rate can be made using passive microwave sensing. While the problem of non-beam filling and the resultant underestimation of precipitation rate is reduced on climatic scales (Prabhakara *et al.*, 1986), the poor temporal sampling of a polar-orbiting microwave sensor leads to considerable uncertainty in the satellite-derived precipitation climatology for these areas (Lovejoy and Austin, 1980). Even using more than one polar-orbiting microwave sensor, such as SMMR and SSM/I, the derivation of a reliable precipitation climatology is complicated by the lack of intercomparability of the channels and their spatial resolutions, and by the different times of satellite overpass. These problems are clearly shared by the attempts to derive a global satellite cloud climatology (Chapter 3). For the lower latitudes, precipitation climatologies based on satellite microwave data should improve significantly with the planned TRMM project and with the deployment of these microwave sensors on operational meteorological platforms in geostationary orbit (e.g., GOES NEXT).

(1) *Oceans* The role of the data-sparse tropical Pacific Ocean as the site of the El Niño Southern Oscillation (ENSO), has made it a prime subject for satellite rainfall climatologies (Arkin and Ardanuy, 1989). Earlier studies (e.g., Ramage, 1975; Kilonsky and Ramage, 1976) determined the locations of HRCs in visible imagery since these are the sites of most active convective precipitation. Strong variations in HRC frequencies were found to occur in association with the major 1982–3 ENSO event. Increases in HRC occurrence for December 1972 relative to December 1971 were observed close to the equator in the central Pacific, in association with the positive anomalies of SST. However, the latter association was not everywhere the same. ESMR-5 data were used by Kidder and Vonder Haar (1977) to derive tropical oceanic rain rates for the period spanning the peak of the 1972–3 ENSO (December–February). From the twice-daily data, they determined both the occurrence of precipitation and three classes of rain rate (0.25–1.0 mm h^{-1}; 1.0–2.5 mm h^{-1}; > 2.5 mm h^{-1}. In their analysis, the mid-Pacific precipitation maximum is substantially greater than climatology, and was associated with ENSO. The results compare favorably with those of Ramage (1975) for December 1972. Kidder and Vonder Haar also found good

spatial consistency for their three classes of rain rate. Comparison of the noon and midnight orbital data permitted an assessment of the diurnal variation of precipitation (Table 14). They found little evidence of a strong diurnal pattern for the oceans, although there is a tendency for the heavier events to occur near local noon.

Table 14 Percentage of the total of noon and midnight precipitation events occurring near local noon in oceanic regions between 20 °N and 30 °S during the season December 1972 through February 1973

Rainfall category, mm h^{-1}	Precipitation frequencies		
	Dry regions (0–5%)	Wet regions (5–100%)	All regions (0–100%)
Light (0.25–1.00)	47	50	48
Moderate (1.0–2.5)	48	52	49
Heavy (2.5–5.0)	49	54	51
Very heavy (>5.0)	53	61	57
All rain	49	52	50

Source: Kidder and Vonder Haar, 1977.

Oceanic precipitation rates on a global scale were derived from ESMR-5 data for the period December 1972 through February 1975 (Rao and Theon, 1977). Circulation features, such as the ITCZ and the major tropical–extratropical cloud bands and associated storm tracks, were noted. Rao and Theon identified a previously unrecognized rain area in the South Atlantic having annual rainfall of 900 mm and an average rain rate of 0.1 mm h^{-1}. They also examined the 1972–3 ENSO and showed major changes in both near-tropical and extratropical rain rates between that event and precipitation in subsequent years. Meisner and Arkin (1987) examined the diurnal characteristics of tropical precipitation, as shown by three years of GOES data. In common with previous studies, they find a lack of diurnal change over the oceans, except in the convergence zones, where the precipitation peaks occur around local noon. Over land there is a generally stronger diurnal cycle. Moreover, differences in the diurnal cycle of precipitation are apparent between the tropical lowlands of South America, the highlands of central America, and middle latitudes. Interannual variations in the diurnal cycle are noted between regions, with the 1982–3 ENSO being strongly evident.

A three-year seasonal climatology (1979–81) of oceanic precipitation from atmospheric liquid water content was derived from SMMR data by Prabhakara *et al.* (1986), and then compared with rainfall data for the major ENSO of 1982–3 obtained from the same instrument. For the Pacific (Fig. 5.17a,b), the zone of strong convection near the equator is noted; especially in the western Pacific, along with its seasonal change. The dry zones off the west coast of North and South America are also apparent. Comparisons with the March–April–May period of 1983 (during the ENSO) show strong increases in

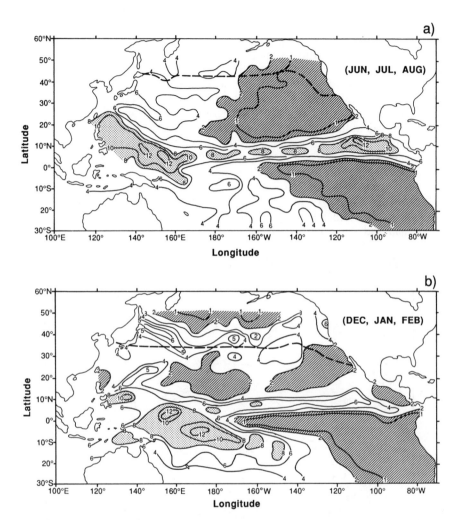

Fig. 5.17 Distribution of mean SMMR-derived rainfall (10^2 mm) for (a) June, July and August 1979–81, and (b) December, January and February over the Pacific Ocean between 50° N and 30° S. Rainfall <200 mm and >800 mm are identified by different shading. (From Prabhakara *et al.*, 1986.)

precipitation in the eastern and central equatorial Pacific, and marked decreases in the western Pacific, relative to the 1979–81 'normals'. There is a close spatial correspondence between the SMMR-derived anomalies and those of satellite OLR (Prabhakara *et al.*, 1986).

Geostationary visible and IR data were used by Arkin and Meisner to develop a three-year (December 1981–November 1984) seasonal climatology of convective precipitation for the Americas and eastern Pacific (the GOES precipitation index: GPI). By utilizing a decrease in the IR threshold temperature from low to high latitudes, they obtained improved estimates when compared with published long-term station rainfall data. Satellite-retrieved precipitation compared well with observations for the tropical oceans; less well over land, where the sparse gauge network apparently led to underestimation of rainfall. The GPI seems to capture well the seasonal variations of precipitation. The influence of the 1982–3 ENSO on GPI precipitation was examined by Arkin and Meisner. Fig. 5.18 shows the difference pattern for summers 1981–2 and 1982–3 over northern South America, and indicates the strong decrease associated with drought over the Nordeste of Brazil and also southern Peru. Conversely, strong precipitation increases occur across southern Brazil and Paraguay, and this was associated with extreme flooding through mid-1983.

(2) *Monsoon regions* Satellite nephanalyses were used by Barrett in climatological studies to estimate precipitation for the monsoon regions of east Asia and Australia/Indonesia (Barrett, 1970). A regression equation

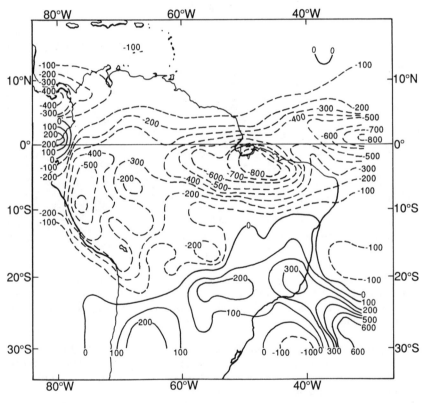

Fig. 5.18 Difference pattern of the GOES precipitation index (GPI) between DJF 1981–2 and DJF 1982–3 for tropical South America. Increases/decreases from 1981–2 to 1982–3 are positive/negative. The contour interval is 100 mm with negative contours dashed. (From Arkin and Meisner, 1987.)

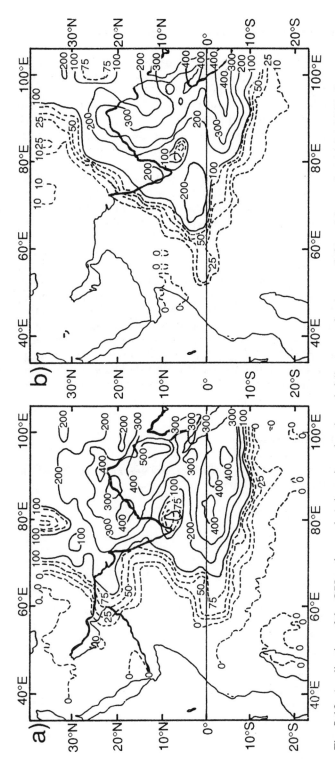

Fig. 5.19 Application of the GPI to the south Asia region, showing rainfall amounts for August (left) and September (right) 1986. Contours are in millimeters. (From Arkin *et al.*, 1989.)

developed from the cloud indexing method was used to compile an estimated precipitation map of the Australia/Indonesia region for July 1966. In a follow-up study, Barrett (1971) related satellite cloud and estimated precipitation fields for July 1966 and January 1967 to standard meteorological charts of pressure and wind. Strong differences in precipitation were noted over the maritime continent between those two months. Barrett also identified 'significant nephsystems', which include lower-latitude vortices and cloud band axes.

Much more recently, Arkin *et al.* (1989) have used the rainfall estimation technique developed for GOES (Arkin and Meisner, 1987) and applied it to data from INSAT; the Indian geostationary platform, for the summer monsoon season (June–September) of 1986. Rainfall estimation using satellite techniques is justified for this region because of the extensive data-void ocean areas that are involved in the monsoon, even though most of India does not lack an adequate rain gauge network. Monthly and full-season estimates of precipitation were derived using the 235 K threshold of the GOES technique. These delineate minima in the OLR. Figure 5.19a,b compares estimated rainfall for August and September 1986. Comparisons of satellite-derived and gauge precipitation for India give a high degree of confidence to the transferability of the GPI technique to other tropical regions. At the same time, Arkin *et al.* (1989) note that the correlations improve for the west coast of the subcontinent if a warmer threshold CTT of 265 K or 270 K is selected. This apparently results from the strongly orographic nature of the precipitation in that region.

Suggested further reading

Barrett, E. C., and Martin, D. W., 1981: *The Use of Satellite Data in Rainfall Monitoring*. Academic Press, London. 340 pp.

6 Monitoring the Earth radiation budget and surface variables important to climate

1 Satellites and the radiation budget

Operational meteorological satellites typically sense in portions of the visible and TIR wavelengths of the EMS. The former represents the portion of the incident solar radiation reflected by clouds, the atmosphere and the Earth's surface (i.e., the planetary albedo). The latter measures the outgoing longwave radiation (OLR) from the Earth's surface and clouds in the window region (Chapter 2).

In cloud-free scenes, the OLR is largely a function of Earth surface temperature: high emittances correspond to high T_s, lower emittances to reduced T_s. The upper limit of the surface temperature is determined not only by the available (net) radiation at the surface but also by the amount of moisture available for evapotranspiration. Highest IR fluxes tend to occur over dry desert surfaces in the subtropics through a highly transmissive atmosphere (low water vapor content; lack of cloud). Where the amount of surface and atmospheric moisture increases, as in the humid tropics, evaporation limits the daytime T_s and the OLR is lower. Away from the tropics and subtropics, the decrease of OLR and T_s appears more zonal compared with the stronger longitudinal variations at lower latitudes that reflect land–sea differences.

The deep and optically thick convective clouds of the ITCZ are characterized by low CTTs and OLR. They also correspond to the most reflective clouds in visible wavelengths (HRCs) and, accordingly, are an indicator of heavy precipitation (Chapter 5). In addition, fields of satellite OLR have been used to indicate atmospheric circulation changes in the tropics and subtropics (e.g., Fig. 1.3), arising from the close association between OLR minima and tropospheric diabatic heating. Climatologically, the regions of minimum OLR are not distributed equally through the tropical zone. They are located predominantly over the maritime continent and undergo within-season variations connected with the 30–60 day oscillation; seasonal variations related to the monsoon circulations; and interannual variations associated with teleconnections to the Walker Circulation (Grossman and Garcia, 1990).

There are some major differences between a radiation budget obtained from conventional measurements at and near the Earth's surface and one made from satellite altitudes (Kandel, 1983). These include the following.

(1) Differences in the spectral response and radiometric calibration of the instruments; also, these may differ *between* satellite systems, making a homogeneous long-term ERB difficult to achieve (Stephens *et al.*, 1981). Most visible wavelength sensors are narrow-band rather than broad-band. Also, sensor 'drift' may produce temporal errors.

(2) Differences in the sensor field of view. If the satellite sensor receives a flux integrated over a wide field of view (WFOV), this incorporates variations related to the Earth's curvature and the increasing thickness of the atmosphere away from nadir (i.e., an *irradiance*: see Kandel, 1983). Use of a narrow field of view (NFOV), such as the NOAA SR, samples the *radiance* for a much smaller area. The effect of reducing the field of view on OLR measurements is to increase the spatial variability (Ramanathan, 1987).

(3) The systematic errors due to temporal sampling. For polar orbiters there is a reduced temporal sampling of ERB components, and this may be especially significant with regard to the diurnal variability of cloud. This arises from the occurrence of only two fixed observations per day for most locations on the Earth (greater frequency at higher latitudes). The local time of the daylight overpass of a polar orbiter is critical for estimates of the surface albedo (Chapter 2). Figure 6.1 shows the effect of different equatorial crossing times of a polar orbiter on estimates of the absorbed solar radiation for January. The latter were calculated using a GCM with interactive clouds. The errors are largest for the early daylight hours and for latitudes of relatively large solar zenith angle.

The situation is improved with the use of geostationary satellites (Arkin *et al.*, 1989); however, even then there is a need to optimize the large amount of data available by selecting an appropriate sampling interval that emphasizes the diurnal variability of ERB components (England and Hunt, 1984). The Earth's curvature away from the sub-satellite point of a geostationary platform may introduce considerable errors.

Notwithstanding these difficulties, satellite data have been able to provide multi-year climatologies of ERB components that facilitate examination of contemporary climate issues, such as the cloud-radiative forcing.

2 Sources for satellite ERB climatologies

Relatively long-term satellite ERB data from visible and IR radiometers (almost 30 years) are available for climatological studies (e.g., Winston, 1967; Raschke *et al.*, 1973; Vonder Haar and Suomi, 1971; Stephens *et al.*, 1981). These are derived from global or near-global measurements of solar irradiance and thermal exitances made on at least a once-daily basis. The main satellite systems used in such climatologies are the following:

(1) the TIROS, ESSA and NOAA series of polar orbiters;
(2) NASA R&D polar orbiters of the Nimbus series; in particular, Nimbus-6 and -7;
(3) development and operational geostationary platforms; notably the ATS/SMS/GOES series (USA), METEOSAT (Europe) and GMS (Japan);
(4) special ERB monitoring programs. These include, most notably, the Earth Radiation Budget Experiment (ERBE), which includes a dedicated platform, the Earth Radiation Budget Satellite (ERBS). In this

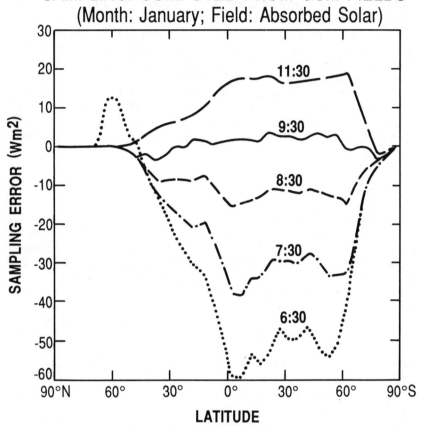

ZONAL MEAN ERROR FOR SUNSYNCHRONOUS SAMPLING: COMPUTED FROM GCM FIELDS (Month: January; Field: Absorbed Solar)

Fig. 6.1 Potential sampling errors in absorbed solar radiation for sun-synchronous (polar-orbiter) spacecraft in January. The curves denote the dependence of the sampling errors on equatorial crossing time of the satellite. The errors are computed from synthetic 'data' obtained from GCM simulations. (From Ramanathan, 1987.)

category also should be included the Heat Capacity Mapping Mission (HCMM), which was able to monitor the diurnal variability of thermal inertia of the Earth's surface.

We now survey the major ERB-monitoring satellites that are providing global-scale climatologies. These emphasize the polar orbiting systems.

(a) The NOAA SR

Operational polar-orbiter satellites of the NOAA series provide narrow-band observations of the reflected shortwave (0.5–0.7 μm) and longwave window (10.5–12.5 μm) fluxes for deriving global satellite radiation climatologies (see, e.g., Gruber, 1977; Gruber and Winston, 1978; Hartmann and Short, 1980; Stephens et al., 1981; Ohring and Gruber, 1983). The SR data were used by Stephens et al. (1981) in a 48-month ERB study that integrated several different satellite sensor data. While the narrow-band visible data have been assumed to be representative of exitances in the full spectral range of visible wavelengths, a regression model converts IR values to those representative of the full window exitances. Moreover, the planetary albedo measurements are assumed to not vary diurnally for this radiation climatology.

(b) The Nimbus-6 and -7

The ERB experiments flown on the Nimbus-6 and -7 polar orbiters (respectively, three years, 1975–8; seven years, 1978–85) have recently provided a 10-year (1975–85) climatology (Bess et al., 1989). These follow the earlier short-term climatologies of Raschke and Bandeen (1970) and Raschke et al. (1973). The data are obtained by wide field of view (WFOV) radiometers for the shortwave and longwave exitances. The regional scale albedo and bidirectional reflectances for the shortwave, and the longwave emission, can be deconvoluted mathematically from these WFOV measurements (Hucek et al., 1987; Ardanuy et al., 1987). Unlike the NOAA SR series, the Nimbus-6 and -7 contain no dedicated longwave radiometer. Rather, the OLR is determined as the difference between the 'total' irradiance across the broad 0.2–50 μm range and the shortwave irradiance for the broad-band visible sensor of 0.2–3.8 μm (Bess et al., 1989). The 10-year Nimbus-6 and -7 climatology facilitates examination of the ERB variations (especially OLR) associated with large-scale climatic anomalies; notably ENSO (Bess et al., 1989). Its value for the detection of an unambiguous CO_2 signal separate from that associated with interannual climate variability has been questioned (Cess, 1990).

(c) The Earth Radiation Budget Experiment (ERBE)

The ERBE consists of a dedicated satellite (ERBS) in near-Earth orbit (600 km) and inclined at 57°, and two sun-synchronous NOAA 9 and 10 platforms (ERBE Science Team, 1986; Smith et al., 1987). Major advantages over previous systems include:

(1) better determination of the *diurnal cycle* of ERB components;
(2) improved accuracy of the measurements, including in-flight calibration of the sensors and cross-comparisons with the different platforms to monitor sensor drift;

(3)   the deployment of scanner (shortwave, longwave, total) and non-scanner (WFOV, medium FOV for short and total irradiance) instruments for purposes of intercomparison; and

(4)   improved determination of the monthly and seasonal variability of ERB components at regional scales (Barkstrom *et al.*, 1989, 1990), and also of the cloud-radiative forcing (Ramanathan *et al.*, 1989a,b). In addition, the ERBE data are of value in general circulation studies; for example, they can be used to estimate the net poleward energy transport that is in response to the latitudinal radiative disequilibrium. Early results on the diurnal variability of OLR indicate a maximum (up to about 70 W m^{-2}) for clear desert regions and a minimum (<5 W m^{-2}) for clear oceans (Harrison *et al.*, 1988). Recent comparisons of the Nimbus-7 ERB with the ERBE for overlapping time periods show close agreement globally in albedo and OLR (Kyle *et al.*, 1990). However, regional-scale and monthly differences between the two data sets are larger, amounting to a few percent.

3 Global-scale ERB climatologies

The main satellite ERB components that can be resolved at a regional scale are: (1) the shortwave planetary albedo; (2) the longwave emittance in the window region; and (3) the available (net) radiation (N). At the top of the atmosphere (TOA), N can be defined thus (after Stephens *et al.*, 1981).

$$N = (1 - \alpha_\mathrm{p})Q - I \qquad (23)$$

where Q is the incoming solar flux,
$\quad\alpha_\mathrm{p}$ is the planetary albedo,
$\quad I$ is the net longwave flux.

The reciprocal of the albedo ($1 - \alpha_\mathrm{p}$) gives the solar radiation absorbed by the clouds, atmosphere and Earth's surface. For TOA measurements, this difference is that between the available solar radiation and the reflected portion. The major temporal and spatial characteristics of the three quantities, derived from the various satellite sources just described, are now presented. More detailed discussion of regional aspects of the ERB climatology, particularly the surface albedo and its association with surface variables important in climate, follows.

(a) Planetary albedo

Figure 6.2 shows the annual cycle of the satellite-derived planetary albedo. Two curves are shown: that based on the 45-month SR record (Ohring and Gruber, 1983), and that derived from the 48-month composite study of Stephens *et al.* (1981). The former shows considerably greater seasonal change. The difference is largest in the Northern Hemisphere winter and smallest in the summer. They imply the importance of changes in Earth–sun

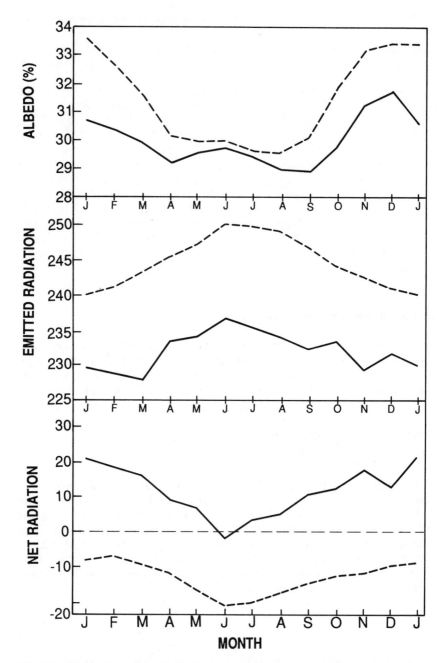

Fig. 6.2 Comparisons of ERB components from two satellite climatologies: planetary albedo, IR flux, and TOA net radiation. That by Stephens *et al.* (1981) is shown by solid lines; that by Ohring and Gruber (1983) is shown by dashed lines. The values in both cases are global or near-global averages. Differences between the two satellite climatologies are explained in the text. (After Stephens *et al.*, 1981; Ohring and Gruber, 1983.)

distance (greatest in northern summer) as well as the role of Northern Hemisphere cryosphere (snow and ice) on global brightness. In addition, differences in the satellite systems and sensor characteristics, and methods of deriving the curves are important influences. For example, it appears that data for all latitude zones were used in the Ohring and Gruber study, while that in Stephens *et al.* (1981) pertains to the area between ±70° latitude. Moreover, the latter are area averages. Despite the differences in the monthly 'global' values of planetary albedo in these two studies, the annual global averages are similar: 0.30 in Stephens *et al.* (1981) compared with 0.314 for the Ohring and Gruber (1983) study. Accurate determination of an albedo depends on the value used for solar irradiance (from the solar 'constant'). The two results for planetary albedo given here are made virtually identical by increasing the solar constant value from 1353 W m^{-2} to 1376 W m^{-2} in the Ohring and Gruber study, as justified from Nimbus-7 ERB data (Ohring and Gruber, 1983). This decreases their planetary albedo to 0.306. A solar irradiance value of about 1365 W m^{-2} is noted from the recent ERBE satellite results (ERBE Science Team, 1986). The satellite-derived planetary albedo values are considerably lower than those obtained from theoretical estimates (Ramanathan, 1987). Furthermore, the annual mean albedo of the Northern and Southern Hemispheres are nearly the same (NH = 0.310, SH = 0.298: Stephens *et al.*, 1981). That this occurs despite the very different land–sea distributions and associated surface albedo contributions implies the importance of the cloud cover component to planetary brightness.

Maps of the annual averaged planetary albedo, and those for DJF and MAM are given in Fig. 6.3a,b,c. The strong symmetry of the two hemispheres is evident in the annual average (Fig. 6.3a). Longitudinally-varying patterns related to land–sea variations in lower latitudes contrast with the more zonal differences in middle and high latitudes. In general, values less than the annual-averaged global planetary albedo of 30% occur over low and subtropical oceans, and result from high sun angles and low surface reflectance. Ocean albedos for cloudless conditions obtained from ERBE data range from 6–10% in low latitudes to 15–20% at high latitudes (Ramanathan *et al.*, 1989a,b). The planetary albedo and its seasonal change increases over land in these latitudes owing to increased frequency of HRCs or, in the case of the deserts, increased surface reflectance and lack of cloud cover. The highest albedos tend to occur in the high latitudes of the winter hemisphere with very low sun angles. The relative importance of these effects can be seen in the maps of DJF (Fig. 6.3b) and JJA (Fig. 6.3d). In middle and higher latitudes of the Northern Hemisphere, the latitudinal gradient of planetary albedo is greatest in winter (Fig. 6.3b). While the annual and semi-annual cycles explain much of the satellite-observed planetary albedo (e.g., in the storm track region of the North Atlantic; over the monsoon regions of south-east Asia and Australia; in the ITCZ of tropical South America; and the Antarctic seasonal sea ice zone), transient effects appear equally or even more important in other regions (Stephens *et al.*, 1981). These include portions of the central and western tropical Pacific, east Asia, and the area west of Drake Passage in the Southern Hemisphere.

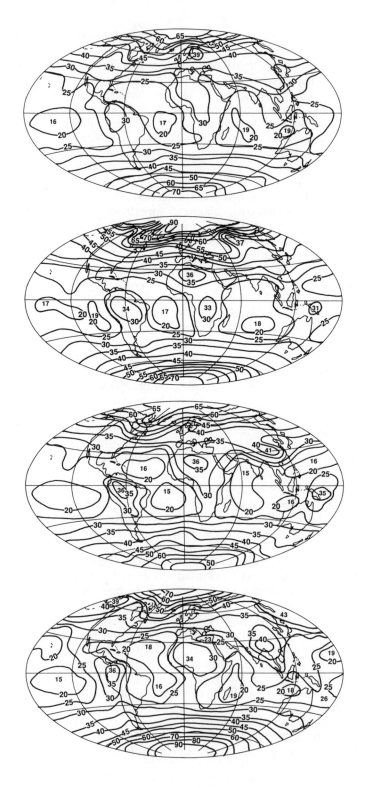

(b) Surface albedo

In the absence of clouds, much of the remainder of the contribution to the planetary albedo is from the surface. Atmospheric aerosols (ash, desert dust) may also be important at certain times and in certain regions. Thus, accuracy in the determination of a surface albedo value depends greatly on the ability to specify the surface type and to identify cloud-free scenes. These procedures are discussed more fully in Section 4c (below). Most satellite cloud retrieval algorithms utilize a combination of visible and infrared information against model-specified clear column reflectances and radiances (see, e.g., Rossow, 1989b; also Chapter 3). Over most surfaces, clouds have a more transient impact on the planetary albedo than do changes in the surface reflectivity. For these surfaces, the minimum brightness recorded over a given time-averaging period (e.g., 3, 5, 15 or 30 days) may be considered representative of the surface albedo. However, it must also be recognized that certain cloud properties, such as persistence, may differ markedly between climatic regions (Rossow et al., 1989b), and this influences the compositing period required to obtain an uncontaminated surface reflectance (ideally, a month or less). The minimum brightness technique has major difficulties over snow and ice surfaces, owing to the similar shortwave reflectance and IR temperature signatures (Chapter 3), and the uncertainties in parameters such as surface emissivity and temperature. Accurate determination of the surface albedo is particularly problematic in areas of unstable, or heterogeneous, surfaces; that is, where conditions are changing rapidly in time or over space. Regions that are particularly prone to this instability and which involve surface albedo changes of the order of about 30–50%, include the marginal sea ice zone and the boundary between snow-covered and snowfree land. Zones characterized by conservative but quite strong spatial gradients of surface albedo are coasts, particularly those in the subtropics.

A detailed global-scale analysis of surface visible reflectances using the NOAA SR has been made for the mid-season months of 1977 by Rossow et al. (1989b). These authors find the global surface albedo to be 10% for the relatively narrow spectral band between 0.5 and 0.7 μm. This result includes the polar regions, but no winter pole results are included owing to darkness. The 10% value contrasts with a planetary albedo of about 30% (Ohring and Gruber, 1983). Excluding the polar regions causes the global annual mean surface albedo to fall to 5%. These results suggest a relatively dark global surface that is dominated by water and vegetated land over most latitudes, and with some effect from desert surfaces, but with strong seasonal contributions from snow and ice cover (Fig. 6.4).

Fig. 6.3 (opposite) Satellite-derived maps of the planetary albedo (%): (a) annual mean; (b) DJF; (c) MAM; (d) JJA. (From Stephens et al., 1981.)

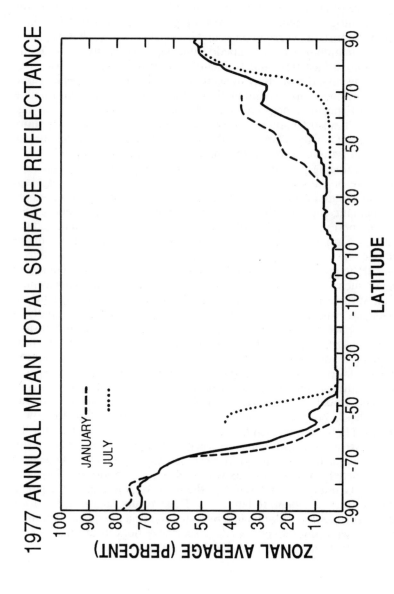

Fig. 6.4 Zonally averaged profile of the surface reflectance for 1977 (January: dashed; July: dotted; annual mean: solid line) based on NOAA-5 SR data. Note the large values and large seasonal change at high latitudes compared with lower latitudes. (From Rossow *et al.*, 1989b.)

(c)   Outgoing longwave radiation (OLR)

Determination of the TOA net radiation (N) budget requires knowledge of the outgoing longwave (thermal emittance) component: equation (23). This may be determined by direct measurement using either a scanner or WFOV sensor, or by subtraction of the shortwave reflectance from the total irradiance over a wide spectral interval (e.g., Nimbus-6, -7). The annual cycle of OLR, as determined from the Ohring and Gruber (1983) and Stephens *et al.* (1981) studies are compared in Fig. 6.2. The consistently larger values of the Ohring and Gruber study are acknowledged by these authors to be in error, probably by about 13 W m^{-2}. They are the apparent cause for consistently negative global values of N in all months, which are not expected from considerations of radiative balance. The annual globally averaged IR flux is 234 W m^{-2} (± 7 W m^{-2}), according to the Stephens *et al.* (1981) climatology. While magnitudes differ, the *patterns* of OLR are broadly similar between the two studies. They show a peak in the Northern Hemisphere summer that is related to the different response of land and ocean to thermal forcing; particularly the role of the hot desert surfaces. The spatial patterns of the OLR for annual-averaged conditions, and the extreme seasons (DJF, JJA) appear in Fig. 6.5a,b,c. Areas of highest emittance for the annual case (Fig. 6.5a) occur over the Sahara and Kalahari deserts, and over portions of the tropical oceans dominated by low-level stratocumulus clouds or largely cloud-free oceans having high SSTs. On the other hand, reduced OLR occurs over tropical land surfaces and the 'maritime continent' since these regions are dominated by high and cold convective clouds. These results are broadly similar to the smoothed fields obtained using Nimbus-6 and -7 for the period 1975–85 (Bess *et al.*, 1989). The more zonal OLR variations over middle and high latitudes are associated with a progressive decrease in temperature. Lowest values of OLR occur over Antarctica.

Comparison of Fig. 6.3a, which shows the annual mean planetary albedo, with Fig. 6.5a suggests an inverse relationship of these two quantities (highly reflective surfaces tend to be cold); except over the deserts, where the opposite is the case (Stephens *et al.*, 1981). Instantaneous flux values from the initial ERBE data show that convective clouds are brighter and colder than suggested by previous studies (ERBE Science Team, 1986). The OLR maps for the extreme seasons (Fig. 6.5b,c) show clearly the strong seasonal change over the low latitudes and subtropics, particularly the monsoon regions (see also Bess *et al.*, 1989). Changes over the lower-latitude oceans are less marked seasonally. The gradient of OLR over middle and higher latitudes tends to be greater in the winter hemisphere. Most of the variance in OLR can be explained by the annual and semi-annual cycles, except in certain lower latitude regions (Stephens *et al.*, 1981).

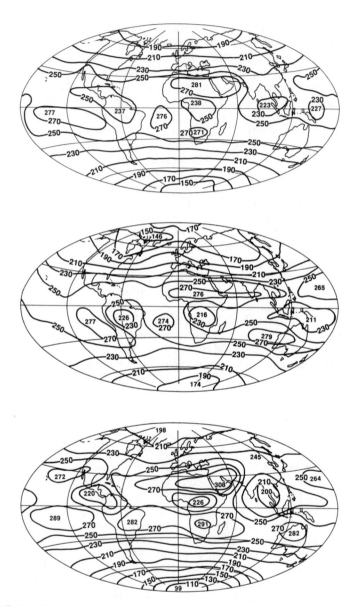

Fig. 6.5 Satellite-derived maps of the outgoing longwave radiaton (emitted flux: W m^{-2}): (a) annual mean; (b) DJF; (c) JJA. (From Stephens *et al.*, 1981.)

(d) Net radiation (N)

The annual cycle of TOA net radiation is plotted in Fig. 6.2. The annual
globally averaged N is 9 ± 10 W m^{-2}. The large absolute error arises from
measurement uncertainty amplified by the calculation of the net flux.
Accordingly, the global N is about 0; positive (negative) equatorward (poleward) of about 37° N and S. The zonal profiles of annual-averaged IR fluxes,
planetary albedo and N are given in Fig. 6.6. The dominant features of
Fig. 6.6 include the inverse relationship of OLR and planetary albedo, and

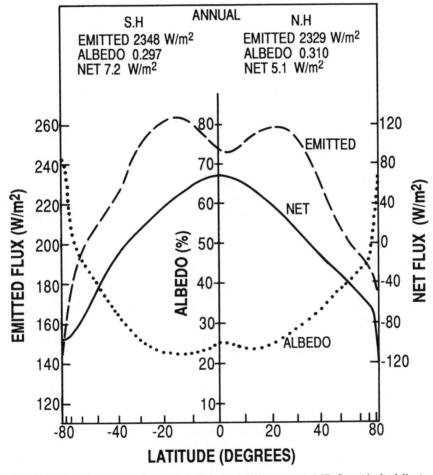

Fig. 6.6 Zonally averaged profiles of the annually averaged IR flux (dashed line),
planetary albedo (dotted line), and net flux (solid line) for the Northern and Southern
Hemispheres. Note the similarity in the net flux curves despite the large differences in
land–sea distribution between hemispheres. (From Stephens *et al.*, 1981.)

the high IR emission (but lower N) in the subtropics compared with the tropics. The slightly higher planetary albedo and slightly lower IR emission in the Northern Hemisphere results in the reduced hemispheric-averaged N.

Seasonal profiles of planetary albedo, IR and N (not shown here) indicate an out-of-phase relationship between the Northern and Southern Hemispheres in the extreme seasons. Satellite-derived spatial patterns of N for the annual averaged and extreme seasons (Fig. 6.7a,b,c) confirm the tropical oceans to be key regions of surplus N; the subtropics as areas having close to 0 or even negative N (deserts); and the pronounced semi-annual reversal in N over monsoon regions. The annual cycle of ERB components in the Asian monsoon region is shown in Fig. 6.8. The figure shows that, despite the large irregular changes of planetary albedo and IR flux, the resulting N curve is very smooth. This is because of the reciprocity of planetary albedo and OLR (see also Fig. 1.2). Similar graphs of the annual march of satellite-derived TOA ERB components can be obtained for other climatic regions (see, e.g., Ohring and Gruber, 1983).

(e) Circulation associations with ERB quantities and their interannual variations

Maps of the mean seasonal and annual satellite-derived OLR (Figs. 6.3, 6.5, 6.7) summarize well the large-scale radiation budget climatology of the Earth–atmosphere system. This arises from:

(1) the close association of total OLR with T_s and CTT;
(2) the reciprocity of OLR with the planetary albedo and, in the tropics especially, the association with the heaviest convective precipitation (see, e.g., Arkin *et al.*, 1989); and
(3) the link with atmospheric column heating and cooling.

Regions of low OLR in the tropics are associated with diabatic heating resulting from the release of latent heat in clouds (see also Fig. 1.3). Conversely, radiational cooling of the atmospheric column, particularly over deserts, helps maintain widespread subsidence in the descending limb of the Hadley Cell, and aridity.

The dominant modes of variation of OLR in the tropics, where the changes are at a maximum (Bess *et al.*, 1989), are:

(1) intraseasonal, associated with the 30–60 day oscillation;
(2) seasonal, associated particularly with the latitudinal movement of the OLR minima between the South Asian and north Australian regions; and
(3) interannual, associated with low-frequency changes in the strength of the Walker Circulation connected with ENSO. Over the maritime continent in northern winter, bursts of tropical convection are evident as OLR minima, and result from surges of cold air of extratropical origin (Chang and Lau, 1980). The OLR variations on all these scales appear closely interrelated. For example, recent studies (Lau and Chan, 1988) suggest that the 30–60 day oscillation may be enhanced prior to the onset of a

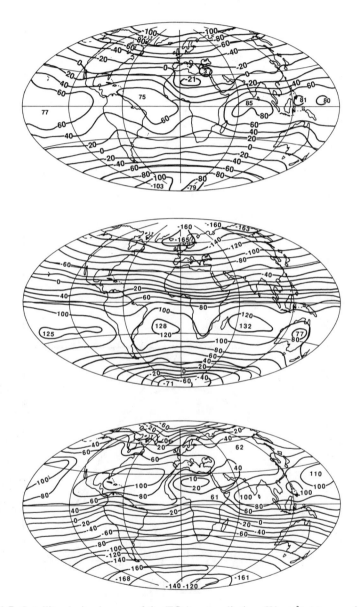

Fig. 6.7 Satellite-derived maps of the TOA net radiation (W m^{-2}): (a) annual mean; (b) DJF; (c) JJA. (From Stephens *et al.*, 1981.)

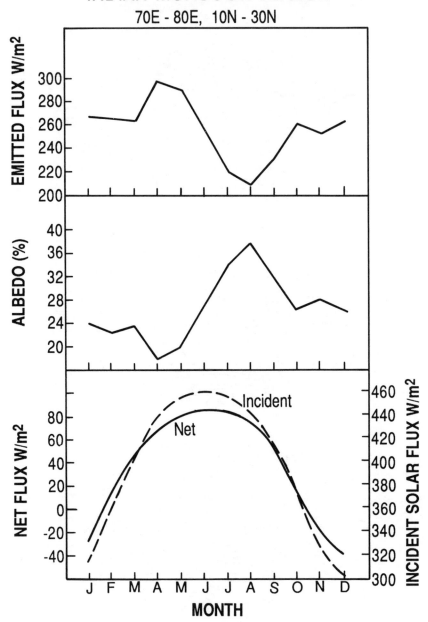

Fig. 6.8 Annual cycle of satellite-observed ERB components: IR flux (top panel); planetary albedo (middle panel), and net radiative flux (lower panel) for the Asian monsoon region (70° E–80° E; 10° N–30° N). Superimposed on the net flux is the annual cycle of the incident solar flux at TOA (dashed curve). (From Stephens *et al.*, 1981.)

major 'warm' ENSO (e.g., that of 1982–3), and suppressed during the decay phase. The persistence over several seasons of a 30–60 day oscillation influence on ENSO is probably facilitated by anomalies of SST and air–sea interaction in the tropical Pacific and Indian Ocean regions (Meehl, 1987; Kiladis and van Loon, 1988). The main OLR features of the two ENSO events occurring in the 1975–85 period covered by the Nimbus-6 and -7 climatology (1976–7, 1982–3) are discussed by Bess *et al.* (1989).

The intraseasonal variations in OLR consist of eastward-propagating pulses (large cloud clusters) that emanate from the western tropical Pacific. These have a period of about 45 days and are maintained by shorter-period synoptic-scale systems (Nakazawa, 1986). Latitude variations in the 30–60 day oscillation are dominantly seasonal, in association with the movement of the zone of maximum heating and of convection. Monthly variability of the IR flux is largest over land compared with ocean at similar latitudes (Stephens *et al.*, 1981). Thus, there is a strong intraseasonal component to the Asian and Australian monsoons, and this is evident in the planetary albedo (Stephens *et al.*, 1981). Diurnally, there is a tendency for greater variability in tropical and subtropical OLR at night over lower latitudes of the Northern Hemisphere in July, at least based on data for one year (Bess, 1986).

(f) Cloud forcing from ERB measurements

The shortwave and longwave radiation streams are strongly modulated in the presence of clouds (Chapter 1, Section 4a). Moreover, cloud–radiation inter-actions influence the zonal and meridional-plane features of the atmospheric circulation, especially in lower latitudes. Given the satellite estimates for globally annual-averaged N of close to 0 (within measurement uncertainty: Stephens *et al.*, 1981), climatologists are seeking to understand the *climate sensitivity* of global cloud cover; that is, the degree to which perturbations in cloud cover influence climate and its variability. It appears that different regional *cloud regimes*, each consisting of particular cloud type and persistence characteristics (e.g., Minnis *et al.*, 1987), differ in terms of their climatic effect. Knowledge of such cloud–climate sensitivity is necessary before reliable estimates can be made of the climate impact of 'greenhouse' (trace) gas-induced warming, and also to assess how climate anomalies, such as drought, may be reinforced by cloud-radiative forcing (Ramanathan *et al.*, 1989a). Similarly, such knowledge should help in the physical and dynamical interpretations of paleo-climate changes, and how the cloud regimes may have been different.

Where radiation observations are available, the cloud–climate sensitivity parameter (equation (1)) can be expressed in a mathematically simple way for an atmospheric column (after Ramanathan *et al.*, 1989a). For a column that contains clouds, the net radiative heating, H, is given as in equation (23); i.e.,

$$H = Q(1 - \alpha) - F. \tag{24}$$

Here F replaces I (longwave emittance in equation (23))
  F is the net longwave emittance
  α is the planetary albedo.

The effect of cloud forcing (C) on H is given as:

$$C = H - H_{clr} \tag{25}$$

where H_{clr} is the net heating in the absence of clouds (clear sky: $C = 0$). In the latter case, $H = H_{clr}$.

Combining equation (25) with (24) gives:

$$C = C_{sw} + C_{lw} \tag{26}$$

where $C_{sw} = Q(\alpha_{clr} - \alpha)$ and $C_{lw} = F_{clr} - F$; α_{clr} is the clear-sky (surface) albedo and F_{clr} is the clear-sky IR emittance. Thus, equation (26) expresses the net effects of the incoming shortwave and back radiation of IR radiation.

Since an imaged region ('domain') is unlikely to be homogeneous in terms of cloud amount (clear or overcast), C can be obtained as a cloud fraction that compares clear with overcast subregions, thus:

$$C_{sw} = Q(\alpha_{clr} - \alpha) = Q\sum_i f_i(\alpha_{clr,i} - \alpha_{0,i}) \tag{27}$$

and

$$C_{lw} = F_{clr} - F = F\sum_i f_i(F_{clr,i} - F_{0,i}) \tag{28}$$

where i is the subregion of the domain,
  f_i is the cloud fraction
  $\alpha_{0,i}$ is the overcast albedo for subregion i
  $F_{0,i}$ is the overcast longwave emittance for subregion i.

The cloud forcing is easier to obtain from the left-hand side of equations (27) and (28) because of the greater spatial homogeneity of fluxes from clear skies than from overcast skies (Ramanathan *et al.*, 1989a,b) and the problems associated with obtaining reliable estimates of cloud fraction (Chapter 3). As noted earlier, however, the task of determining clear-sky values from satellite data is not inconsequential.

Ramanathan *et al.* (1989a,b) and Harrison *et al.* (1990) have determined regional- and global-scale cloud forcing for the mid-season month of April 1985 using clear-sky brightness and longwave scanner data from ERBE. Their results confirm the broad-scale features of surface albedo noted earlier. In addition, the distribution of clear-sky radiative heating (F_{clr}) shows a maximum emission of about 330 W m^{-2} in the tropics, declining to 150 W m^{-2} in polar regions. For the shortwave component of the cloud forcing (C_{sw}), large negative values (albedo enhancement) occur in tropical monsoon regions and deep convective regions, but peak values occur in middle latitudes. The longwave cloud forcing (greenhouse enhancement) shows peak values in the tropics declining towards the poles, and also over the subtropical deserts. The reduction in F_{clr} is at a maximum over deep or multi-layered cloud systems in the tropics and middle latitudes (see also

Ardanuy et al., 1989). ERBE results for high latitude snow and ice surfaces are uncertain at this time; however, results from the Nimbus-7 ERB data (Ardanuy et al., 1989) suggest weak daytime longwave cloud forcing (<10 W m^{-2}) for these regions. The total cloud forcing, or $C = (C_{sw} + C_{lw})$, is close to zero in the tropics, which are dominated by the convective cloud regime. However, C peaks (large negative values) in the low-latitude stratus and stratocumulus cloud decks off the west coasts of continents. It is positive over portions of the western Sahara and Sahel (strong longwave forcing, negligible shortwave forcing due to lack of clouds). The cloud forcing factor is dominantly negative over middle- to high-latitude oceans and weakly positive over certain land masses. Hemispheric differences tend to manifest the seasonal cycle, whereby the longwave and shortwave cloud forcings approximately cancel in the winter hemisphere but the albedo enhancement dominates in the summer hemisphere (Harrison et al., 1990). The overall effect is to reduce the equator-to-pole radiative heating gradient during the spring and summer.

The differences in net cloud forcing between the two low-latitude regimes (convective cloud, west coast stratocumulus) can be appreciated from examination of Fig. 6.9. These compare, for each region, the interannual changes in monthly means of OLR (i.e., ΔF) with those for the planetary albedo ($\Delta \alpha$). It is assumed that changes in the surface albedo are negligible over ocean surfaces. Note the steeper slope of the least squares regression line for the convective cloud regime compared with that for marine stratus (Fig. 6.9). This corresponds with the much greater forcing of C_{lw} in the former case. C_{lw} changes much less for a large change in $\Delta \alpha$ in the stratus cloud regime. Cloud forcing tends to be positive (greenhouse enhancement exceeds shortwave reflectance) for cirrus clouds over ocean areas (Harrison et al., 1990). Differences in the *longwave* cloud forcing for cloud level classes (clear, low, middle, high) have been identified using the METEOSAT TIR and water vapor channels (Schmetz et al., 1990). The biggest impact is noted for high-level clouds in the tropical convective regime (15–20 W m^{-2}); the smallest for low clouds in the marine stratocumulus and mid-latitudes (5–10 W m^{-2}).

The results of Ramanathan et al. (1989a) and Harrison et al. (1990) show that, globally, the shortwave forcing may dominate over the longwave enhancement effect of clouds. This amounts to between about 14–21 W m^{-2}, at least between April 1985 and January 1986; or an average of 17 W m^{-2}. The direction of the forcing is consistent with some previous satellite-based studies (see, e.g., Hartmann and Short, 1980; Ohring and Clapp, 1980; Ohring et al., 1981) but not with others. The study by Cess (1976) suggested a minimal cloud-radiative forcing for the globe as a whole. In the ERBE studies the net cooling is shown to exceed by about 3–5 times the warming predicted by most GCMs for a doubled CO_2 scenario (Ramanathan et al., 1989a).

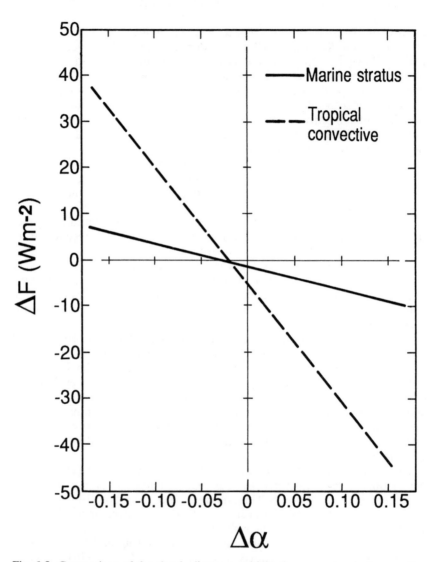

Fig. 6.9 Comparison of the cloud–climate sensitivity for ocean tropical convective (dashed line) and marine stratus (solid line). The figure shows the best fit regression lines of the interannual changes in monthly means of longwave radiation plotted against interannual changes of albedo. Note the larger change in longwave forcing for a given change in albedo for the convective compared with stratus regimes. (After Ohring and Gruber, 1983.)

4 Monitoring surface influences on the albedo and on climate

Important energy exchanges take place at the Earth's surface and these help drive climate. Satellites can be used to monitor both the surface conditions and the incoming (solar) and outgoing (reflected solar, IR) energy fluxes, as well as the exchanges of latent and sensible heat (Diak and Stewart, 1989; Sellers *et al.*, 1990). It seems appropriate, therefore, to discuss in some detail the satellite monitoring of surface conditions on sub-regional scales in the context of the surface albedo and its retrieval from measurements of the Earth irradiance (planetary albedo).

(a) Surface effects and the surface albedo

The surface albedo, in conjunction with the cloud forcing, determines the surface net radiation and the partitioning of that energy into sensible and latent heating (equation 10). Accurate information on the surface albedo at synoptic scales is crucial to an understanding of climate variability and change; most notably the occurrence and enhancement of drought (Norton *et al.*, 1979; Henderson-Sellers, 1980) and the impact and potential impact of humans through desertification and deforestation (Charney *et al.*, 1977; Otterman, 1981, 1989).

Of particular relevance from the satellite perspective is the association between surface albedo and surface temperature. Land surfaces in semi-arid regions tend to exhibit a positive correlation between these two variables: increased albedo is associated with high ground temperatures in which stressed, dead or otherwise sparse vegetation occur with much bare dry ground exposed (Vukovich *et al.*, 1987; Jackson and Idso, 1975). This apparent contradiction, whereby the reduction in the absorbed solar radiation is associated with higher T_s, can be explained largely on the basis of soil moisture and the partitioning of the available energy into the sensible and latent heat fractions. The presence of a dry soil leads to an increase in the sensible heat exchange and a reduction in the evaporation (and latent heat). Accordingly, an increase in the surface albedo in semi-arid regions caused by overgrazing or other removal of vegetation, may be associated with reduced cloud development and precipitation (Anthes, 1984). Inclusion of the soil moisture status in a GCM determines strongly the model sensitivity to both a projected CO_2 increase (see, e.g., Kellogg and Zhao, 1988) and the regional-scale persistence of drought (Oglesby and Erickson, 1989). On smaller scales (say, 100 km × 300 km), convective cloud development has been observed to occur first over high albedo wheat stubble in Oklahoma during times of weak synoptic flow and when the ambient air is dry (Rabin *et al.*, 1990; also Fig. 3.2). The enhanced sensible heat flux promotes convective cloud development over the wheat areas, but clouds are suppressed over cooler lakes and areas dominated by growing vegetation (higher latent heat fluxes). Part of the explanation for these differences in the surface albedo–cloud relationship between climate zones probably involves scale considerations. For example, a recent analysis of surface–atmosphere interaction for the Midwest US in summer (Brinegar, 1990) finds that, in the drought year of 1988, convective cloud days were much more frequent over the forested surfaces than the highly stressed, higher-albedo and higher-temperature cropped surfaces. The region considered is considerably larger than the wheat stubble areas of the Rabin et al. study.

Not all semi-arid surfaces exhibit the positive association of surface albedo and T_s. Satellite and ground truth studies of a fenced-off portion ('exclosure') of the northern Sinai, in which anthropogenic pressures were virtually eliminated (Otterman, 1977, 1981), suggests the opposite: lower shortwave reflectance and higher radiant temperatures over the exclosure (Otterman and Tucker, 1985). Here, the sensible heat flux dominates over the albedo reduction, especially at large solar zenith angles, in connection with reductions in the soil heat flux and longwave radiation over the rough dry vegetation of the exclosure. Accordingly, it has been suggested that removal of desert fringe vegetation to expose bright bare soil surfaces in semi-arid and arid regions may reduce convective precipitation on regional scales since the reduced sensible heat flux may not be sufficient to carry atmospheric moisture to the convective condensation level. Conversely, the planting of vegetation in bands having scale widths of about 100 km could enhance precipitation by promoting moist convection of magnitude $10\ cm\ s^{-1}$ to extend to heights of 1 km or higher, under appropriate synoptic conditions (Anthes, 1984). Norton et al. (1979) found some tendency for higher surface albedo to follow reduced precipitation over the southern Sahara, Sahel and tropical rainy areas of West Africa for the period 1967–74 (Fig. 6.10). There was also the suggestion of the albedo influencing the precipitation of the following year, but this relationship was tentative. In an extension of that satellite study, Courel et al. (1984) used METEOSAT and Landsat data to show that the dry season albedo actually decreased between 1973 and 1979 in the western Sahel despite continued drought. This seems to contradict the Charney (1975) hypothesis of an increase in albedo as drying of the surface proceeds, and suggests that the albedo-cloud and precipitation feedback is not a simple one. Courel et al. suggest that the reduced albedo may arise from an accumulation of dead vegetation or recovery of the grass cover even as the total amount of living vegetation decreases. A similar mechanism has been proposed for the northern Sinai by Otterman (1977).

The sensitivity of climate to albedo change can be estimated through use of various types of climate model (see, e.g., Charney, 1975; Charney et al., 1977). However, as input to those models, reliable albedo information on spatial scales commensurate with those used in the model is required, and with the elimination of time-dependent influences. Changing the land surface albedo boundary conditions in a GCM to values derived using satellite information (Preuss and Geleyn, 1980) reveals a small effect on T_s for a 10-day integration. However, these authors note the possibility of larger effects for climate-scale integrations. Henderson-Sellers and Wilson (1983) argue for a required accuracy of surface albedo data of ±0.05 (refer to Table 1). Satellite investigations for limited regions (see, e.g., Pinty et al., 1985) suggest present accuracies of about 0.10 for typical land surface values at GCM grid scales. Accordingly, much effort is being directed towards the retrieval of

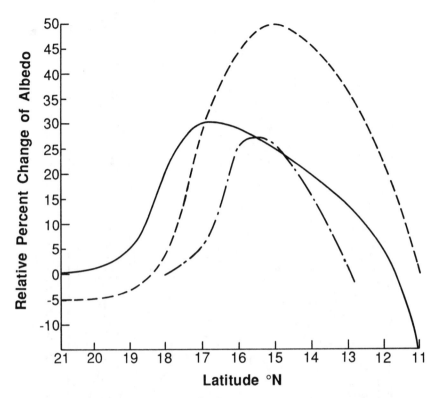

Fig. 6.10 Change in the mean latitudinal seasonal surface albedo for West Africa for the period July 2, 1974–September 21, 1974 (solid line) and September 15, 1969– January 6, 1970 (dashed line). Values from Rockwood and Cox for July 2 and September 20, 1974 are shown by the dot-dashed line. (From Norton *et al.*, 1979.)

surface albedos using satellites, and this forms a key element of such programs as ISLSCP. Satellites clearly play a role in determining both the surface albedo and also the incident solar radiation at the Earth's surface (irradiance).

Preuss and Geleyn (1980) used the Nimbus-3 minimum planetary albedo values of Raschke *et al.* (1973) to infer global-scale spectrally integrated (direct plus diffuse) surface albedos on an annual basis for input to GCMs. The effects of the intervening atmosphere were modeled for broad-scale (climatic) differences in surface type. The surface albedo was determined from its linear association with the minimum planetary albedo. Figure 6.11 shows the Nimbus-3 surface albedos determined from this physically based retrieval method. While some cloud contamination occurs (e.g., over China and adjacent ocean) the gradients between ice-covered and ocean surfaces in higher latitudes of both hemispheres are considered realistic. This map may be compared with that derived by Stephens *et al.* (1981: Fig. 6.3) which

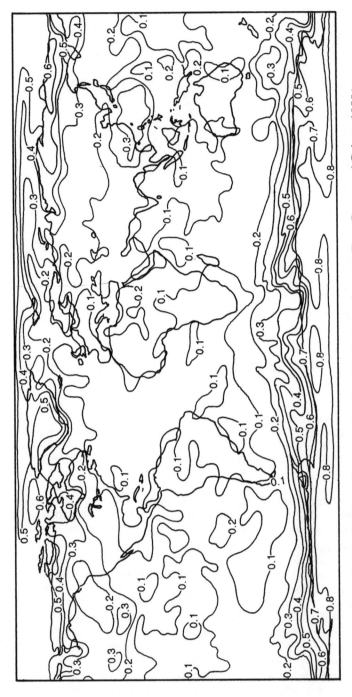

Fig. 6.11 Satellite-derived integrated surface albedos (annual mean). (From Preuss and Geleyn, 1980.)

shows considerably less spatial variability. Other albedo climatologies (e.g., those of Vonder Haar and Suomi, 1971 and Raschke *et al.*, 1973), while highly useful, are representative of the planetary albedo (TOA flux) and do not consider explicitly the surface albedo. Figure 6.11, and regional-scale satellite studies (e.g., Rockwood and Cox, 1978; Norton *et al.*, 1979; Courel *et al.*, 1984; Pinty and Szejwach, 1985; Pinty *et al.*, 1985; Gutman *et al.*, 1989a,b), emphasize the importance of cover type to the retrieved surface reflectances and the need for its accurate specification for deriving the surface albedo.

(b) Retrieval and characterization of surface types

(1) *Snowfree land surfaces* Land surfaces are characterized by a high degree of spatial heterogeneity of surface cover. These types have associated differences in emissivity, thermal properties and reflectance characteristics (Matthews and Rossow, 1987). Table 15 is an example of a surface cover classification developed using aircraft and satellite data for north-west Africa (Rockwood and Cox, 1978). The table shows strong meso-scale changes in surface albedo associated with the occurrence of different cover types. Accordingly, the characterization of surface types is a current concern of energy budget studies.

Satellite studies (e.g., Gutman, 1990a) show that the soil moisture and vegetation (status, leaf area density, photosynthetic activity) are not in-dependent. Naturally-vegetated surfaces can pull moisture from deeper within the soil and may show signs of stress later than adjacent arable land. Thus, there is a close relationship between meteorologic drought indicators and satellite-based indices of vegetation activity (see, e.g., Walsh, 1987). Photosynthetically active vegetation typically has a reflectance less than 20% in the narrow-band visible (0.5–0.7 μm) but a much higher reflectance (up to 60%) in the near-IR (0.7–1.3 μm): refer to Fig. 1.15. These portions of the EMS correspond, respectively, with AVHRR channels 1 and 2, or Landsat MSS bands 4 or 5 and 7 (Brest and Goward, 1987). A normalized difference vegetation index (NDVI) (see, e.g., Tarpley *et al.*, 1984) can be derived, which capitalizes on the spectral sensitivities of different land cover types. The procedure involves difference-ratioing the simultaneous NOAA channel 1 and 2 counts, thus:

$$\text{NDVI} = \frac{\text{channel 2} - \text{channel 1}}{\text{channel 2} + \text{channel 1}} \tag{29}$$

The NDVI is an integrated function of photosynthesis, leaf area and evapotranspiration (Running and Nemani, 1988). It is a bounded ratio that ranges between -1.0 and 1.0. Clouds, water and snow have negative NDVI since they are more reflective in visible than near-IR wavelengths; soil and rocks have broadly similar reflectances, giving NDVI close to 0. Only active vegetation has positive NDVI, being typically between about 0.1 to 0.6. Values at the higher end of the range indicate increased photosynthetic activity and greater density of the canopy (Tarpley *et al.*, 1984).

Table 15 Surface classification system

Class	Description	Surface albedo α_{SFC} mean	Standard deviation σ	Surface albedo α_{SFC} range
0	Swampland, oceans–coastal swamps, rivers, smooth oceans; >50% water with small solar zenith angle	0.09	0.015	0.0–0.10
1	Dense forest–uniform dark vegetation; 10% light vegetation or soils	0.15	0.017	0.10–0.16
2	Moderate forest—mostly dark vegetation; 30% light vegetation or soils	0.17	0.021	0.16–0.21
3	Mixed vegetation–evenly mixed surface; 50% dark vegetation; 50% light vegetation or soils	0.22	0.032	0.21–0.26
4	Savanna–mostly low, light grasses, cultivated fields, some dark scrub; 70% light dry grasses; 30% dark scrub or rock	0.28	0.025	0.26–0.31
5	Mixed desert–dry, light surface, colored soils and rock outcroppings; 50% light soils; 50% scrub, light grasses	0.33	–	0.31–0.36
6	Moderate desert–sparse vegetation, mostly light sands, scrub; 30% low scrub, light grasses, rock	0.39	0.016	0.36–0.42
7	Desert–uniform sand surface; little variation in surface over large areas; 10% low scrub or colored rock	0.43	0.021	0.42–1.0

Source: Rockwood and Cox, 1978.

Figure 6.12 is an NDVI image for the Midwest US obtained on a clear day in June 1988. The darker tones in the cropped areas of Illinois, Indiana and Ohio indicate stressed conditions; the whiter tones in the surrounding areas show enhanced biophysical activity. The latter corresponds with naturally-vegetated (mostly forest) surfaces. Limitations of the NDVI, some of which are pertinent to Fig. 6.12, include the following.

(1) Exposed soil early in the growing season can reduce NDVI values even where vegetation may be actively growing. Thus, part of the reason for the stressed signature of the cropped areas may result from the presence of bare ground, especially given the 1.1 km nominal resolution of the AVHRR used here (Gutman *et al.*, 1989b).
(2) Differences in relative moisture stress between cropped and natural surfaces are partly related to root depth.

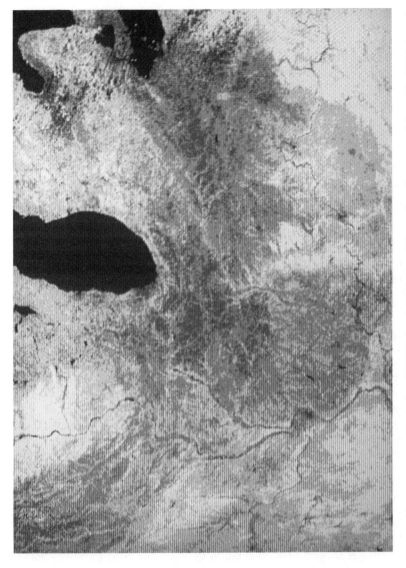

Fig. 6.12 Normalized difference vegetation index (NDVI) for the Midwest US on the afternoon of June 10, 1988, based on data from the NOAA-9 AVHRR. Dark (light) areas correspond with low (high) band ratios and indicate stressed cropped (less-stressed forest) vegetation.

231

(3) There is an antecedent moisture component present in the NDVI. Light antecedent precipitation may completely change the NDVI values but not significantly influence the soil moisture status.

The impact of these effects may be seen in Fig. 6.13, which is an NDVI image for the same region as in Fig. 6.12, only for mid-August 1988. The strong spatial differences in NDVI values have been eliminated, even though widespread soil moisture deficits continue. These are likely the result of the reduction of exposed soil during the growing season and the occurrence of antecedent precipitation in some areas. For studies of land surface climate anomalies, it is clearly desirable to compare NDVI for similar times across years.

For large-scale land surface studies, including the detection of phenological changes, NDVI images may be constructed based on the composite minimum brightness technique to ensure cloud-free reflectances (Chapter 3). These may extend over a week or a month. The maximum NDVI value in a given 4 km resolution cell of the AVHRR GAC (global area coverage) data is retained for construction of weekly or monthly global vegetation index (GVI) maps. This procedure aims both to reduce regional cloud contamination effects, and to account for changing viewing angles of the satellite by favoring pixels close to nadir, where atmospheric backscatter related to scan angle is lower (Thomas and Henderson-Sellers, 1987). The latter effect is greater in channel 1. Cloud cover is not inconsequential given the mid-afternoon orbit of the NOAA polar orbiter. The extended compositing period for monthly GVI would be expected to minimize the probability of cloud contamination; however, regional-scale studies (e.g., Gutman, 1987; Thomas and Henderson-Sellers, 1987) reveal that this is often not the case. Also, there is a regional dependence to the GVI values which are not transferable to the same vegetation types at different locations (Thomas and Henderson-Sellers, 1987). Problems with the GVI data set and their potential solution have been considered at a recent workshop of satellite and climate model users (Gutman, 1990b). Recommendations include the formulation of new algorithms and reprocessing of the satellite record; improved cloud screening techniques; the application of bidirectional reflectance models to the satellite-sensed radiances, and corrections for atmospheric attenuation due to water vapor and aerosols. Studies of the GVI product have proven useful for delimiting continental biomes and for comparing growing seasons between continents (see, e.g, Goward et al., 1987), as well as in studies of the global carbon cycle (Tucker et al., 1986). They are particularly useful for monitoring primary productivity in sensitive marginal areas, such as the Sahel, and as input to crop yield models (Rao et al., 1984; Yates et al., 1986). Recent studies show a high correlation between the NDVI and vegetation activity derived from the horizontally and vertically polarized T_B differences of the 37 GHz SMMR channel, at least for arid and semi-arid regions (Choudhury et al., 1987; Choudhury, 1989). There, the vegetation tends to be sparse and the differences in T_H and T_V tend to be larger. The NDVI appears to be better applied to biomes having a higher density of vegetation (Townshend et al., 1989).

Fig. 6.13 Similar to Fig. 6.12, except for August 21, 1988. Note the elimination of strong spatial differences in NDVI compared with June 10. There is some cloud contamination.

Land cover classification from satellite data (e.g., Townshend *et al.*, 1987) is increasingly viewed as a means to determine biophysical parameters (albedo, T_s, soil moisture, evaporation, and green foliage) important in climate, rather than as an end in itself (Goward, 1990a). Matthews and Rossow (1987) derived global seasonal minimum (surface) reflectance maps for the 0.6 μm channel of the NOAA-5 SR, and compared them with digital inventories of vegetation and soils derived from other sources (Matthews, 1983). They find that the changes in reflectance related to vegetation type are relatively small and limit classification to five basic land cover types at these large scales. These types are: forest/woodland; grassland; shrubland; un-vegetated (desert) and ice. At regional scales, some distinctions within vegetation classes are observed, but Matthews and Rossow (1987) note the need to corroborate these with ground truth.

(2) *Snow cover* Snow cover is important in climate variability (Chapter 2, Section 3d), and also hydrology; the latter with regard to the prediction of runoff from snow melt. Climatically significant parameters associated with snow cover include its high albedo in most solar wavelengths; its large areal extent; its low surface temperature; and its high emissivity in IR wavelengths.

While synoptic measurements of snow depth and snow-water equivalent constitute relatively recent operational data, there is a long-term satellite record of snow-cover extent from the ESSA and NOAA polar orbiters and the SMS/GOES platforms that dates back to late 1966 (Matson and Wiesnet, 1981). Initial analyses utilized CMB data since the changes in snow-cover extent are generally more conservative than cloud effects. The NOAA/NESS Northern Hemisphere snow cover maps have been produced weekly at the scale of 1:62 500 000 and classify snow reflectivity subjectively according to three classes (low, moderate, high). Recently, separation into these reflectivity classes has ceased. The data are now in digital form.

Robock and Scialdone (1986) compared different Northern Hemisphere snow-cover data sets derived from satellite and conventional observations. They find a tendency for CMB charts to overestimate snow and ice extent at the southward margin of the Arctic boundary relative to the NOAA/NESS data, and to underestimate snow cover in densely populated and forested areas. These authors conclude that the NOAA/NESS snow-cover data are the best currently available. There is evidence of a trend towards increasing annual snow-cover extent for North America and Eurasia, although the latter is influenced, at least in part, by inconsistencies in the mapping of the Himalayan snow cover (Matson and Wiesnet, 1981). Seasonal differences are also evident for the 1967–80 period. Snow-cover extent changes in Eurasia dominate the Northern Hemisphere record, giving it a greater feedback potential. These longer-term trends in both snow and sea ice extent have been used as indicators of recent climate change (cf. Kukla and Gavin, 1981; Zwally *et al.*, 1983). The conflicting interpretations appear to be largely a function of the length of the satellite snow and ice record.

(3) *Sea ice* The influence of polar sea ice on the planetary albedo is obviously limited to that time of the year when the sun is above the horizon

(Fig. 3.3). During the polar night, sea ice exerts a major impact on the high-latitude energy budget by influencing the exchanges of latent and sensible heat between the underlying ocean and the atmosphere. Spatial discontinuities in these fluxes are at a maximum at the ice–ocean margin, and in association with the open-water areas of leads and polynyas. In winter, these fluxes are typically of the order of 400 W m^{-2} for sensible heat and 130 W m^{-2} for latent heat (Andreas et al., 1979). Polynya size may be self-limiting due to the formation of new ice with heat loss from the open water (Pease, 1987). Barry et al. (1989) estimate, from DMSP visible image analysis, that leads exceeding 300 m across (±50 m) account for about 20% of the total area of open water/young ice in the Arctic. Mapping of these leads for the Beaufort Sea shows that lead orientations correlate broadly with geostrophic wind direction, in association with synoptic circulations. Airborne lidar measurements in the near-IR (1.06 μm) reveal plumes of moisture and ice crystals emanating from open-water areas in the pack ice of the eastern Canadian Arctic (Barry et al., 1989). Some of these plumes are observed to penetrate the surface-based temperature inversion, indicating that strong heat and moisture fluxes may extend to altitudes of at least 4 km. The large thermal contrasts in ice-covered to ice-free or partial ice cover have long been used to monitor open-water areas in the winter pack using TIR data for cloud-free conditions (see, e.g., Streten, 1974; Dey, 1980, 1981).

Satellite passive microwave data (ESMR, SMMR) provide all-weather surveillance (Chapter 2). The T_B are a function of the ice-water concentration and the ice thickness. The Nimbus-5 ESMR showed that the ice-water concentration increases across the ice margin by about 60% over distances of the order of 100–150 km (Carsey, 1982), and that the margin is narrower in winter than in summer. Similarly, the SMMR has shown that short-term fluctuations in the width of the MIZ occur during summer (Campbell et al., 1987).

Comparisons between aircraft-derived microwave and SMMR summer ice concentrations compare favorably, despite the differences in resolution (Campbell et al., 1987). The biggest differences occur for very low and very high ice concentrations (Burns et al., 1987). The ice-edge position recorded by the SMMR changes according to the degree of compaction of the ice. Under diffuse ice-edge conditions, the actual ice edge correlates best with the 30% SMMR ice concentration isopleth. Where compaction has occurred, then the 40–50% SMMR ice concentration correlates best with the actual ice edge. Away from strong ice concentration gradients, however, the aircraft- and satellite-microwave ice concentrations agree within the range 8–20 km (Campbell et al., 1987). Similarly, comparisons between SMMR passive and SIR-B active microwave observations of sea ice concentrations in the Weddell Sea (Antarctica) show strong agreement (Martin et al., 1987). A test of the SSM/I retrieval algorithm for ice concentration has been performed by Steffen and Schweiger (1990). Comparisons of the satellite passive microwave product with Landsat and AVHRR image data for cases in the Arctic reveals that the larger errors (of a few percent) occur in the fall compared with the spring.

In summer, the radiative and thermal contrasts between ice-covered and ice-free ocean become less important than the albedo effect. In particular, the onset of spring melt and its interannual variations affect the shortwave absorption for the summer and, accordingly, the seasonal and annual surface energy budgets. Knowledge of the timing of the spring melt, its impact on the surface albedo change, and interactions with clouds and radiation fluxes, is essential to a fuller understanding of the role of polar regions in global climate (see, e.g., Barry et al., 1984, 1987; Kukla and Robinson, 1988). It is during the spring melt, also, that ice concentration differences sensed by the SMMR and by Landsat increase to about 10% (Steffen and Maslanik, 1988).

Robinson et al. (1987) and Scharfen et al. (1987) have used the high-resolution DMSP and AVHRR visible imagery to map patterns of snow melt on the Arctic pack ice for four spring and summer seasons (1977, 1979, 1984, 1985). Four tonal and textural classes of melt having characteristic mean albedos were identified, and adjusted to account for average summer cloud cover and for ice concentration changes. In general, the melting is observed to progress into the central Arctic from the Kara/Barents seas and from the southern Beaufort and Chukchi seas. The Arctic Basin albedo decreases from about 0.75–0.80 in early May to about 0.35–0.45 in late July and early August. However, there is considerable between-year variability in the time of onset and duration of the melt, and also the albedo, as shown in Table 16. Snow melt onset is negatively related to the surface air temperature anomalies for the zone 65–85° N in the early part of the melt season, as would be expected from energy budget considerations. Barry et al. (1989) estimate that the variability of the surface net radiation of the central Arctic resulting from changes in the timing of snow melt on the sea ice may be of the order of 160 MJ m^{-2} for the season. Hypothetically at least, this is sufficient to melt 0.66 m of saline ice. The occurrence and interannual variability of the surface melt signatures can also be monitored using satellite passive microwave data, such as the 18 and 37 GHz channels of the SMMR (Anderson, 1987).

Table 16 Average surface albedo over the Arctic Basin for each month in the four study years

Month/year	1977	1979	1984	1985
May	0.73	0.77	0.73	0.76
June	0.58	0.66	0.61	0.65
July	0.43	0.44	0.48	0.48
August*	0.36	0.40	0.42	0.42

* For the period August 1–17.
Source: Robinson et al., 1987.

(c) Satellite retrieval of a clear-sky surface albedo

The method used to ensure cloud-free ('uncontaminated') conditions is crucial to the reliable determination of surface reflectances from satellite altitudes. The most commonly used technique is that of saving the minimum brightness pixel value over a given time period (e.g., a week or month) and assuming that this represents the true surface reflectance over a range of viewing conditions (e.g., the GVI). In this way, it is hoped that transient cloud effects (higher reflectances except over snow) are eliminated, but this technique has been shown to be sensitive to factors other than real changes in surface conditions. These include: the use of non-perfect or noisy radiometers (Henderson-Sellers and Wilson, 1983); the time interval for compositing the minimum brightness; and the positions of cumuliform clouds (even quite large ones) with respect to a given pixel and its neighbours. Thus, the ability to detect clouds is improved with higher resolution data (e.g., Landsat: Courel *et al.*, 1984) and makes the use of a single image as representative of cloud-free conditions for a given month or even a season more acceptable (e.g., Figs. 6.12, 6.13). In addition, cirrus clouds may take on the spectral characteristics of the underlying surface making their identification difficult (Henderson-Sellers and Wilson, 1982).

In such analyses, a reference target or area can be selected that is known to undergo minimal change in reflectance with time for a range of sun angles (e.g., the Sahara Desert). Obviously, the situation for eliminating cloud effects is more problematic for meteorological satellites with their coarser spatial resolution, since broken cloud fields are typical on these scales (see, e.g., Coakley and Kobayashi, 1989). Simulations with ATS data at $36 \, km^2$ resolution over North Africa (Norton *et al.*, 1979) have shown that, beyond about 10% cumulus cloud coverage, the reflectance is no longer representative of the range of brightness values commonly assigned to the surface. Cloud detection algorithms (Chapter 3) also perform differently according to the albedo of the underlying surface (e.g., ocean versus land, desert versus forest vegetation), and as a function of the relative azimuth (Taylor and Stowe, 1984). Thus, the minimum brightness in the presence of clouds exhibits different characteristics for snow-covered (high albedo) and snow-free land surfaces, and presupposes a knowledge of the type and state of the underlying surface (Matthews and Rossow, 1987). It is very difficult to extract reliable minimum brightness values on climatic time scales (i.e., greater than about 5–7 days) in regions of rapidly varying surface conditions, such as the snow-melt margin, and in areas of persistent cloud cover.

(d) Satellite sensor influences on albedo determination

The true albedo is the shortwave reflectance integrated across all solar wavelengths, over all viewing angles and over time (Pinker, 1985). As such, it is only imperfectly measured by satellites in specific (usually narrow) wavelength bands; often at fixed local times and in narrow solid angles. As indicated earlier, actively growing vegetation has a peak reflectance in the near-IR (about 0.7–1.3 μm) but lower reflectance in the visible (0.3–0.7 μm) and mid-IR (about 1.4–2.0 μm). Thus, omitting the mid-IR will give albedo values that are too high for vegetated surfaces (Brest and Goward, 1987), while omitting this band over bare semi-arid and arid surfaces may result in underestimated full-spectral albedos there (Otterman and Fraser, 1976). Accordingly, various weighting schemes may be used on these narrow-band reflectances, based on the known wavelength dependence of different surfaces and the contribution of different wavelengths to the total albedo (see, e.g., Otterman, 1977). Norton *et al.* (1979) used ATS-3 satellite data in the 0.48–0.58 μm ('green') channel to study variations in surface reflectance for the Sahel in drought and non-drought years of the 1970s. They derived a function of the covariance of model solar spectrum and green channel albedo for different surface types, and showed that a large albedo change accompanies the transition from normal wet to normal dry season or even early drought. However, there is little albedo change from vegetation in drought conditions to bare ground. Norton *et al.* note that the observed albedo change should be multiplied by a factor of about 0.6 to bring it closer to the full solar spectrum albedo.

Estimates of the surface reflectance on synoptic scales have made use of the NOAA SR sensor (Chen and Ohring, 1984; Matthews and Rossow, 1987; Gutman *et al.*, 1989a,b). The visible channel is sensitive to radiation generally between about 0.5–0.9 μm but with a peak in the 0.5–0.7 μm range. This omits much of the sensitivity to growing vegetation. In a study of the influence of satellite sensor response on estimates of cloud–climate sensitivity, Shine *et al.* (1984) compared the narrow-band NOAA SR with the broad-band Nimbus sensors. They show that the influence of cloudiness variations on the planetary albedo is somewhat dependent on the choice of satellite sensor, and can be overestimated by a factor of up to 1.38 at low latitudes for the SR. Following the previous findings of Briegleb and Ramanathan (1982) on the spectral dependence of the clear-sky planetary albedo, Chen and Ohring (1984) demonstrate that the 0.5–0.7 μm band is reasonably representative of the broad-band planetary albedo for zonal averages. Moreover, the surface albedo is linearly related to the planetary albedo, such that the surface albedo explains 98% of the latter. Model simulations of narrow-band and broad-band clear-sky planetary albedo for a range of SZAs (Pinker and Ewing, 1986) reveal a generally greater impact of the surface albedo compared with the atmospheric conditions. Broad-band surface albedos have been approximated from NOAA SR channels 1 (0.58–0.68 μm) and 2 (0.73–1.1 μm) reflectances using a regression formula for vegetated surfaces (Gutman *et al.*, 1989a). Pinty *et al.* (1985) assumed that surface albedos derived from the 0.4–1.1 μm channel of METEOSAT are equivalent to those integrated over the entire solar spectrum.

(e) Geometric effects on surface albedo

The bidirectional reflectance and albedo is dependent not only on the surface type and spectral resolution of the sensor but also on geometric influences. These are the solar zenith angle and the relative azimuth (sun–target–sensor) effects (Chapter 2). Satellite measurements of surface albedo can be examined for zenith angle dependence related to time of day using data from geostationary platforms. These can be used to provide corrections to albedos determined from polar orbiters with their fixed local time sampling of ERB components and to help in the formulation of appropriate sampling strategies (England and Hunt, 1984). Using METEOSAT data, Saunders and Hunt (1980) showed the diurnal variations of albedo (Earth surface, clouds) and outgoing IR flux for north-west Africa and the adjacent ocean (Fig. 6.14). The albedo remains relatively constant over the desert but increases markedly over the ocean with increasing SZA (morning, evening). Multiple reflections from cumulus clouds and changes in the structure of high clouds are responsible for the different diurnal signatures represented in Fig. 6.14.

In a study of clear-sky planetary albedo for the US Great Plains, Gutman *et al.* (1989a) applied the Taylor and Stowe (1984) Nimbus bidirectional reflectance model for clear land to AVHRR radiances. This procedure smooths the albedo variations, except in late fall and early winter when SZAs are large. It also helps remove albedo discontinuities caused by adjacent orbits of a polar orbiter (i.e., changes in the relative azimuth angle). When differences in surface vegetation types and seasonal changes in sun angle are accounted for, the albedo variability is reduced still further, especially for winter.

A simpler technique for estimating the hemispheric (integrated) surface albedo from NOAA-9 AVHRR data that does not require application of a bidirectional reflectance model has been advanced by Gutman (1988) and applied to a range of surface types for the US Great Plains (Gutman *et al.*, 1989b). For NOAA-9, the sun–target–sensor geometry exhibits a nine-day repeat cycle. The effect of this change on the TOA albedo for NOAA-7, which has a similar repeat cycle, is illustrated for the Libyan Desert and Arabian Sea in Fig. 6.15 (Darnell *et al.*, 1988). Gutman *et al.* assigned clear-sky reflectances according to the appropriate day in the nine-day cycle; compositing (averaging) for all observations within that day number, and then averaging over the nine geometry configurations (the compositing interval). This procedure gives a reasonable approximation to the hemispheric surface reflectance. The surface albedo is derived by performing a narrow-band to broad-band conversion that includes the near-IR and by applying a simple atmospheric correction. The only remaining source of error with this technique (assuming that reflectances are for totally clear-sky conditions) involves the seasonal change in solar zenith angle. However, the latter is considered to be significant only for large angles, and not applicable to growing season conditions for which the studies were initiated. Preliminary comparisons between the surface albedo results of this method and those developed using the bidirectional model of Taylor and Stowe (1984) for vegetated land are reported as encouraging (Gutman, 1988).

Robock and Kaiser (1985) examined the influence of snow and cloud cover on the bidirectional reflectance for different land surface cover types using the narrow-band NOAA SR. A range of conditions was examined, including overcast skies with and without snow cover present, and clear skies with and without snow cover. These authors find that concurrent cloud and snow cover do not produce higher reflectance values than does cloud cover alone, at least for optically thick clouds. Similarly, overcast conditions tend to

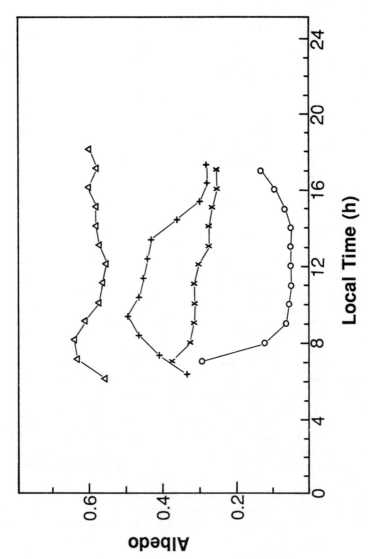

Fig. 6.14 The diurnal variability of albedo as measured on one day (November 5, 1978) by METEOSAT over four surface/cloud types. These are: X, over land; solid circles over sea; +, over low cumulus cloud; solid triangle, over high cumulonimbus cloud. Absolute values of the albedo have an uncertainty of 0.05 but relative uncertainties will be much smaller. (From Saunders and Hunt, 1980.)

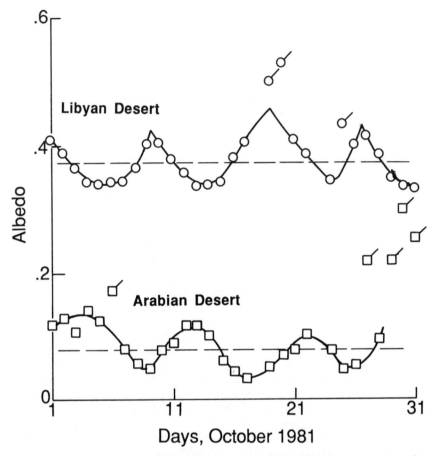

Fig. 6.15 Illustration of the nine-day repeat cycle in TOA albedo at two locations for NOAA-7. Flagged symbols indicate cloudy conditions. Horizontal dashed lines represent clear-sky albedos for the month. (From Darnell *et al.*, 1988.)

give higher reflectances than clear-sky, snow-covered conditions. There was relatively little angular dependence of the reflectivities on relative azimuth angle, even given the quite high latitudes and the day-to-day change in sun–target–satellite geometry of the NOAA polar orbiters.

(f) Atmospheric influences on satellite-derived albedo

Reflected shortwave irradiances received at a satellite sensor need to be corrected for atmospheric effects (Chapter 2). Radiation transfer theory can be used to account for the effects of Rayleigh scatter and gaseous absorption by ozone in short wavelengths and by water vapor in the mid-IR; however, the effects of aerosols (especially dust) on the absorption, scatter and transmittance of solar radiation is more problematic (Pinker, 1985). Methods to correct for these effects include:

(1) model simulations, which require information on the concurrent atmospheric state;
(2) concomitant aircraft and satellite observations; and
(3) the extrapolation of surface albedos to regional scales from a few 'representative' surface sites, in which the atmospheric transmittance is assumed either to not change or to change little spatially (Pinty and Szejwach, 1985).

The planetary albedo can be determined from satellites as the difference between the shortwave radiation incident at the top of the atmosphere and the upwelling (reflected) shortwave radiation (see, e.g., Raschke *et al.*, 1973; Saunders and Hunt, 1980). Assuming clear-sky conditions and a constant surface reflectance, the satellite-sensed brightness is the net effect of the contributions of reflection, absorption, scatter and transmission by atmospheric gases and aerosols. The relative contributions are wavelength dependent (Briegleb and Ramanathan, 1982) and, hence, values for the planetary albedo are influenced by the satellite sensor spectral resolution. The atmospheric effects of Rayleigh scatter, which is more important between 0.2–0.5 μm than between 0.7–4.0 μm; and atmospheric absorption, which shows the opposite tendency for the same wavelengths, can be modeled with varying degrees of sophistication (see, e.g., Rockwood and Cox, 1978; Chen and Ohring, 1984). Previously developed models of atmospheric transmission, scattering and absorption of solar radiation, in which the surface is considered Lambertian and the radiation unpolarized, were used by Otterman and Fraser (1976) to examine the relationship of Earth surface reflectivities to the observed planetary albedo in the four Landsat MSS bands. These were for two water vapor and turbidity levels, at several SZAs, and applied to selected arid and semi-arid surfaces. The authors report little difference between the surface and system reflectivities for ground reflectivities in the range 0.2–0.5 μm; however, much larger differences are reported for the 0.9 μm wavelength (MSS-7). There, the system reflectance can be significantly lower than the surface reflectance owing to the influence of water vapor absorption. In its simplest form (Rockwood and Cox, 1978) the clear-sky system albedo comprises the fractional reflectivity (ρ), absorptivity (a) and transmissivity (τ) in the 0.285–2.8 μm band, or:

$$\rho_{SYS} \equiv H_1\!\uparrow/H_1\!\downarrow \equiv (\rho H_1\!\downarrow + \tau H\!\uparrow_{SFC})/H_1\!\downarrow \tag{30}$$

where $H_1\!\downarrow$ refers to radiation incident at the top of the model atmosphere, $H_1\!\uparrow$ is the radiant exitance at the top of the model, and $H\!\uparrow_{SFC}$ is the surface reflectance (lower level of model). Thus, the effective atmospheric absorptivity is:

$$a_{AT} \equiv (H_1\!\downarrow - H_1\!\uparrow - H\!\downarrow_{SFC} + H\!\uparrow_{SFC})/H_1\!\downarrow \equiv (aH_1\!\downarrow + aH\!\uparrow_{SFC})/H_1\!\downarrow$$

$$\tag{31}$$

where $H\downarrow_{SFC}$ is the incident radiation at the surface. The effective atmospheric transmissivity is:

$$\tau_{AT} \equiv H\downarrow_{SFC}/H_1\downarrow \equiv (\tau H_1\downarrow + \rho H\uparrow_{SFC})/H_1\downarrow. \tag{32}$$

To get at the surface albedo in this model (or $\alpha_{SFC} = H\uparrow_{SFC}/H\downarrow_{SFC}$) there is an assumed conservation of solar energy incident at the top of the model ($H_1\downarrow$) and as reflected by the surface ($H\uparrow_{SFC}$), so that:

$$H\uparrow_{SFC} = \tau H_1\downarrow \alpha_{SFC}. \tag{33}$$

By manipulation and substitution one obtains the budget equation:

$$1 = \rho_{SYS} + a_{AT} + \tau_{AT}(1 - \alpha_{SFC}) \tag{34}$$

and this can be rearranged to give the system albedo from satellite brightness observations:

$$\rho_{SYS} = 1 - a_{AT} - \tau_{AT}(1 - \alpha_{SFC}) \tag{35}$$

In the Rockwood and Cox (1978) satellite model, concurrent observations of the atmospheric transmissivity (τ_{AT}) and absorptivity (a_{AT}) are required (e.g., from aircraft). To derive the surface albedo, equation (35) is rearranged to give:

$$\alpha_{SFC} = 1 - [(1 - \rho_{SYS} - a_{AT})/\tau_{AT}] \tag{36}$$

Rockwood and Cox were able to classify surface types over north-west Africa (refer to Table 15) and to derive maps of surface albedo at the mesoscale for the period July–September, 1974 (GATE). In addition, they were able to show temporal changes in surface albedo for the different types and in the latitudinal gradient over the wet season (Table 17). In lieu of collocated satellite and aircraft observations, some 'representative' values of τ_{AT} and α_{AT} may be used with the assumption of only limited temporal or spatial variability. This assumption is least appropriate in the case of aerosol effects (especially dust) on the atmospheric optical depth.

Table 17 Percentage of the area ($15–20°$ N by $5–15°$ W) attributable to each surface class denoted in Table 15

Surface class	α_{SFC} range	Percentage of total area	
		2 July	20 September
7	0.42–1.0	30	29
6	0.36–0.42	24	13
5	0.31–0.36	24	15
4	0.26–0.31	17	14
3	0.21–0.26	4	20
2	0.16–0.21	0	0
1	0.10–0.16	0	0
0	0–0.10	0	0

Source: Rockwood and Cox, 1978.

Pinty and Szejwach (1985) describe a method for obtaining surface albedo on the scale of a 'climatic region' (West Africa) and in the absence of detailed ground truth. The method relies on the ability of a geostationary satellite (METEOSAT) to monitor the diurnal variation of surface radiances in the 0.4–1.1 μm range, and on the spatial correlation of atmospheric conditions and surface albedo at a reference site with surrounding areas. These areas are, ideally, in the same climatic region. The method also assumes that any spatial gradient in the atmospheric transmittance is not large enough to cause the albedos extrapolated from a reference site to exceed a given error. Thus, for an albedo determination 1000 km removed from the site and a desired relative accuracy of 25%, the transmittance gradient must not exceed 3%/100 km (or 5%/100 km for a 25% relative accuracy at 500 km distance). Pinty and Szejwach find, for two days in February and July 1979, that the spatial variability of retrieved surface albedos is 0.18–0.6, and is highest over the deserts. Both the spatial pattern and seasonal change in surface albedo are in general agreement with previous studies for this region (e.g., Rockwood and Cox, 1978; Norton *et al.*, 1979; Courel *et al.*, 1984).

Suggested further reading

Carter, W. D., Arking, A., McCormick, M. P., and Raschke, E. (eds.), 1987: Remote Sensing: Earth's Surface and Atmosphere (special issue). *Adv. Space Res.*, 7: (3). 252 pp.

Henderson-Sellers, A. (ed.), 1984: *Satellite Sensing of a Cloudy Atmosphere: Observing the Third Planet*. Taylor and Francis, London and Philadelphia. 340 pp.

Ohring, G. and Gruber, A., 1983: Satellite Radiation Observations and Climate Theory. In: *Theory of Climate*, Vol. 25 of Advances in Geophysics (ed. Barry Saltzman). Academic Press, New York. pp. 237–304.

Epilogue:
Satellites in the climatological future

The increasing awareness of environmental degradation, and potential climatic alterations as a result of human activities, is clearly extending the contribution of meteorological satellites far beyond that for which they were originally intended: the surveillance of cloud and moisture patterns to aid weather analysis and prediction over data-remote areas. Satellite information has had a profound impact on the monitoring of climate and the Earth's surface characteristics, as well as in the development of contemporary theories of climate. This trend can only continue as we approach the twenty-first century.

Satellites with more diverse sensors will be deployed over the next few years. These will emphasize particularly the higher frequencies of the microwave region and will, moreover, finally extend the successful results of the 1978 Seasat mission that orbited active (radar) sensors. While new generations of geostationary and sun-synchronous polar orbiters with microwave sensors will be deployed, so too, increasingly, will satellites that have dedicated mission objectives and for which non-conventional orbits are required; as in the monitoring of the diurnal variability of clouds and precipitation. The return of the US Space Shuttle to duty, and possible future deployment of similar manned vehicles by other countries and consortia, will have an impact in the area of Earth monitoring.

A particularly exciting prospect from the standpoint of Earth monitoring for global change (Eos) is the proposed manned US Space Station in low-latitude orbit, and the unmanned near-polar orbiting platforms of the Japanese and Europeans. Highly advanced sensors in hitherto little-explored spectral regions and with ever increasing resolution capabilities, are planned for these platforms. The data from these, it is anticipated, will represent a major leap forward for satellite climatological studies. At that time, it will be important to remember that those advances built upon a long line of scientific and technological accomplishments; a sampling of the former having been reviewed in this monograph on remote sensing research in climatology.

References

Ackerman, S. A., and Cox, S. K., 1981: Comparison of satellite and all-sky camera estimates of cloud cover during GATE. *J. Appl. Met.*, *20:* 581–7.

Adang, T. C., and Gall, R. L., 1989: Structure and dynamics of the Arizona monsoon boundary. *Mon. Wea. Rev.*, *117:* 1423–38.

Adler, R. F., and Fenn, D. D., 1979: Thunderstorm intensity as determined from satellite data. *J. Appl. Met.*, *18:* 502–17.

Adler, R. F., Markus, M. J., and Fenn, D. D., 1985: Detection of severe Midwest thunderstorms using geosynchronous satellite data. *Mon. Wea. Rev.*, *113:* 769–81.

Agee, E. M., 1987: Mesoscale cellular convection over the ocean. *Dynam. Atmos. Oceans*, *10:* 317–41.

Albrecht, B. A., 1990: Aerosols, cloud microphysics, and fractional cloudiness. *Science*, *245:* 1227–30.

Alfultis, M. A., and Martin, S., 1987: Satellite passive microwave studies of the Sea of Okhotsk ice cover and its relation to oceanic processes, 1978–1982. *J. Geophys. Res.*, *92:* 13 013–28.

Alishouse, J. C., 1983: Total precipitable water and rainfall distributions from SEASAT Scanning Multichannel Microwave Radiometer. *J. Geophys. Res.*, *88:* 1929–35.

Alishouse, J. C., Ferraro, R. R., and Fiore, J. V., 1990: Inference of oceanic rainfall properties from the Nimbus-7 SMMR. *J. Appl. Meteor.*, *29:* 551–60.

Allison, L. J., Steranka, J., Cherrix, G. T., and Hilsenrath, E., 1972: Meteorological applications of the Nimbus 4 Temperature-Humidity Infrared Radiometer, 6.7 micron channel data. *Bull., Amer. Meteor. Soc.*, *53:* 526–35.

Allison, L. J., Rodgers, E. B., Wilheit, T. T., and Fett, R. W., 1974: Tropical cyclone rainfall as measured by the Nimbus 5 electrically scanning microwave radiometer. *Bull., Amer. Meteor. Soc.*, *55:* 1074–89.

Anderson, M. R., 1987: The onset of spring melt in first-year ice regions of the Arctic as determined from Scanning Multichannel Microwave Radiometer data for 1979 and 1982. *J. Geophys. Res.*, *92:* 13 153–63.

Anderson, R. K., Ashman, J. P., Bittner, F., Farr, G. R. M., Ferguson, E. W., Oliver, V. J., and Smith, A. H., 1969; 1973: Application of meteorological satellite data in analysis and forecasting. ESSA Tech. Rept. NESC 51. (Also issued as Air Weather Service Tech. Rept. No. 212). National Environmental Satellite Center, ESSA, Washington, DC [AD697-033].

Andersson, T., and Nilsson, S., 1990: Topographically induced convective snowbands over the Baltic Sea and their precipitation distribution. *Weather and Forecasting*, *5:* 299–312.

Andreas, E. L., Paulson, C. A., Williams, R. M., Lindsay, R. W., and Businger, J. A., 1979: The turbulent heat flux from Arctic leads. *Bound. Layer Meteor.*, *17:* 57–91.

Angell, J. K., Korshover, J., and Cotton, G. F., 1984: Variation in United States cloudiness and sunshine, 1950–1982. *J. Clim. Appl. Meteor.*, *23:* 752–61.

Anthes, R. A., 1984: Enhancement of convective precipitation by mesoscale variations in vegetative covering in semiarid regions. *J. Clim. Appl. Meteor.*, *23:* 541–54.

Ardanuy, P. E., Kyle, H. L., and Chang, H-D., 1987: Evolution of the Southern Oscillation as observed by the Nimbus-7 ERB experiment. *Mon. Wea. Rev.*, *115:* 2615–25.

Ardanuy, P. E., Stowe, L. L., Gruber, A., Weiss, M., and Long, C. S., 1989: Longwave cloud forcing as determined from Nimbus 7 observations. *J. Climate*, 2: 766–99.

Arkin, P. A., 1979: The relationship between fractional coverage of high cloud and rainfall accumulations during GATE over the B-scale array. *Mon. Wea. Rev.*, *107:* 1382–7.

Arkin, P. A., and Ardanuy, P. E., 1989: Estimating climatic-scale precipitation from space: a review. *J. Climate*, 2: 1229–38.

Arkin, P. A., and Meisner, B. N., 1987: The relationship between large-scale convective rainfall and cold cloud over the western hemisphere during 1982–84. *Mon. Wea. Rev.*, *115:* 51–74.

Arkin, P. A., and Webster, P. J., 1985: Annual and interannual variability of tropical–extratropical interaction: an empirical study. *Mon. Wea. Rev.*, *113:* 1510–23.

Arkin, P. A., Krishna Rao, A. V. R., and Kelkar, R. R., 1989: Large-scale precipitation and outgoing longwave radiation from INSAT-1B during the 1986 Southwest monsoon season. *J. Climate*, 2: 619–28.

Arking, A., 1964: Latitudinal distribution of cloud cover from TIROS III photographs. *Science*, *143:* 569–72.

Arking, A., and Childs, J. D., 1985: Retrieval of cloud cover parameters from multispectral satellite images. *J. Clim. Appl. Meteor.*, *24:* 322–33.

Augustine, J. A., and Howard, K. W., 1988: Mesoscale Convective Complexes over the United States during 1985. *Mon. Wea. Rev.*, *116:* 685–701.

Barker, H. W., and Davies, J. A., 1989: Surface albedo estimates from Nimbus-7 ERB data and a two-stream approximation of the radiative transfer equation. *J. Climate*, 2: 409–18.

Barkstrom, B. R., 1984: The Earth Radiation Budget Experiment (ERBE). *Bull., Amer. Meteor. Soc.*, *65:* 1170–85.

Barkstrom, B., Harrison, E., Smith, G., Green, R., Kibler, J., Cess, R., and the ERBE Science Team, 1989: Earth Radiation Budget Experiment (ERBE) archival and April 1985 results. *Bull., Amer. Meteor. Soc.*, *70:* 1254–62.

Barkstrom, B. R., Harrison, E. F., and Lee, R. B., 1990: Earth Radiation Budget Experiment, preliminary seasonal results. *EOS (Trans. Amer. Geophys. Union)*, *71:* 297, 304–5.

Barnes, J. C., 1966: Note on the use of satellite observations to determine average cloudiness over a region. *J. Geophys. Res.*, *71:* 6137–40.

Barnett, T., Graham, N., Cane, M., Zebiak, S., Dolan, S., O'Brien, J., and Legler, D., 1988: On the prediction of the El Niño event of 1986–1987. *Science*, *241:* 192–6.

Barnston, A. G., and Thomas, J. L., 1983: Rainfall measurement accuracy in FACE: a comparison of gage and radar rainfalls. *J. Clim. Appl. Meteor.*, *22:* 2038–52.

Barr, S., Lawrence, M. B., and Sanders, F., 1966: TIROS vortices and large-scale vertical motion. *Mon. Wea. Rev.*, *94:* 675–96.

Barrett, E. C., 1970: The estimation of monthly rainfall from satellite data. *Mon. Wea. Rev.*, *98:* 322–27.

Barrett, E. C., 1971: The tropical Far East: ESSA satellite evaluations of high season climatic parameters. *J. Geog.*, *137:* 535–55.

Barrett, E. C., 1973: Forecasting daily rainfall from satellite data. *Mon. Wea. Rev.*, *101:* 215–22.

Barrett, E. C., 1974: *Climatology from Satellites*. Methuen, London and New York. 418pp.

Barrett, E. C., 1979: The use of weather satellite data in the evaluation of national water resources, with special reference to the Sultanate of Oman, In: Rycroft, M. J. (ed.). *Space Research*, Vol. 19, Pergamon Press. pp. 41–6.

Barrett, E. C., and Grant, C. K., 1979: Relations between frequency distributions of cloud over the United Kingdom based on conventional observations and imagery from Landsat 2. *Weather, 34:* 416–23.

Barrett, E. C., and Harris, R., 1977: Satellite infra-red nephanalysis. *Meteor. Mag., 106:* 11–26.

Barrett, E. C., Kidd, C., Bailey, J. O., and Collier, C. G., 1990: The Great Storm of 15/16 October 1987: passive microwave evaluations of associated rainfall and marine wind speeds. *Meteorol. Mag., 119:* 177–87.

Barrett, E. C., and Martin, D. W., 1981: *The Use of Satellite Data in Rainfall Monitoring.* Academic Press. 340pp.

Barrett, E. C., D'Souza, G., Power, C. H., and Kidd, C., 1988: Toward trispectral satellite rainfall monitoring algorithms, In: Theon, J., and Fugono, N. (eds.). *Tropical Rainfall Measurements.* A. Deepak Publ. pp. 285–92.

Barry, R. G., and Maslanik, J., 1989: Arctic sea ice characteristics and associated atmosphere-ice interactions in summer inferred from SMMR data and drifting buoys: 1979–1984. *GeoJournal, 18:* 35–44.

Barry, R. G., Crane, R. G., Schweiger, A., and Newell, J., 1987: Arctic cloudiness in spring from satellite imagery. *J. Climatol., 7:* 423–51.

Barry, R. G., Henderson-Sellers, A., and Shine, K. P., 1984: Climate sensitivity and the marginal cryosphere, In: *Climate Processes and Climate Sensitivity,* Geophys. Monog. 29, Maurice Ewing Vol. 5. Amer. Geophys. Union, Washington, DC. pp. 221–37.

Barry, R. G., Miles, M. W., Cianflone, R. C., Scharfen, G., and Schnell, R. C., 1989: Characteristics of Arctic sea ice from remote-sensing data and their relationship to atmospheric processes. *Ann. Glaciol., 12:* 9–15.

Barton, I. J., 1983a: Dual channel satellite measurements of sea surface temperature. *Quart. J. Royal Meteor. Soc., 109:* 365–78.

Barton, I. J., 1983b: Upper level cloud climatology from an orbiting satellite. *J. Atmos. Sci., 40:* 435–47.

Bates, T. S., Charlson, R. J., and Gammon, R. H., 1987: Evidence for the climatic role of marine biogenic sulphur. *Nature, 329:* 319–21.

Bean, S. J., and Somerville, P. N., 1981: Some new worldwide cloud-cover models. *J. Appl. Meteor., 20:* 223–8.

Beer, T., 1980: Satellite microwave estimation of tropical cyclone rainfall: a case study of cyclone Joan. *Austral. Meteor. Mag., 28:* 155–66.

Bell, I. D., 1986: The Northwest Australian cloud band, In: Preprint volume, *Second International Conference on Southern Hemisphere Meteorology,* December 1–5, 1986: Wellington, New Zealand. Amer. Meteor. Soc., Boston, Mass. pp. 42–5.

Bellon, A., Lovejoy, S., and Austin, G. L., 1980: Combining satellite and radar data for the short-range forecasting of precipitation. *Mon. Wea. Rev., 108:* 1554–66.

Bess, T. D., 1986: Variability of Earth-emitted radiation from one year of Nimbus-6 ERB data. *J. Atmos. Sci., 43:* 1445–53.

Bess, T. D., Smith, G. L., and Charlock, T. P., 1989: A ten-year monthly data set of outgoing longwave radiation from Nimbus-6 and Nimbus-7 satellites. *Bull., Amer. Meteor. Soc., 70:* 480–9.

Bjerknes, J., Allison, L. J., Kreins, E. R., Godshall, F. A., and Warnecke, G., 1969: Satellite mapping of the Pacific tropical cloudiness. *Bull., Amer. Meteor. Soc., 50:* 313–22.

Blackmer, Jr., R. H., and Serebreny, S. M., 1968: Analysis of maritime precipitation using radar data and satellite cloud photographs. *J. Appl. Meteor., 7:* 122–31.

Bolle, H-J., 1985: Assessment of thin cirrus and low cloud over snow by means of the maximum likelihood method. *Adv. Space Res., 5:* 169–75.

Bonatti, J. P., and Rao, V. B., 1987: Moist baroclinic instability in the development of

North Pacific and South American intermediate-scale disturbances. *J. Atmos. Sci.*, *44:* 2657–67.

Booth, A. L., and Taylor, V. R., 1969: Meso-scale archive and computer products of digitized video data from ESSA satellites. *Bull., Amer. Meteor. Soc.*, *50:* 431–8.

Böttger, H., Eckardt, M., and Katergiannakis, U., 1975: Forecasting extratropical storms with hurricane intensity using satellite information. *J. Appl. Meteor.*, *14:* 1259–65.

Boucher, R. J., and Newcomb, R. J., 1962: Synoptic interpretations of some TIROS vortex patterns: a preliminary cyclone model. *J. Appl. Meteor.*, *1:* 127–36.

Bowman, K. P., 1988: Global trends in total ozone. *Science*, *239:* 48–50.

Bowman, K. P., 1990: Evolution of the total ozone field during the breakdown of the Antarctic circumpolar vortex. *J. Geophys. Res.*, *95:* 16529–43.

Brakke, T. W., and Kanemasu, E. T., 1981: Insolation estimation from satellite measurements of reflected radiation. *Remote Sensing Envt.*, *11:* 157–67.

Brest, C. L., and Goward, S. N., 1987: Deriving albedo measurements from narrow band satellite data. *Int. J. Remote Sensing*, *8:* 351–67.

Brewer, P. G., Bruland, K. W., Eppley, R. W., and McCarthy, J. J., 1986: The Global Ocean Flux Study (GOFS): status of the U.S. GOFS Program. *EOS (Trans. Amer. Geophys. Union)*, *67:* 827–32.

Briegleb, B., and Ramanathan, V., 1982: Spectral and diurnal variations in clear sky planetary albedo. *J. Appl. Meteor.*, *21:* 1160–71.

Brinegar, R., 1990: Midwest U.S. land surface–atmosphere interactions from satellite analysis, summers 1987 and 1988. Masters thesis, Department of Geography, Indiana University.

Bristor, C. L., and Ruzecki, M. A., 1960: TIROS 1 photographs of the Midwest storm of April 1, 1960. *Mon. Wea. Rev.*, *88:* 315–26.

Bristor, C. L., Callicott, W. M., and Bradford, R. E., 1966: Operational processing of satellite cloud pictures by computer. *Mon. Wea. Rev.*, *94:* 515–27.

Bromwich, D. H., 1989: Satellite analyses of Antarctic katabatic wind behavior. *Bull., Amer. Meteor. Soc.*, *70:* 738–49.

Brown, S. C., 1970: Simulating the consequences of cloud cover on Earth-viewing space missions. *Bull., Amer. Meteor. Soc.*, *51:* 126–31.

Browning, K. A., 1979: The FRONTIERS plan: a strategy for using radar and satellite imagery for very-short-range precipitation forecasting. *Meteor. Mag.*, *108:* 161–84.

Browning, K. A., 1985: Conceptual models of precipitation systems. *Meteor. Mag.*, *114:* 293–319.

Browning, K. A., and Hill, F. F., 1985: Mesoscale analysis of a polar trough interacting with a polar front. *Quart. J. Royal Meteor. Soc.*, *111:* 445–62.

Browning, K. A., Eccleston, A. J., and Monk, G. A., 1985: The use of satellite and radar imagery to identify persistent shower bands downwind of the North Channel. *Meteor. Mag.*, *114:* 325–31.

Bryson, R. A., and Wendland, W. M., 1975: Climatic effects of atmospheric pollution, In: Singer, S. F. (ed.) *The Changing Global Environment*. Reidel Publ., Holland. pp. 139–47.

Budd, W. F., 1982: The role of Antarctica in southern hemisphere weather and climate. *Austral. Meteor. Mag.*, *30:* 265–72.

Bunting, J. T., Hawkins, R. S., d'Entremont, R. P., 1983: R and D nephanalysis at the Air Force Geophysical Laboratory, In: *Preprints volume, Fifth Conference on Atmospheric Radiation*. Baltimore, MD. Amer. Meteor. Soc., Boston, Mass. pp. 272–5.

Burfeind, C. R., Weinman, J. A., and Barkstrom, B. R., 1987: A preliminary computer pattern analysis of satellite images of mature extratropical cyclones. *Mon. Wea. Rev.*, *115:* 556–63.

Burns, B. A., Cavalieri, D. J., Keller, M. R., Campbell, W. J., Grenfell, T. C., Maykut, G. A., and Gloersen, P., 1987: Multisensor comparison of ice-concentration estimates in the marginal ice zone. *J. Geophys. Res.*, *92:* 6843–56.

Burtt, T. G., and Junker, N. W., 1976: A typical rapidly developing extratropical cyclone as viewed in SMS-II imagery. *Mon. Wea. Rev.*, *104:* 489–90.

Businger, S., 1985: The synoptic climatology of polar low outbreaks. *Tellus, 37A:* 419–32.

Businger, S., 1987: The synoptic climatology of polar-low outbreaks over the Gulf of Alaska and the Bering Sea. *Tellus, 39A:* 307–25.

Businger, S., and Reed, R. J., 1989: Cyclogenesis in cold air masses. *Weather and Forecasting, 4:* 133–56.

Businger, S., and Hobbs, P. V., 1987: Mesoscale structures of two comma cloud systems over the Pacific Ocean. *Mon. Wea. Rev.*, *115:* 1908–28.

Cahalan, R. F., 1983: Climatological statistics of cloudiness, In: Preprint volume, *Fifth Conference on Atmospheric Radiation, October 31–November 4, 1983*: Baltimore, MD. Amer. Meteor. Soc. pp. 206–13.

Cahalan, R. F., and Joseph, J. H., 1989: Fractal statistics of cloud fields. *Mon. Wea. Rev.*, *117:* 261–72.

Cahalan, R. F., Short, D. A., and North, G. R., 1982: Cloud fluctuation statistics. *Mon. Wea. Rev.*, *110:* 26–43.

Campbell, W. J., Gloersen, P., Josberger, E. G., Johannessen, O. M., Guest, P. S., Mognard, N., Schuchman, R., Burns, B. A., Lannelongue, N., and Davidson, K. L., 1987: Variations of mesoscale and large-scale sea ice morphology in the 1984 Marginal Ice Zone Experiment as observed by microwave remote sensing. *J. Geophys. Res.*, *92:* 6805–24.

Campbell, W. J., Weeks, W. F., Ramseier, R. O., and Gloersen, P., 1975: Geophysical studies of floating ice by remote sensing. *J. Glaciol.*, *15:* 305–27.

Carleton, A. M., 1979: A synoptic climatology of satellite-observed extratropical cyclone activity for the Southern Hemisphere winter. *Arch. Meteor. Geophys. Bioklim.*, *B27:* 265–79.

Carleton, A. M., 1980: Polynya development in the Cape Thompson-Point Hope region, Alaska. *Arct. Alp. Res.*, *12:* 205–14.

Carleton, A. M., 1981a: Climatology of the 'instant occlusion' phenomenon for the Southern Hemisphere winter. *Mon. Wea. Rev.*, *109:* 177–81.

Carleton, A. M., 1981b: Ice–ocean–atmosphere interactions at high southern latitudes in winter from satellite observation. *Austral. Meteor. Mag.*, *29:* 183–95.

Carleton, A. M., 1981c: Monthly variability of satellite-derived cyclonic activity for the Southern Hemisphere winter. *J. Climatol.*, *1:* 21–38.

Carleton, A. M., 1983: Variations in Antarctic sea ice conditions and relationships with Southern Hemisphere cyclonic activity, winters 1973–77. *Arch. Meteor. Geophys. Bioklim.*, *B32:* 1–22.

Carleton, A. M., 1984a: Synoptic sea ice-atmosphere interactions in the Chukchi and Beaufort seas from Nimbus-5 ESMR data. *J. Geophys. Res.*, *89:* 7245–58.

Carleton, A. M., 1984b: Cloud-cryosphere interactions, In: Henderson-Sellers, A. (ed.) *Satellite Sensing of a Cloudy Atmosphere: Observing the Third Planet.* Taylor and Francis, London and Philadelphia. pp. 289–325.

Carleton, A. M., 1985a: Synoptic and satellite aspects of the southwestern U.S. summer 'monsoon'. *J. Climatol.*, *5:* 389–402.

Carleton, A. M., 1985b: Satellite climatological aspects of the 'polar low' and 'instant occlusion'. *Tellus, 37A:* 433–50.

Carleton, A. M., 1985c: Synoptic cryosphere-atmosphere interactions in the Northern Hemisphere from DMSP image analysis. *Int. J. Remote Sensing, 6:* 239–61.

Carleton, A. M., 1986: Synoptic-dynamic character of 'bursts' and 'breaks' in the Southwest U.S. summer precipitation singularity. *J. Climatol., 6:* 605–23.

Carleton, A. M., 1987a: Summer circulation climate of the American Southwest, 1945–84. *Annals, Assoc. Amer. Geogr., 77:* 619–34.

Carleton, A. M., 1987b: Satellite-derived attributes of cloud vortex systems and their application to climate studies. *Remote Sensing Envt., 22:* 271–96.

Carleton, A. M., 1989: Antarctic sea-ice relationships with indices of the atmospheric circulation of the Southern Hemisphere. *Climate Dynamics, 3:* 207–20.

Carleton, A. M., and Carpenter, D. A., 1989: Intermediate-scale sea ice-atmosphere interactions over high southern latitudes in winter. *GeoJournal, 18:* 87–101.

Carleton, A. M., and Carpenter, D. A., 1990: Satellite climatology of 'polar lows' and broadscale climatic associations for the Southern Hemisphere. *Int. J. Climatol., 10:* 219–46.

Carleton, A. M., and Lamb, P. J., 1986: Jet contrails and cirrus cloud: a feasibility study employing high resolution satellite imagery. *Bull., Amer. Meteor. Soc., 67:* 301–9.

Carleton, A. M., and Whalley, D., 1988: Eddy transport of sensible heat and the life history of synoptic systems: a statistical analysis for the Southern Hemisphere winter. *Meteorol. Atmos. Physics, 38:* 140–52.

Carleton, A. M., Carpenter, D. A., and Weser, P. J., 1990: Mechanisms of interannual variability of the Southwest U.S. summer rainfall maximum. *J. Climate, 3:* 999–1015.

Carlson, T. N., 1980: Airflow through midlatitude cyclones and the comma cloud pattern. *Mon. Wea. Rev., 108:* 1498–509.

Carlson, T. N., Dodd, J. K., Benjamin, S. G., and Cooper, J. N., 1981: Satellite estimation of the surface energy balance, moisture availability and thermal inertia. *J. Appl. Meteor., 20:* 67–87.

Carsey, F. D., 1982: Arctic sea ice distribution at end of summer 1973–1976 from satellite microwave data. *J. Geophys. Res. , 87:* 5809–35.

Cavalieri, D. J., and Parkinson, C. L., 1981: Large-scale variations in observed Antarctic sea ice extent and associated atmospheric circulation. *Mon. Wea. Rev., 109:* 2323–36.

Cavalieri, D. J., and Parkinson, C. L., 1987: On the relationship between atmospheric circulation and the fluctuations in the sea ice extents of the Bering and Okhotsk seas. *J. Geophys. Res., 92:* 7141–62.

Cavalieri, D. J., Martin, S., and Gloersen, P., 1983: Nimbus 7 SMMR observations of the Bering Sea ice cover during March 1979. *J. Geophys. Res., 88:* 2743–54.

Cess, R. D., 1976: Climate change: an appraisal of atmospheric feedback mechanisms employing zonal climatology. *J. Atmos. Sci., 33:* 1831–43.

Cess, R. D., 1990: Interpretation of an 8-year record of Nimbus-7 wide-field-of-view infrared measurements. *J. Geophys. Res., 95:* 16653–7.

Cess, R. D., Briegleb, B. P., and Lian, M. S., 1982: Low-latitude cloudiness and climate feedback: comparative estimates from satellite data. *J. Atmos. Sci., 39:* 53–9.

Cess, R. D., Potter, G. L., Blanchet, J. P., Boer, G. J., Ghan, S. J., Kiehl, J. T., LeTreut, H., Li, Z-X., Liang, X-Z., Mitchell, J. F. B., Morcrette, J-J., Randall, D. A., Riches, M. R., Roeckner, E., Schlese, U., Slingo, A., Taylor, K. E., Washington, W. M., Wetherald, R. T., and Yagai, I., 1989: Interpretation of cloud-climate feedback as produced by 14 atmospheric general circulation models. *Science, 245:* 513–16.

References

Chang, A. T. C., and Wilheit, T. T., 1979: Remote sensing of atmospheric water vapor, liquid water, and wind speed at the ocean surface by passive microwave techniques from the Nimbus 5 satellite. *Radio Sci.*, *14:* 793–802.

Chang, A. T. C., Foster, J. L., and Hall, D. K., 1990: Satellite sensor estimates of northern hemisphere snow volume. *Int. J. Remote Sensing*, *11:* 167–71.

Chang, C-P., and Lau, K. M. W., 1980: Northeasterly cold surges and near-equatorial disturbances over the Winter MONEX area during December 1974. Part II: Planetary-scale aspects. *Mon. Wea. Rev.*, *108:* 298–312.

Chang, H. D., Hwang, P. H., Wilheit, T. T., Chang, A. T. C., Staelin, D. H., and Rosenkranz, P. W., 1984: Monthly distributions of precipitable water from the NIMBUS 7 SMMR data. *J. Geophys. Res.*, *89:* 5328–34.

Changnon, S. A., 1981: Midwestern cloud, sunshine and temperature trends since 1901: possible evidence of jet contrail effects. *J. Appl. Meteor.*, *20:* 496–508.

Charalambides, S., Hunt, G. E., Rycroft, M. J., Murgatroyd, R. J., and Limbert, W. S., 1985: Studies of the radiation budget anomalies over Antarctica during 1974–1983, and their possible relationship to climatic variations. *Adv. Space Res.*, *5:* 127–32.

Charney, J. G., 1975: Dynamics of deserts and droughts in the Sahel. *Quart. J. Royal Meteor. Soc.*, *101:* 193–202.

Charney, J., Quirk, W. J., Chow, S-H., and Kornfield, J., 1977: A comparative study of the effects of albedo change on drought in semi-arid regions. *J. Atmos. Sci.*, *34:* 1366–85.

Chen, H. S., 1985: *Space Remote Sensing Systems: An Introduction.* Academic Press, Inc. 257pp.

Chen, T. S., and Ohring, G., 1984: On the relationship between clear-sky planetary and surface albedos. *J. Atmos. Sci.*, *41:* 156–8.

Chesters, D., Robinson, W. D., and Uccellini, L. W., 1987: Optimized retrievals of precipitable water from the VAS 'split window'. *J. Clim. Appl. Meteor.*, *26:* 1059–66.

Chesters, D., Uccellini, L. W., and Robinson, W. D., 1983: Low-level water vapor fields from the VISSR Atmospheric Sounder (VAS) 'split window' channels. *J. Clim. Appl. Meteor.*, *22:* 725–43.

Chou, S-H., and Atlas, D., 1982: Satellite estimates of ocean-air heat fluxes during cold air outbreaks. *Mon. Wea. Rev.*, *110:* 1434–50.

Choudhury, B. J., 1989: Monitoring global land surface using Nimbus-7 37 GHz data. Theory and examples. *Int. J. Remote Sensing*, *10:* 1579–605.

Choudhury, B. J., Tucker, C. J., Golus, R. E., and Newcomb, W. W., 1987: Monitoring vegetation using Nimbus-7 scanning multichannel microwave radiometer's data. *Int. J. Remote Sensing*, *8:* 533–8.

Christian, H. J., Blakeslee, R. J., and Goodman, S. J., 1989: The detection of lightning from geostationary orbit. *J. Geophys. Res.*, *94:* 13329–37.

Clapp, P. F., 1964: Global cloud cover for seasons using TIROS nephanalysis. *Mon. Wea. Rev.*, *92:* 495–507.

Clayton, K., 1974: Synoptic observations at a global scale. *J. Brit. Interplanet. Soc.*, *27:* 23–8.

Coakley, J. A., and Bretherton, F. P., 1982: Cloud cover from high resolution scanner data: detecting and allowing for partially filled fields of view. *J. Geophys. Res.*, *87:* 4917–32.

Coakley, Jr., J. A., and Kobayashi, T., 1989: Broken cloud biases in albedo and surface insolation derived from satellite imagery data. *J. Climate*, *2:* 721–30.

Coakley, Jr., J. A., Bernstein, R. L., and Durkee, P. A., 1987: Effect of ship-stack effluents on cloud reflectivity. *Science*, *237:* 1020–2.

Cogley, J. G., and Henderson-Sellers, A., 1984: Effects of cloudiness on the high-latitude surface radiation budget. *Mon. Wea. Rev.*, *112:* 1017–32.

Collier, C. G., Fair, C. A., and Newsome, D. H., 1988: International weather-radar networking in western Europe. *Bull., Amer. Meteor. Soc.*, *69:* 16–21.

Conover, J. H., 1962: Cloud interpretation from satellite altitudes. Research note 81, U.S. Air Force Cambridge Research Laboratories, Cambridge, Mass. 55pp.

Coulmann, S., Bakan, S., and Hinzpeter, H., 1986: A cloud climatology for the South Atlantic derived from METEOSAT I images. *Tellus, 38A:* 453–61.

Courel, M. F., Kandel, R. S., and Rasool, S. I., 1984: Surface albedo and the Sahel drought. *Nature, 307:* 528–31.

Cox, S. K., 1971: Cirrus clouds and the climate. *J. Atmos. Sci.*, *28:* 1513–15.

Cox, S. K., McDougal, D. S., Randall, D. A., and Schiffer, R. A., 1987: FIRE—the First ISCCP Regional Experiment. *Bull., Amer. Meteor. Soc.*, *68:* 114–18.

Crane, R. G., 1983: Atmosphere–sea ice interactions in the Beaufort/Chukchi Sea and in the European sector of the Arctic. *J. Geophys. Res.*, *88:* 4505–23.

Crane, R. G., and Anderson, M. R., 1984: Satellite discrimination of snow/cloud surfaces. *Int. J. Remote Sensing, 5:* 213–23.

Crane, R. G., and Anderson, M. R., 1989: Spring melt patterns in the Kara/Barents Sea: 1984. *GeoJournal, 18:* 25–33.

Crane, R. G., and Barry, R. G., 1984: The influence of clouds on climate with a focus on high latitude interactions: review paper. *J. Climatol.*, *4:* 71–93.

Crane, R. G., Barry, R. G., and Zwally, H. J., 1982: Analysis of atmosphere-sea ice interactions in the Arctic Basin using ESMR microwave data. *Int. J. Remote Sensing, 3:* 259–76.

Curry, J. A., 1988: Arctic cloudiness in spring from satellite imagery: some comments. *J. Climatol.*, *8:* 533–38.

Curry, J. A., Ardeel, C. D., and Tian, L., 1990: Liquid water content and precipitation characteristics of stratiform clouds as inferred from satellite microwave measurements. *J. Geophys. Res.*, *95:* 16659–71.

Darnell, W. L., Gupta, S. K., and Staylor, W. F., 1983: Downward longwave radiation at the surface from satellite measurements. *J. Clim. Appl. Meteor.*, *22:* 1956–60.

Darnell, W. L., Staylor, W. F., Gupta, S. K., and Denn, F. M., 1988: Estimation of surface insolation using sun-synchronous satellite data. *J. Climate, 1:* 820–35.

Davis, N. E., 1981: Meteosat looks at the general circulation. III. Tropical-extratropical interaction. *Weather, 36:* 168–73.

Davis, N. E., 1982: METEOSAT water vapor channel and a low-latitude vortex. *Bull., Amer. Meteor. Soc.*, *63:* 747–50.

Davis, P. A., Major, E. R., and Jacobowitz, H., 1984: An assessment of NIMBUS-7 ERB shortwave scanner data by correlative analysis with narrowband CZCS data. *J. Geophys. Res.*, *89:* 5077–88.

Davison, J., and Harrison, D. E., 1990: Comparison of Seasat scatterometer winds with tropical Pacific observations. *J. Geophys. Res.*, *95:* 3403–10.

DeGrand, J. Q., Carleton, A. M., and Lamb, P. J., 1990: A mid-season climatology of jet condensation trails from high resolution satellite data, In: *Proceedings of the Seventh Conference on Atmospheric Radiation*, held at San Francisco, Calif.: July 23–7, 1990. Amer. Meteor. Soc., Boston, Mass. pp. 309–11.

DelBeato, R., and Barrell, S. L., 1985: Rain estimation in extratropical cyclones using GMS imagery. *Mon. Wea. Rev.*, *113:* 747–55.

Desbois, M., Sèze, G., and Szejwach, G., 1982: Automatic classification of clouds on METEOSAT imagery: application to high-level clouds. *J. Appl. Meteor.*, *21:* 401–12.

References

Detwiler, A., and Cho, H., 1982: Reduction of residential heating and cooling requirements possible through atmospheric seeding and ice-forming nuclei. *J. Appl. Meteor.*, *21:* 1045–7.

Detwiler, A., and Pratt, R., 1984: Clear-air seeding: opportunities and strategies. *J. Wea. Modif.*, *16:* 46–60.

Dey, B., 1980: Applications of satellite thermal infrared images for monitoring North Water during the periods of polar darkness. *J. Glaciol.*, *25:* 425–38.

Dey, B., 1981: Monitoring winter sea ice dynamics in the Canadian Arctic with NOAA-TIR images. *J. Geophys. Res.*, *86:* 3223–35.

Dey, C. H., 1989: The evolution of objective analysis methodology at the National Meteorological Center. *Weather and Forecasting*, *4:* 297–312.

Diak, G. R., and Gautier, C., 1983: Improvements to a simple physical model for estimating insolation from GOES data. *J. Clim. Appl. Meteor.*, *22:* 505–8.

Diak, G. R., and Stewart, T. R., 1989: Assessment of surface turbulent fluxes using geostationary satellite surface skin temperatures and a mixed layer planetary boundary layer scheme. *J. Geophys. Res.*, *94:* 6357–73.

Dickson, R. R., 1973: An active low-latitude storm track across the United States. *Mon. Wea. Rev.*, *101:* 461–6.

Doneaud, A. A., Miller, Jr., J. R., Johnson, L. J., Vonder Haar, T. H., and Laybe, P., 1987: The Area-Time-Integral technique to estimate convective rain volumes over areas applied to satellite data—a preliminary investigation. *J. Clim. Appl. Meteor.*, *26:* 156–69.

Douglas, A. V., and Englehart, P. J., 1981: On a statistical relationship between autumn rainfall in the central equatorial Pacific and subsequent winter precipitation in Florida. *Mon. Wea. Rev.*, *109:* 2377–82.

Dowling, D. R., and Radke, L. F., 1990: A summary of the physical properties of cirrus clouds. *J. Appl. Meteor.*, *29:* 970–8.

Downey, W. K., Tsuchiya, T., and Schreiner, A. J., 1981: Some aspects of a northwestern Australian cloudband. *Austral. Meteor. Mag.*, *29:* 99–113.

Drosdowsky, W., and Holland, G. J., 1987: North Australian cloud lines. *Mon. Wea. Rev.*, *115:* 2645–59.

Durran, D. R., and Weber, D. B., 1988: An investigation of the poleward edges of cirrus clouds associated with midlatitude jet streams. *Mon. Wea. Rev.*, *116:* 702–14.

Duvel, J. P., 1988: Analysis of diurnal, interdiurnal and interannual variations during Northern Hemisphere summers using METEOSAT infrared channels. *J. Climate.*, *1:* 471–84.

Duynkerke, P. G., 1989: The diurnal variation of a marine stratocumulus layer: a model sensitivity study. *Mon. Wea. Rev.*, *117:* 1710–25.

Dvorak, V. F., 1975: Tropical cyclone intensity analysis and forecasting from satellite imagery. *Mon. Wea. Rev.*, *103:* 420–30.

Dziewonski, A. M., 1989: Mission to Planet Earth, concept of a global observing system. *EOS (Trans. Amer. Geophys. Union)*, *70:* 242, 244–8.

Ebert, E., 1987: A pattern recognition technique for distinguishing surface and cloud types in the polar regions. *J. Clim. Appl. Meteor.*, *26:* 1412–27.

Ebert, E. E., 1989: Analysis of polar clouds from satellite imagery using pattern recognition and a statistical cloud analysis scheme. *J. Appl. Meteor.*, *28:* 382–99.

Egorova, I. R., 1970: Features of atmospheric fronts in the Southern Hemisphere from satellite observations, In: Vetlov, I. P. and Morskoi, G. I. (eds.) *Problems of Satellite Meteorology*. Trudy No. 36. Israel Program for Scientific Translations, Jerusalem, 1970. pp. 35–8.

Eigenwillig, N., and Fischer, H., 1982: Determination of midtropospheric wind vectors by tracking pure water vapor structures in METEOSAT water vapor image sequences. *Bull., Amer. Meteor. Soc., 63:* 44–58.

Emery, W. J., and Schluessel, P., 1989: Global differences between skin and bulk sea surface temperatures. *Eos (Trans. Amer. Geophys. Union), 70:* 210–11.

England, C. F., and Hunt, G. E., 1984: A study of the errors due to temporal sampling of the Earth's radiation budget. *Tellus, 36B:* 303–16.

ERBE Science Team, 1986: First data from the Earth Radiation Budget Experiment (ERBE). *Bull., Amer. Meteor. Soc., 67:* 818–24.

Erickson, C. O., and Winston, J. S., 1972: Tropical storm, mid-latitude, cloud-band connections and the autumnal buildup of the planetary circulation. *J. Appl. Meteor., 11:* 23–36.

Ese, T., Kanestrom, I., and Pedersen, K., 1988: Climatology of polar lows over the Norwegian and Barents Sea. *Tellus, 40A:* 248–55.

Eyre, J. R., 1981: Meteosat water vapor imagery. *Meteor. Mag., 110:* 345–53.

Feldman, G., Ng, C., Esaias, W., McClain, C., Elrod, J., Maynard, N., Endres, D., Evans, R., Brown, J., Walsh, S., Carle, M., and Podesta, G., 1989: The Oceanography Report: Ocean color, availability of the global data set. *Eos (Trans. Amer. Geophys. Union), 70:* 634–5, 640–1, 644.

Ferrare, R. A., Fraser, R. S., and Kaufman, Y. J., 1990: Satellite measurements of large-scale air pollution: measurements of forest fire smoke. *J. Geophys. Res., 95:* 9911–25.

Fitch, M. J., and Carleton, A. M., 1990: Antarctic meso-cyclone regimes from satellite and conventional data (abstract), In: *Annals Geophysicae,* European Geophysical Society, special issue: XV General Assembly, Copenhagen, Denmark: 23–27 April 1990. pp. 177–8.

Fleming, J. R., and Cox, S. K., 1974: Radiative effects of cirrus clouds. *J. Atmos. Sci., 31:* 2182–8.

Flitcroft, I. D., Milford, J. R., and Dugdale, G., 1989: Relating point to area average rainfall in semiarid West Africa and the implications for rainfall estimates derived from satellite data. *J. Appl. Meteor., 28:* 252–66.

Flores, A. L., and Carlson, T. N., 1987: Estimation of surface moisture availability from remote temperature measurements. *J. Geophys. Res., 92:* 9581–5.

Forbes, G. S., and Lottes, W. D., 1985: Classification of mesoscale vortices in polar airstreams and the influence of the large-scale environment on their evolutions. *Tellus, 37A:* 132–55.

Forbes, G. S., and Merritt, J. H., 1984: Mesoscale vortices over the Great Lakes in wintertime. *Mon. Wea. Rev., 112:* 377–81.

Foster, J. L., Hall, D. K., and Chang, A. T. C., 1987: Remote sensing of snow. *EOS (Trans. Amer. Geophys. Union), 68:* 681–4.

Fritz, S., and Rao, P. K., 1967: On the infrared transmission through cirrus clouds and the estimation of relative humidity from satellites. *J. Appl. Meteor., 6:* 1088–96.

Gadgil, S., and Guruprasad, A., 1990: An objective method for the identification of the Intertropical Convergence Zone. *J. Climate., 3:* 558–67.

Garand, L., 1988: Automated recognition of oceanic cloud patterns. Part I: Methodology and application to cloud climatology. *J. Climate, 1:* 20–39.

Garand, L., 1989: Two automated methods to derive probability of precipitation fields over oceanic areas from satellite imagery. *J. Appl. Meteor., 28:* 913–24.

Garand, L., Weinman, J. A., and Moeller, C. C., 1989: Automated recognition of oceanic cloud patterns. Part II: Detection of air temperature and humidity anomalies above the ocean surface from satellite imagery. *J. Climate, 2:* 356–66.

Garcia, O., 1981: A comparison of two satellite rainfall estimates for GATE. *J. Appl. Meteor., 20:* 430–8.

References

Garcia, O., 1985: *Atlas of Highly Reflective Clouds for the Global Tropics: 1971–83.* NOAA-ERL, Boulder, CO, 365pp.

Gautier, C., 1982: Mesoscale insolation variability derived from satellite data. *J. Appl. Meteor.*, *21:* 51–8.

Gautier, C., and Katsaros, K. B., 1984: Insolation during STREX. 1. Comparisons between surface measurements and satellite estimates. *J. Geophys. Res.*, *89:*11779–88.

Gautier, C., Diak, G., and Masse, S., 1980: A simple physical model to estimate incident solar radiation at the surface from GOES satellite data. *J. Appl. Meteor.*, *19:* 1005–12.

Gloersen, P., 1984: Sea surface temperatures from Nimbus-7 SMMR radiances. *J. Clim. Appl. Meteor.*, *23:* 336–40.

Gloersen, P., and Campbell, W. J., 1988: Variations in the Arctic, Antarctic, and global sea ice covers during 1978–1987 as observed with the Nimbus 7 Scanning Multichannel Microwave Radiometer. *J. Geophys. Res.*, *93:* 10666–74.

Gloersen, P., and Salomonson, V. V., 1975: Satellites—new global observing techniques for ice and snow. *J. Glaciol.*, *15:* 373–89.

Gloersen, P., Zwally, H. J., Chang, A. T. C., Hall, D. K., Campbell, W. J., and Ramseier, R. O., 1978: Time-dependence of sea-ice concentration and multiyear ice fraction in the Arctic Basin. *Bound.-Layer Meteor.*, *13:* 339–59.

Gloersen, P., Cavalieri, D. J., Chang, A. T. C., Wilheit, T. T., Campbell, W. J., Johannessen, O. M., Katsaros, K. B., Kunzi, K. F., Ross, D. B., Staelin, D., Windsor, E. P. L., Barath, F. T., Gudmandsen, P., Langham, E., and Ramseier, O., 1984: A summary of results from the first NIMBUS 7 SMMR observations. *J. Geophys. Res.*, *89:* 5335–44.

Godshall, F. A., 1968: Intertropical Convergence Zone and mean cloud amount in the tropical Pacific Ocean. *Mon. Wea. Rev.*, *96:* 172–5.

Goodman, A. H., Henderson-Sellers, A., and McGuffie, K., 1990: Climatological contingency probabilities of clouds. *Int. J. Climatol.*, *10:* 565–89.

Gordon, H. R., Clark, D. K., Mueller, J. L., and Hovis, W. A., 1980: Phytoplankton pigments from the Nimbus-7 Coastal Zone Color Scanner: comparisons with surface measurements. *Science*, *210:* 63–6.

Goward, S. N., 1990: Experiences and perspectives in compiling long-term remote sensing data sets on landscapes and biospheric processes. *GeoJournal*, *20.2:* 107–14.

Goward, S. N., Dye, D., Kerber, A., and Kalb, V., 1987: Comparison of North and South American biomes from AVHRR observations. *Geocarto Intl.*, *1:* 27–39.

Graham, N. E., and White, W. B., 1988: The El Niño cycle: a natural oscillator of the Pacific Ocean–atmosphere system. *Science*, *240:* 1293–302.

Gray, T. I., and Clapp, J. F., 1978: An interaction between low- and high-latitude cloud bands as recorded by GOES-1 imagery. *Bull., Amer. Meteor. Soc.*, *59:* 808–9.

Griffith, C. G., 1987: Comparisons of gauge and satellite rain estimates for the central United States during August 1979. *J. Geophys. Res.*, *92:* 9551–66.

Griffith, C. G., and Woodley, W. L., 1973: On the variation with height of the top brightness of precipitating convective clouds. *J. Appl. Meteor.*, *12:* 1086–9.

Griffith, C. G., Augustine, J. A., and Woodley, W. L., 1981: Satellite rain estimation in the U.S. High Plains. *J. Appl. Meteor.*, *20:* 53–66.

Griffith, C. G., Woodley, W. L., Grube, P. G., Martin, D. W., Stout, J., and Sikdar, D. N., 1978: Rain estimation from geosynchronous satellite imagery – visible and infrared studies. *Mon. Wea. Rev.*, *106:* 1153–71.

Griffith, K. T., Cox, S. K., and Knollenberg, R. G., 1980: Infrared radiative properties of tropical cirrus clouds inferred from aircraft measurements. *J. Atmos. Sci.*, *37:* 1077–87.

Grody, N. C., Gruber, A., and Shen, W. C., 1980: Atmospheric water content over the tropical Pacific derived from the Nimbus-6 Scanning Microwave Radiometer. *J. Appl. Meteor.*, *19:* 986–96.

Grossman, R. L., and Garcia, O., 1990: The distribution of deep convection over ocean and land during the Asian summer monsoon. *J. Climate*, *3:* 1032–44.

Gruber, A., 1973: Estimating rainfall in regions of active convection. *J. Appl. Meteor.*, *12:* 110–18.

Gruber, A., 1977: Determination of the Earth-atmosphere radiation budget from NOAA satellite data. *NOAA Tech. Rept., NESS 76*, Washington, DC. 28pp.

Gruber, A., and Jacobowitz, H., 1985: The longwave radiation estimated from NOAA polar orbiting satellites: an update and comparisons with Nimbus-7 ERB results. *Adv. Space Physics*, *5:* 111–20.

Gruber, A., and Winston, J. S., 1978: Earth-atmosphere radiative heating based on NOAA scanning radiometer measurements. *Bull., Amer. Meteor. Soc.*, *59:* 1570–3.

Gruber, A., Herman, L., and Kreuger, A. F., 1971: The use of satellite cloud motions for estimating the circulation over the tropics. *Mon. Wea. Rev.*, *99:* 739–43.

Gutman, G., 1987: The derivation of vegetation indices from AVHRR data. *Int. J. Remote Sensing*, *8:* 1235–43.

Gutman, G., 1988: A simple method for estimating monthly mean albedo of land surfaces from AVHRR data. *J. Clim. Appl. Meteor.*, *27:* 973–88.

Gutman, G. G., 1990a: Towards monitoring droughts from space. *J. Climate*, *3:* 282–95.

Gutman, G. G., 1990b: Review of the Workshop on the Use of Satellite-Derived Vegetation Indices in Weather and Climate Prediction Models. *Bull., Amer. Meteor. Soc.*, *71:* 1458–63.

Gutman, G., Gruber, A., Tarpley, D., and Taylor, R., 1989a: Application of angular models to AVHRR data for determination of the clear-sky planetary albedo over land surfaces. *J. Geophys. Res.*, *94:* 9959–70.

Gutman, G., Ohring, G., Tarpley, D., and Ambroziak, R., 1989b: Albedo of the U.S. Great Plains as determined from NOAA-9 AVHRR data. *J. Climate*, *2:* 608–17.

Guymer, L. B., 1978: Operational application of satellite imagery to synoptic analysis in the Southern Hemisphere. Tech. Rept. No. 29, Commonwealth Bureau of Meteorology, Dept. of Science, Melbourne, Australia. 87pp.

Halpern, P., 1984: Ground level solar energy estimates using Geostationary Operational Environmental Satellite measurements and realistic model atmospheres. *Remote Sensing Envt.*, *15:* 47–61.

Hanson, K. J., Vonder Haar, T. H., and Suomi, V. E., 1967: Reflection of sunlight to space and absorption by the Earth and atmosphere over the United States during spring 1962. *Mon. Wea. Rev.*, *95:* 354–62.

Harrison, E. F., Brooks, D. R., Minnis, P., Wielicki, B. A., Staylor, W. F., Gibson, G. G., Young, D. F., Denn, F. M., and the ERBE Science Team, 1988: First estimates of the diurnal variation of longwave radiation from multiple-satellite Earth Radiation Budget Experiment. *Bull., Amer. Meteor. Soc.*, *69:* 1144–51.

Harrison, E .P., Minnis, P., Barkstrom, B. R., Ramanathan, V., Cess, R. D., and Gibson, G. G., 1990: Seasonal variation of cloud radiative forcing derived from the Earth Radiation Budget Experiment. *J. Geophys. Res.*, *95:* 18 687–703.

Harrison, M. S. J., 1984: A generalized classification of South African summer rain-bearing synoptic systems. *J. Climatol.*, *4:* 547–60.

References

Hartmann, D. L., and Short, D. A., 1980: On the use of earth radiation budget statistics for studies of clouds and climate. *J. Atmos. Sci.*, *37:* 1233–50.

Hastenrath, S., 1990: The relationship of Highly Reflective Clouds to tropical climate anomalies. *J. Climate*, *3:* 353–65.

Hayden, C. M., Smith, W. L., and Woolf, H. M., 1981: Determination of moisture from NOAA polar orbiting satellite sounding radiances. *J. Appl. Meteor.*, *20:* 450–66.

Heinemann, G., 1990: Mesoscale vortices in the Weddell Sea region (Antarctica). *Mon. Wea. Rev.*, *118:* 779–93.

Helfert, M. R., and Lulla, K. P., 1990: Mapping of continental-scale biomass burning and smoke palls over the Amazon Basin as observed from the Space Shuttle. *Photogramm. Eng. Remote Sensing*, *56:* 1367–73.

Henderson-Sellers, A., 1978: Surface type and its effect on cloud cover: a climatological investigation. *J. Geophys. Res.*, *83:* 5057–62.

Henderson-Sellers, A., 1980: Albedo changes – surface surveillance from satellites. *Climatic Change*, *2:* 275–81.

Henderson-Sellers, A., 1986: Layer cloud amounts for January and July 1979 from 3-D nephanalysis. *J. Clim. Appl. Meteor.*, *25:* 118–32.

Henderson-Sellers, A., 1989: North American total cloud amount variations this century. *Palaeogeog., Palaeoclimatol., Palaeoecol. (Global and Planetary Change Section)*, *75:* 175–94.

Henderson-Sellers, A., and Hughes, N. A., 1985: 1979 3D-nephanalysis global total cloud amount climatology. *Bull., Amer. Meteor. Soc.*, *66:* 626–7.

Henderson-Sellers, A., and Wilson, M. F., 1982: High resolution planetary albedos – values and variability. *Remote Sensing Envt*, *12:* 479–84.

Henderson-Sellers, A., and Wilson, M. F., 1983: Surface albedo data for climatic modeling. *Revs. Geophys. Space Physics*, *21:* 1743–78.

Henderson-Sellers, A., Hughes, N. A., and Wilson, M., 1981: Cloud cover archiving on a global scale: a discussion of principles. *Bull., Amer. Meteor. Soc.*, *62:* 1300–7.

Henderson-Sellers, A., Seze, G., Drake, F., and Desbois, M., 1987: Surface-observed and satellite-retrieved cloudiness compared for the 1983 ISCCP Special Study Area in Europe. *J. Geophys. Res.*, *92:* 4019–33.

Holligan, P. M., Viollier, M., Harbour, D. S., Camus, P., Champagne-Philippe, M., 1983: Satellite and ship studies of coccolithophore production along a continental shelf edge. *Nature*, *304:* 339–42.

Hopkins, M. M., 1967: An approach to the classification of meteorological satellite data. *J. Appl. Meteor.*, *6:* 164–78.

Houghton, D. D., and Suomi, V. E., 1978: Information content of satellite images. *Bull., Amer. Meteor. Soc.*, *59:* 1614–17.

Houze, R. A., Rutledge, S. A., Biggerstaff, M. I., and Smull, B. F., 1989: Interpretation of Doppler weather radar displays of midlatitude mesoscale convective systems. *Bull., Amer. Meteor. Soc.*, *70:* 608–19.

Hovis, W. A., Clark, D. K., Anderson, F., Austin, R. W., Wilson, W. H., Baker, E. T., Ball, D., Gordon, H. R., Mueller, J. L., El-Sayed, S. Z., Sturm, B., Wrigley, R. C., and Yentsch, C. S., 1980: Nimbus-7 Coastal Zone Color Scanner: system description and initial imagery. *Science*, *210:* 60–3.

Huang, C-H., Panofsky, H. A., and Schwalb, A., 1967: Some relationships between synoptic variables and satellite radiation data. *Mon. Wea. Rev.*, *95:* 483–6.

Huang, H-J., and Vincent, D. G., 1985: Significance of the South Pacific Convergence Zone in energy conversions of the Southern Hemisphere during FGGE, 10–27 January 1979. *Mon. Wea. Rev.*, *113:* 1359–71.

Hueck, R. R., Kyle, H. L., and Ardanuy, P. E., 1987: Nimbus-7 earth radiation budget wide field of view climate data set improvement. 1. The earth albedo from deconvolution of shortwave measurements. *J. Geophys. Res.*, *92:* 4107–23.

Hughes, N. A., 1984: Global cloud climatologies: a historical review. *J. Clim. Appl. Meteor.*, *23:* 724–51.

Hughes, N. A., and Henderson-Sellers, A., 1985: Global 3D-nephanalysis of total cloud amount: climatology for 1979. *J. Clim. Appl. Meteor.*, *24:* 669–86.

Hunt, G. E., Saunders, R. W., Rumball, D. A., and Marriage, N., 1981: Some quantitative meteorological measurements from geostationary satellites. *Weather*, *36:* 96–104.

Hwang, P. H., Stowe, L. L., Yeh, H. Y. M., Kyle, H. L., and the Nimbus-7 Cloud Data Processing Team, 1988: The Nimbus-7 Global Cloud Climatology. *Bull., Amer. Meteor. Soc.*, *69:* 743–52.

Ingram, W. J., Wilson, C. A., and Mitchell, J. F. B., 1989: Modeling climate change: an assessment of sea ice and surface albedo feedbacks. *J. Geophys. Res.*, *94:* 8609–22.

Jackson, R. D., and Idso, S. B., 1975: Surface albedo and desertification. *Science, 189:* 1012–13.

Jasperson, W. H., Nastrom, G. D., Davis, R. E., and Holdeman, J. D., 1985: Variability of cloudiness at airline cruise altitudes from GASP measurements. *J. Clim. Appl. Meteor.*, *24:* 74–82.

Johannessen, O. M., 1987: Introduction: summer marginal ice zone experiments during 1983 and 1984 in Fram Strait and the Greenland Sea. *J. Geophys. Res.*, *92:* 6716–18.

Jones, A. S., and Vonder Haar, T. H., 1990: Passive microwave remote sensing of cloud liquid water over land regions. *J. Geophys. Res.*, *95:* 16673–83.

Joseph, J. H., and Cahalan, R. F., 1990: Nearest neighbour spacing of fair weather cumulus clouds. *J. Appl. Meteor.*, *29:* 793–805.

Justus, C. G., Paris, M. V., and Tarpley, J. D., 1986: Satellite-measured insolation in the United States, Mexico, and South America. *Remote Sensing Envt.*, *20:* 57–83.

Kakar, R. K., 1983: Retrieval of clear sky moisture profiles using the 183 GHz water vapor line. *J. Clim. Appl. Meteor.*, *22:* 1282–9.

Kandel, R., 1983: Satellite observation of the Earth radiation budget. *Beitr. Phys. Atmosph.*, *56:* 322–40.

Kastner, M., Fischer, H., and Bolle, H-J., 1980: Wind determination from Nimbus 5 observations in the 6.3 μm water vapor band. *J. Appl. Meteor.*, *19:* 409–18.

Katsaros, K. B., and Lewis, R. M., 1986: Mesoscale and synoptic scale features of North Pacific weather systems observed with the Scanning Multichannel Microwave Radiometer on Nimbus 7. *J. Geophys. Res.*, *91:* 2321–30.

Katsaros, K. B., Bhatti, I., McMurdie, L. A., and Petty, G. W., 1989: Identification of atmospheric fronts over the ocean with microwave measurements of water vapor and rain. *Weather and Forecasting, 4:* 449–60.

Kaufman, Y. J., and Joseph, J. H., 1982: Determination of surface albedos and aerosol extinction characteristics from satellite imagery. *J. Geophys. Res.*, *87:* 1287–99.

Kaufman, Y. J., Tucker, C. J., and Fung, I., 1990: Remote sensing of biomass burning in the tropics. *J. Geophys. Res.*, *95:* 9927–39.

Kedem, B., Chiu, L. S., and North, G. R., 1990: Estimation of mean rain rate: application to satellite observations. *J. Geophys. Res.*, *95:* 1965–72.

Kellogg, W. W., and Zhao, Z-c., 1988: Sensitivity of soil moisture to doubling of carbon dioxide in climate model experiments. Part I: North America. *J. Climate, 1:* 348–66.

Kelly, G. A. M., 1978: Interpretation of satellite cloud mosaics for Southern Hemisphere analysis and reference level specification. *Mon. Wea. Rev., 106:* 870–89.

References

Kelly, G. A. M., Forgan, B. W., Powers, P. E., and LeMarshall, J. F., 1982: Mesoscale observations from a polar-orbiting satellite vertical sounder, In: Browning, K. A. (ed.) *Nowcasting*. Academic Press. pp. 107–21.

Key, J., and Barry, R. G., 1989: Cloud cover analysis with Arctic AVHRR data. 1. Cloud detection. *J. Geophys. Res.*, *94:* 18521–35.

Key, J. R., Maslanik, J. A., and Barry, R. G., 1989: Cloud classification from satellite data using a fuzzy sets algorithm: a polar example. *Int. J. Remote Sensing*, *10:* 1823–42.

Key, J., Maslanik, J. A., and Schweiger, A. J., 1989: Classification of merged AVHRR and SMMR Arctic data with neural networks. *Photogramm. Eng. Remote Sensing*, *55:* 1331–8.

Kidder, S. Q., and Vonder Haar, T. H., 1977: Seasonal oceanic precipitation frequencies from Nimbus 5 microwave data. *J. Geophys. Res.*, *82:* 2083–6.

Kidder, S. Q., and Vonder Haar, T. H., 1990: On the use of satellites in Molniya orbits for meteorological observations of middle and high latitudes. *J. Atmos. Oceanic Technol.*, *7:* 517–22.

Kidder, S. Q., and Wu, H-T., 1984: Dramatic contrast between low clouds and snow cover in daytime 3.7 μm imagery. *Mon. Wea. Rev.*, *112:* 2345–6.

Kiehl, J. T., and Ramanathan, V., 1990: Comparison of cloud forcing derived from Earth Radiation Budget Experiment with that simulated by the NCAR Community Climate Model. *J. Geophys. Res.*, *95:* 11679–98.

Kiladis, G. N., and van Loon, H., 1988: The Southern Oscillation. Part VII: Meteorological anomalies over the Indian and Pacific sectors associated with the extremes of the oscillation. *Mon. Wea. Rev.*, *116:* 120–36.

Kilonsky, B. J., and Ramage, C. S., 1976: A technique for estimating tropical open-ocean rainfall from satellite observations. *J. Appl. Meteor.*, *15:* 972–5.

King, P., Yip, T-C., and Steenbergen, J. D., 1989: RAINSAT: a one-year evaluation of a bispectral method for the analysis and short-range forecasting of precipitation areas. *Weather and Forecasting*, *4:* 210–21.

Klassen, W., and VanDenberg, W., 1985: Evapotranspiration derived from satellite-observed surface temperatures. *J. Clim. Appl. Meteor.*, *24:* 412–24.

Kletter, L., 1972: Globale Beobachtung signifikanter Wirbelstrukturen. *Arch. Meteor. Geophys. Bioklim.*, *A21:* 353–72.

Kodama, Y., and Asai, T., 1988: Large-scale cloud distributions and their seasonal variations as derived from GMS-IR observations. *J. Meteor. Soc. Japan*, *66:* 87–101.

Koenig, G., Liou, K-N., and Griffin, M., 1987: An investigation of cloud/radiation interactions using three-dimensional nephanalysis and earth radiation budget data bases. *J. Geophys. Res.*, *92:* 5540–54.

Kogan, F. N., 1987: On using smoothed vegetation time series for identifying near-optimal climatic conditions, In: Preprints volume, *Tenth Conference on Probability and Statistics in Atmospheric Sciences*, October 6–8, 1987: Edmonton, Alberta, Canada. Amer. Meteor. Soc., Boston. pp. 285–6.

Kondrat'ev, K. Y., Borisenkov, E. P., and Morozkin, A. A., 1970: *Interpretation of Observation Data from Meteorological Satellites*. Israel Program for Scientific Translations, Jerusalem. 370pp.

Kornfield, J., and Hasler, A. F., 1969: A photographic summary of the Earth's cloud cover for the year 1967. *J. Appl. Meteor.*, *8:* 687–700.

Kornfield, J., Hasler, A. F., Hanson, K. J., and Suomi, V. E., 1967: Photographic cloud climatology from ESSA III and V computer produced mosaics. *Bull., Amer. Meteor. Soc.*, *48:* 878–83.

Kousky, V. E., 1987: Seasonal climate summary. The global climate for December

1986–February 1987: El Niño returns to the tropical Pacific. *Mon. Wea. Rev.*, *115:* 2822–38.

Kruspe, G., and Bakan, S., 1990: The atmospheric structure during episodes of open cellular convection observed in KonTur 1981. *J. Geophys. Res.*, *95:* 1973–84.

Kuettner, J., 1959: The band structure of the atmosphere. *Tellus*, *11:* 267–94.

Kuhn, P. M. 1970: Airborne observations of contrail effects on the thermal radiation budget. *J. Atmos. Sci.*, *27:* 937–42.

Kuhnel, I., 1989: Tropical-extratropical cloudband climatology based on satellite data. *Int. J. Climatol.*, *9:* 441–63.

Kukla, G. J., and Gavin, J., 1981: Summer ice and carbon dioxide. *Science*, *214:* 497–503.

Kukla, G. J., and Robinson, D. A., 1988: Variability of summer cloudiness in the Arctic Basin. *Meteorol. Atmos. Physics*, *39:* 42–50.

Kuo, K. S., Welch, R. M., and Sengupta, S. K., 1988: Structural and textural characteristics of cirrus clouds observed using high spatial resolution LANDSAT imagery. *J. Appl. Meteor.*, *27:* 1242–60.

Kyle, H. L., Mecherikunnel, A., Ardanuy, P., Penn, L., Groveman, B., Campbell, G. G., and Vonder Haar, T. H., 1990: A comparison of two major Earth radiation budget data sets. *J. Geophys. Res.*, *95:* 9951–70.

Lacis, A. A., and Hansen, J. E., 1974: A parameterization for the absorption of solar radiation in the Earth's atmosphere. *J. Atmos. Sci.*, *31:* 118–33.

Lander, M. A., 1990: Evolution of the cloud pattern during the formation of tropical cyclone twins symmetrical with respect to the equator. *Mon. Wea. Rev.*, *118:* 1194–202.

Lau, K. M., and Chan, P. H., 1988: Intraseasonal and interannual variations of tropical convection: a possible link between the 40–50 day oscillation and ENSO? *J. Atmos. Sci.*, *45:* 506–21.

Lau, K. M., and Lim, H., 1984: On the dynamics of equatorial forcing of climate teleconnections. *J. Atmos. Sci.*, *41:* 161–76.

Lau, K. M., Chang, C. P., and Chan, P. H., 1983: Short-term planetary-scale interactions over the tropics and midlatitudes. Part II: Winter-MONEX period. *Mon. Wea. Rev.*, *111:* 1372–88.

Lau, N. C., and Lau, K. M., 1984: The structure and energetics of midlatitude disturbances accompanying cold-air outbreaks over East Asia. *Mon. Wea. Rev.*, *112:* 1309–27.

Lee, J. E., and Johnson, S. D., 1985: Expectancy of cloudless photographic days in the contiguous United States. *Photogramm. Eng. Remote Sensing*, *LI:* 1883–91.

Lee, T. F., 1989a: Dust tracking using composite visible/IR images: a case study. *Weather and Forecasting*, *4:* 258–63.

Lee, T. F., 1989b: Jet contrail identification using the AVHRR infrared split window. *J. Appl. Meteor.*, *28:* 993–5.

Lee, R., and Taggart, C. I., 1969: A procedure for satellite cloud photo-interpretation, and appearance of clouds from satellite altitudes, In: *Satellite Meteorology*, Australian Bureau of Meteorology, Melbourne. pp. 17(e)–(f).

Lee, T-H., Chesters, D., and Mostek, A., 1983: The impact of conventional surface data upon VAS regression retrievals in the lower troposphere. *J. Clim. Appl. Meteor.*, *22:* 1853–74.

Leese, J. A., 1962: The role of advection in the formation of vortex cloud patterns. *Tellus*, *14:* 409–21.

Lenschow, D. H., and Agee, E. M., 1974: The Air Mass Transformation Experiment (AMTEX): preliminary results from 1974 and plans for 1975. *Bull., Amer. Meteor. Soc.*, *36:* 1228–35.

References

Lethbridge, M., 1967: Precipitation probability and satellite radiation data. *Mon. Wea. Rev.*, *95:* 487–90.

Liebmann, B., 1987: Observed relationships between large-scale tropical convection and the tropical circulation on subseasonal time scales during Northern Hemisphere winter. *J. Atmos. Sci.*, *44:* 2543–61.

Lighthill, Sir J., and Pearce, R. P. (eds.), 1981: *Monsoon Dynamics*. Cambridge University Press.

Lim, H., and Chang, C-P., 1983: Dynamics of teleconnections and Walker Circulations forced by equatorial heating. *J. Atmos. Sci.*, *40:* 1897–915.

Liou, K-N., 1986: Influence of cirrus clouds on weather and climate processes: a global perspective. *Mon. Wea. Rev.*, *114:* 1167–99.

Liou, K-N., and Ou, S-C., 1989: The role of cloud microphysical processes in climate: an assessment from a one-dimensional perspective. *J. Geophys. Res.*, *94:* 8599–607.

Liou, K-N., Ou, S-C., Takano, Y., Valero, F. P. J., and Ackerman, T. P., 1990: Remote sounding of the tropical cirrus cloud temperature and optical depth using 6.5 and 10.5 μm radiometers during STEP. *J. Appl. Meteor.*, *29:* 716–26.

Liu, W. T., 1987: 1982–1983 El Niño Atlas Nimbus 7 Microwave Radiometer Data. *JPL Publ. 87-5*, NASA/JPL, Calif. Institute of Technol., Pasadena. 68pp.

Lo, R. C., and Johnson, D. R., 1971: An investigation of cloud distribution from satellite infrared radiation data. *Mon. Wea. Rev.*, *99:* 599–605.

Locatelli, J. D., Hobbs, P. V., and Werth, J. A., 1982: Mesoscale structures of vortices in polar air streams. *Mon. Wea. Rev.*, *110:* 1417–33.

Lovejoy, S., 1982: Area-perimeter relation for rain and cloud areas. *Science, 216:* 185–7.

Lovejoy, S., and Austin, G. L., 1979: The delineation of rain areas from visible and IR satellite data for GATE and mid-latitudes. *Atmos.-Ocean*, *17:* 77–92.

Lovejoy, S., and Austin, G. L., 1980: The estimation of rain from satellite-borne microwave radiometers. *Quart. J. Royal Meteor. Soc.*, *106:* 255–76.

Lovejoy, S., and Schertzer, D., 1990: Multifractals, universality classes and satellite and radar measurements of cloud and rain fields. *J. Geophys. Res.*, *95:* 2021–34.

Lovelock, J. E., 1989: Geophysiology, the science of Gaia. *Revs. Geophys.*, *27:* 215–22.

Machta, L., and Carpenter, T., 1971: Trends in high cloudiness at Denver and Salt Lake City, In: Matthews, W. H., Kellogg, W. W., and Robinson, G. D. (eds.) *Man's Impact on the Climate*. MIT Press. pp. 401–5.

Maddox, R. A., 1980: Mesoscale convective complexes. *Bull., Amer. Meteor. Soc.*, *61:* 1374–87.

Malberg, H., 1973: Comparison of mean cloud cover obtained by satellite photographs and ground-based observations over Europe and the Atlantic. *Mon. Wea. Rev.*, *101:* 893–7.

Manabe, S., 1975: Cloudiness and the radiative convective equilibrium, In: Singer, S. F. (ed.) *The Changing Global Environment*. Reidel Publ., pp. 175–6.

Martin, D. W., Goodman, B., Schmidt, T. J., and Cutrim, E. C., 1990: Estimates of daily rainfall over the Amazon Basin. *J. Geophys. Res.*, *95:* 17043–50.

Martin, F. L., and Salomonson, V. V., 1970: Statistical characteristics of subtropical jet-stream features in terms of MRIR observations from Nimbus II. *J. Appl. Meteor.*, *9:* 508–20.

Martin, S., Holt, B., Cavalieri, D. J., and Squire, V., 1987: Shuttle Imaging Radar B (SIR-B) Weddell Sea ice observations: a comparison of SIR-B and Scanning Multichannel Microwave Radiometer ice concentrations. *J. Geophys. Res.*, *92:* 7173–9.

Matson, M., and Wiesnet, D. R., 1981: New data base for climate studies. *Nature, 289:* 451–6.

Matthews, E., 1983: Global vegetation and land use: new high-resolution data bases for climate studies. *J. Clim. Appl. Meteor.*, *22:* 474–87.

Matthews, E., and Rossow, W. B., 1987: Regional and seasonal variations of surface reflectance from satellite observations at 0.6 μm. *J. Clim. Appl. Meteor.*, *26:* 170–202.

McAnelly, R. L., and Cotton, W. R., 1989: The precipitation life cycle of Mesoscale Convective Complexes over the central United States. *Mon. Wea. Rev.*, *117:* 784–808.

McBean, G. A., 1989: The World Climate Research Programme, In: Berger, A., Dickinson, R. E., and Kidson, J. W. (eds.) *Understanding Climate Change*, Geophysical Monograph 52, IUGG Vol. 7. Amer. Geophys. Union, Washington, DC. pp. 3–8.

McClain, E. P., 1966: On the relation of satellite viewed cloud conditions to vertically integrated moisture fields. *Mon. Wea. Rev.*, *94:* 509–14.

McClain, E. P., Pichel, W. G., Walton, C. C., Ahmad, Z., and Sutton, J., 1983: Multichannel improvements to satellite-derived global sea surface temperatures. *Adv. Space Res.*, *2:* 43–7.

McGinnigle, J. B., 1988: The development of instant occlusions in the North Atlantic. *Meteor. Mag.*, *117:* 325–41.

McGinnigle, J. B., 1990: Numerical weather prediction model performance on instant occlusion developments. *Meteorol. Mag.*, *119:* 149–63.

McGuffie, K., and Henderson-Sellers, A., 1986: Technical note: Illustration of the influence of shadowing on high latitude information derived from satellite imagery. *Int. J. Remote Sensing*, *7:* 1359–65.

McGuffie, K., and Henderson-Sellers, A., 1988: Observations of oceanic cloudiness. *EOS (Trans. Amer. Geophys. Union)*, *69:* 713–15.

McGuffie, K., and Henderson-Sellers, A., 1989: Almost a century of 'imaging' clouds over the whole-sky dome. *Bull., Amer. Meteor. Soc.*, *70:* 1243–53.

McGuffie, K., and Robinson, D. A., 1988: Examination of USAF nephanalysis performance in the marginal cryosphere region. *J. Climate*, *1:* 1124–37.

McGuffie, K., Barry, R. G., Schweiger, A., Robinson, D. A., and Newell, J., 1988: Intercomparison of satellite-derived cloud analyses for the Arctic Ocean in spring and summer. *Int. J. Remote Sensing*, *9:* 447–67.

McGuffie, K., Henderson-Sellers, A., and Goodman, A. H., 1989: Regional analysis of 3D (three-dimensional) Nephanalysis total cloud amounts for July 1983. *Int. J. Remote Sensing*, *10:* 1395–422.

McGuirk, J. P., Thompson, A. H., and Smith, N. R., 1987: Moisture bursts over the tropical Pacific Ocean. *Mon. Wea. Rev.*, *115:* 787–98.

McGuirk, J. P., and Ulsh, D. J., 1990: Evolution of tropical plumes in VAS water vapor imagery. *Mon. Wea. Rev.*, *118:* 1758–66.

McMurdie, L. A., and Katsaros, K. B., 1985: Atmospheric water distribution in a midlatitude cyclone observed by the Seasat Scanning Multichannel Microwave Radiometer. *Mon. Wea. Rev.*, *113:* 584–98.

McMurdie, L. A., Levy, G., and Katsaros, K. B., 1987: On the relationship between scatterometer-derived convergences and atmospheric moisture. *Mon. Wea. Rev.*, *115:* 1281–94.

Meehl, G. A., 1987: The annual cycle and interannual variability in the tropical Pacific and Indian Ocean regions. *Mon. Wea. Rev.*, *115:* 27–50.

Meisner, B. N., and Arkin, P. A., 1987: Spatial and annual variations in the diurnal cycle of large-scale tropical convective cloudiness and precipitation. *Mon. Wea. Rev.*, *115:* 2009–32.

Menzel, W. P., Schmit, T. J., and Wylie, D. P., 1990: Cloud characteristics over central

References

Amazonia during GTE/ABLE 2B derived from multispectral visible and infrared spin scan radiometer atmospheric sounder observations. *J. Geophys. Res.*, *95:* 17039–42.

Miller, D. B., 1971: Automated production of global cloud-climatology based on satellite data, In: U.S. Naval Academy, Air Weather Service USAF Tech. Report 242, Proceedings of the Sixth Automated Weather Support Technical Exchange Conference. pp. 291–306.

Miller, D. B., and Feddes, R. G., 1971: *Global Atlas of Relative Cloud Cover, 1969–70, Based on Photographic Signals from Meteorological Satellites*. U.S. Dept. of Commerce and U.S. Air Force, Washington, DC. 237pp.

Milner, S., 1986: NEXRAD, the coming revolution in radar storm detection and warning. *Weatherwise, 39:* 72–85.

Minnis, P., and Harrison, E. F., 1984a: Diurnal variability of regional cloud and clear-sky radiative parameters derived from GOES-data. Part II: November 1978 cloud distributions. *J. Clim. Appl. Meteor., 23:* 1012–31.

Minnis, P., and Harrison, E. F., 1984b: Diurnal variability of regional cloud and clear-sky radiative parameters derived from GOES data. Part I: Analysis method. *J. Clim. Appl. Meteor., 23:* 993–1011.

Minnis, P., and Harrison, E. F., 1984c: Diurnal variablility of regional cloud and clear-sky radiative parameters derived from GOES data. Part III: November 1978 radiative parameters. *J. Clim. Appl. Meteor., 23:* 1032–51.

Minnis, P., and Wielicki, B. A., 1988: Comparison of cloud amounts derived using GOES and Landsat data. *J. Geophys. Res., 93:* 9385–403.

Minnis, P., Harrison, E. F., and Gibson, G. G., 1987: Cloud cover over the equatorial eastern Pacific derived from July 1983 International Satellite Cloud Climatology Project data using a hybrid bispectral threshold method. *J. Geophys. Res., 92:* 4051–73.

Mitchell, J. F. B., Senior, C. A., and Ingram, W. J., 1989: CO_2 and climate: a missing feedback? *Nature, 341:* 132–4.

MIZEX Group, 1986: MIZEX East 83/84: the summer Marginal Ice Zone Program in the Fram Strait/Greenland Sea. *EOS (Trans. Amer. Geophys. Union), 67:* 513–17.

Morel, P., Desbois, M., and Szejwach, G., 1978: A new insight into the troposphere with the water vapor channel of Meteosat. *Bull., Amer. Meteor. Soc., 59:* 711–14.

Moser, W., and Raschke, E., 1984: Incident solar radiation over Europe estimated from METEOSAT data. *J. Clim. Appl. Meteor., 23:* 166–70.

Motell, C. E., and Weare, B. C., 1987: Estimating tropical Pacific rainfall using digital satellite data. *J. Clim. Appl. Meteor., 26:* 1436–46.

Mueller, B. M., and Fuelberg, H. E., 1990: A simulation and diagnostic study of water vapor image dry bands. *Mon. Wea. Rev., 118:* 705–22.

Muench, H. S., 1981: Short-range forecasting of cloudiness and precipitation through extrapolation of GOES imagery. *AFGL-TR-81-0218, Environmental Research Papers, No. 750*: Met. Division, Air Force Geophysics Lab., Hanscom AFB, Mass. 46pp.

Mullen, S. L., 1979: An investigation of small synoptic-scale cyclones in polar air streams. *Mon. Wea. Rev., 107:* 1636–47.

Muramatsu, T., 1983: Diurnal variation of satellite-measured T_{BB} area distribution and eye diameter of mature typhoons. *J. Meteor. Soc. Japan, 61:* 77–90.

Nagle, R. E., and Serebreny, S. M., 1962: Radar precipitation echo and satellite cloud observations of a maritime cyclone. *J. Appl. Meteor., 1:* 279–95.

Nakazawa, T., 1986: Main features of 30–60 day variations as inferred from 8-year OLR data. *J. Meteor. Soc. Japan, 64:* 777–86.

Nakazawa, T., 1988: Tropical super clusters within intraseasonal variations over the western Pacific. *J. Meteor. Soc. Japan, 66:* 823–39.

Namias, J., 1962: Influences of abnormal surface heat sources and sinks on atmospheric behavior, In: *Proc. Int. Symp. Numerical Weather Prediction,* Tokyo: Meteor. Soc. Japan. pp. 615–27.

Nemani, R. R., and Running, S. W., 1989: Estimation of regional surface resistance to evapotranspiration from NDVI and thermal-IR AVHRR data. *J. Appl. Meteor., 28:* 276–84.

Nicholson, S. E., 1989: Remote sensing of land surface parameters of relevance to climate studies. *Progr. Phys. Geog., 13:* 1–12.

Niebauer, H. J., and Day, R. H., 1989: Causes of interannual variability in the sea ice cover of the eastern Bering Sea. *GeoJournal, 18:* 45–59.

Nitta, T., 1986: Long-term variations of cloud amount in the western Pacific region. *J. Meteor. Soc. Japan, 64:* 373–90.

Njoku, E. G., 1985: Satellite-derived sea surface temperature: workshop comparisons. *Bull., Amer. Meteor. Soc., 66:* 274–81.

Njoku, E. G., and Swanson, L., 1983: Global measurements of sea surface temperature, wind speed and atmospheric water content from satellite microwave radiometry. *Mon. Wea. Rev., 111:* 1977–87.

Njoku, E. G., Barnett, T. P., Laurs, R. M., and Vastano, A. C., 1985: Advances in satellite sea surface temperature measurement and oceanographic applications. *J. Geophys. Res., 90:* 11573–86.

Nogues-Paegle, J., and Mo, K. C., 1988: Transient response of the Southern Hemisphere subtropical jet to tropical forcing. *J. Atmos. Sci., 45:* 1493–1508.

North, G. R., 1987: Sampling studies for satellite estimation of rain, In: Preprint volume, *Tenth Conference on Probability and Statistics in Atmospheric Sciences,* October 6–8, 1987: Edmonton, Alberta, Amer. Meteor. Soc., Boston. pp. 129–35.

Norton, C. C., Mosher, F. R., and Hinton, B., 1979: An investigation of surface albedo variations during the recent Sahel drought. *J. Appl. Meteor., 18:* 1252–62.

Oglesby, R. J., and Erickson III, D. J., 1989: Soil moisture and the persistence of North American drought. *J. Climate, 2:* 1362–80.

Ohring, G., and Clapp, P., 1980: The effect of changes in cloud amount on the net radiation at the top of the atmosphere. *J. Atmos. Sci., 37:* 447–54.

Ohring, G., Clapp, P. F., Heddinghaus, T. R., and Kreuger, A. F., 1981: The quasi-global distribution of the sensitivity of the Earth-atmosphere radiation budget due to clouds. *J. Atmos. Sci., 38:* 2539–41.

Ohring, G., Gallo, K., Gruber, A., Planet, W., Stowe, L., and Tarpley, J. D., 1989: Climate and global change, characteristics of NOAA satellite data. *EOS (Trans. Amer. Geophys. Union), 70:* 889, 894, 901.

Ohring, G., and Gruber, A., 1983: Satellite radiation observations and climate theory. *Adv. Geophys., 25:* 237–304.

Oliver, V. J., 1969: Tropical storm classification system, In: *Satellite Meteorology,* Australian Bureau of Meteorology, Melbourne. pp. 27–9.

Oliver, V. J., and Anderson, R. K., 1969: Circulation in the tropics as revealed by satellite data. *Bull., Amer. Meteor. Soc., 50:* 702–7.

O'Sullivan, F., Wash, C. H., Stewart, M., and Motell, C. E., 1990: Rain estimation from infrared and visible GOES satellite data. *J. Appl. Meteor., 29:* 209–23.

Otterman, J., 1977: Monitoring surface albedo change with Landsat. *Geophys. Res. Lett., 4:* 441–4.

Otterman, J., 1981: Satellite and field studies of man's impact on the surface in arid regions. *Tellus, 33:* 68–77.

References

Otterman, J., 1989: Enhancement of surface-atmosphere fluxes by desert-fringe vegetation through reduction of surface albedo and of soil heat flux. *Theoret. Appl. Climatol.*, *40:* 67–79.

Otterman, J., and Fraser, R. S., 1976: Earth-atmosphere system and surface reflectivities in arid regions from LANDSAT MSS data. *Remote Sensing Envt.*, *5:* 247–66.

Otterman, J., and Tucker, C. J., 1985: Satellite measurements of surface albedo and temperatures in semi-desert. *J. Clim. Appl. Meteor.*, *24:* 228–35.

Parikh, J., 1977: A comparative study of cloud classification techniques. *Remote Sensing Envt.*, *6:* 67–81.

Parkinson, C. L., 1989: On the value of long-term satellite passive microwave data sets for sea ice/climate studies. *GeoJournal*, *18:* 9–20.

Parkinson, C. L., and Cavalieri, D. J., 1989: Arctic sea ice 1973–1987: seasonal, regional, and interannual variability. *J. Geophys. Res.*, *94:* 14499–523.

Parkinson, C. L., and Gratz, A. J., 1983: On the seasonal sea ice cover of the Sea of Okhotsk. *J. Geophys. Res.*, *88:* 2793–802.

Parkinson, C. L., Comiso, J.C., Zwally, H. J., Cavalieri, D. J., and Gloersen, P., 1987: Seasonal and regional variations of Northern Hemisphere sea ice as illustrated with satellite passive-microwave data for 1974. *Ann. Glaciol.*, *9:* 119–26.

Pathak, P. N., 1987: Empirical analysis of passive microwave observations from Bhaskara-II SAMIR and remote sensing of atmospheric water vapor and liquid water. *J. Clim. Appl. Meteor.*, *26:* 3–17.

Pease, C. H., 1987: The size of wind-driven coastal polynyas. *J. Geophys. Res.*, *92:* 7049–59.

Pease, S. R., Lyons, W. A., Keen, C. S., and Hjelmfelt, M., 1988: Mesoscale spiral vortex embedded within a Lake Michigan snow squall band: high resolution satellite observations and numerical model simulations. *Mon. Wea. Rev.*, *116:* 1374–80.

Perry, M. D., 1990: A satellite climatology of Mesoscale Convective Systems for the southwestern United States. Masters thesis, Department of Geography, Indiana University, Bloomington.

Petersen, R. A., and Hoke, J. E., 1989: The effect of snow cover on the Regional Analysis and Forecast System (RAFS) low-level forecasts. *Weather and Forecasting*, *4:* 253–7.

Petersen, R. A., Uccellini, L. W., Mostek, A., and Keyser, D. A., 1984: Delineating mid- and low-level water vapor patterns in pre-convective environments using VAS moisture channels. *Mon. Wea. Rev.*, *112:* 2178–98.

Petty, G. W., and Katsaros, K. B., 1990: Precipitation observed over the South China Sea by the Nimbus-7 Scanning Multichannel Microwave Radiometer during Winter MONEX. *J. Appl. Meteor.*, *29:* 273–87.

Phillips, N., McMillin, L., Gruber, A., and Wark, D., 1979: An evaluation of early operational temperature soundings from TIROS-N. *Bull., Amer. Meteor. Soc.*, *58:* 1188–97.

Picon, L., and Desbois, M., 1990: Relation between METEOSAT water vapor radiance fields and large-scale tropical circulation features. *J. Climate*, *3:* 865–76.

Pinker, R. T., 1985: Determination of surface albedo from satellites. *Adv. Space Res.*, *5:* 333–43.

Pinker, R. T., and Ewing, J. A., 1985: Modeling surface solar radiation: model formulation and validation. *J. Clim. Appl. Meteor.*, *24:* 389–401.

Pinker, R. T., and Ewing, J. A., 1986: Effect of surface properties on the narrow to broadband spectral relationship in clear sky satellite observations. *Remote Sensing Envt.*, *20:* 267–82.

Pinty, B., and Szejwach, G., 1985: A new technique for inferring surface albedo from satellite observations. *J. Clim. Appl. Meteor.*, *24:* 741–50.

Pinty, B., Szejwach, G., Stum, J., 1985: Surface albedo over the Sahel from METEOSAT radiances. *J. Clim. Appl. Meteor.*, *24:* 108–13.

Platt, C. M. R., 1983: On the bispectral method for cloud parameter determination from satellite VISSR data: separating broken cloud and semitransparent cloud. *J. Clim. Appl. Meteor.*, *22:* 429–39.

Platt, C. M. R., 1989: The role of cloud microphysics in high-cloud feedback effects on climate change. *Nature*, *341:* 428–9.

Platt, C. M. R., Scott, J. C., and Dilley, A. C., 1987: Remote sounding of high clouds. Part VI: Optical properties of midlatitude and tropical cirrus. *J. Atmos. Sci.*, *44:* 729–47.

Platt, T., and Sathyendranath, S., 1988: Oceanic primary production: estimation by remote sensing at local and regional scales. *Science*, *241:* 1613–20.

Poc, M. M., and Roulleau, M., 1983: Water vapor fields deduced from METEOSAT-1 water vapor channel data. *J. Clim. Appl. Meteor.*, *22:* 1628–36.

Poc, M. M., Roulleau, M., Scott, N. A., and Chedin, A., 1980: Quantitative studies of Meteosat water-vapor channel data. *J. Clim. Appl. Meteor.*, *19:* 868–76.

Powell, G. L., Brazel, A. J., and Pasqualetti, M. J., 1984: New approach to estimating solar radiation from satellite imagery. *Prof. Geogr.*, *36:* 227–33.

Prabhakara, C., Chang, H. D., and Chang, A. T. C., 1982: Remote sensing of precipitable water over the oceans from Nimbus 7 microwave measurements. *J. Appl. Meteor.*, *21:* 59–68.

Prabhakara, C., Dalu, G., Lo, R. C., and Nath, N. R., 1979: Remote sensing of seasonal distribution of precipitable water vapor over the oceans and the inference of boundary-layer structure. *Mon. Wea. Rev.*, *107:* 1388–401.

Prabhakara, C., Fraser, R. S., Dalu, G., Wu, M-L. C., and Curran, R. J., 1988: Thin cirrus clouds: seasonal distribution over oceans deduced from Nimbus-4 IRIS. *J. Appl. Meteor.*, *27:* 379–99.

Prabhakara, C., Short, D. A., and Vollmer, B. E., 1985: El Niño and atmospheric water vapor: observations from Nimbus 7 SMMR. *J. Clim. Appl. Meteor.*, *24:* 1311–24.

Prabhakara, C., Short, D. A., Wiscombe, W., and Fraser, R. S., 1986: Rainfall over oceans inferred from Nimbus 7 SMMR: application to 1982–83 El Niño. *J. Clim. Appl. Meteor.*, *25:* 1464–74.

Preuss, H. J., and Geleyn, J. F., 1980: Surface albedos derived from satellite data and their impact on forecast models. *Arch. Meteor. Geophys. Bioklim.*, *A29:* 345–56.

Price, J. C., 1982: On the use of satellite data to infer surface fluxes at meteorological scales. *J. Appl. Meteor.*, *21:* 1111–22.

Purdom, F. W., 1984: Using satellite imagery to understand mesoscale weather development, In: Preprint volume, *Conference on Satellite/Remote Sensing and Applications*, Clearwater Beach, Fla., June 25–29, 1984. Amer. Meteor. Soc., Boston. pp. 106–11.

Puri, K., and Miller, M. J., 1990: The use of satellite data in the specification of convective heating for diabatic initialization and moisture adjustment in numerical weather prediction models. *Mon. Wea. Rev.*, *118:* 67–93.

Rabin, R. M., Stadler, S. J., Wetzel, P. J., Stensrud, D. J., and Gregory, M., 1990: Observed effects of landscape variability on convective clouds. *Bull., Amer. Meteor. Soc.*, *71:* 272–80.

Radke, L. F., Coakley, Jr., J. A., and King, M. D., 1989: Direct and remote sensing observations of the effects of ships on clouds. *Science*, *246:* 1146–9.

References

Ramage, C. S., 1975: Preliminary discussion of the meteorology of the 1972–73 El Niño. *Bull., Amer. Meteor. Soc.*, *56:* 234–42.

Ramanathan, V., 1987: The role of earth radiation budget studies in climate and general circulation research. *J. Geophys. Res.*, *92:* 4075–95.

Ramanathan, V., Cess, R. D., Harrison, E. F., Minnis, P., Barkstrom, B. R., Ahmad, E., and Hartmann, D., 1989a: Cloud-radiative forcing and climate: results from the Earth Radiation Budget Experiment. *Science*, *243:* 57–63.

Ramanathan, V., Barkstrom, B. R., and Harrison, E. F., 1989b: Climate and the Earth's radiation budget. *Phys. Today*, *42:* 22–32.

Ramond, D., Corbin, H., Desbois, M., Szejwach, G., and Waldteufel, P., 1981: The dynamics of polar jet streams as depicted by the METEOSAT water vapor channel radiance field. *Mon. Wea. Rev.*, *109:* 2164–76.

Randel, D. L., Vonder Haar, T. H., and Campbell, G. G., 1985: Hemispheric differences in the Earth Radiation Budget derived from Nimbus-7 related to climate studies. *Adv. Space Res.*, *5:* 99–103.

Rao, M. S. V., and Theon, J. S., 1977: New features of global climatology revealed by satellite-derived oceanic rainfall maps. *Bull., Amer. Meteor. Soc.*, *58:* 1285–8.

Rao, P. K., Smith, W. L., and Koffler, R., 1972: Global sea-surface temperature distribution determined from an environmental satellite. *Mon. Wea. Rev.*, *100:* 10–14.

Rao, P. K., Holmes, S. J., Anderson, R. K., Winston, J. S., and Lehr, P. E. (eds.), 1990: *Weather Satellites: Systems, Data, and Environmental Applications*. Amer. Meteor. Soc., Boston. 503pp.

Rao, R. K., Tarpley, J. D., Scofield, R. A., and Moses, J. F., 1984: Meteorological satellite data useful for agroclimate, In: Preprints volume, *Conference on Satellite/Remote Sensing and Applications*. Clearwater Beach, Fla., June 25–29, 1984. Amer. Meteor. Soc., Boston. pp. 15–21.

Raphael, C., and Hay, J. E., 1984: An assessment of models which use satellite data to estimate solar irradiance at the Earth's surface. *J. Clim. Appl. Meteor.*, *23:* 832–44.

Raschke, E., and Bandeen, W. R., 1967: A quasi-global analysis of tropospheric water vapor content from TIROS IV radiation data. *J. Appl. Meteor.*, *6:* 468–81.

Raschke, E., and Bandeen, W. R., 1970: The radiation balance of the Planet Earth from radiation measurements of the satellite Nimbus II. *J. Appl. Meteor.*, *9:* 215–38.

Raschke, E., Vonder Haar, T. H., Bandeen, W. R., and Pasternak, M., 1973: The annual radiation balance of the Earth-atmosphere system during 1969–70 from Nimbus 3 measurements. *J. Atmos. Sci.*, *30:* 341–64.

Raschke, E., Rossow, W., and Schiffer, R., 1987: The International Satellite Cloud Climatology Project – preliminary results and its potential aspects. *Adv. Space Res.*, *7:* (3) 137–45.

Rasmussen, E., 1981: An inestigation of a polar low with a spiral cloud structure. *J. Atmos. Sci.*, *38:* 1785–92.

Rasmussen, E., and Lystad, M., 1987: The Norwegian Polar Lows Project: a summary of the International Conference on Polar Lows, 20–23 May 1986, Oslo, Norway. *Bull., Amer. Meteor. Soc.*, *68:* 801–16.

Rasmussen, L., 1989: Greenland winds and satellite imagery. *Vejret* (Journal of Danish Meteor. Soc.).

Rasool, S. I., 1964: Cloud heights and nighttime cloud cover from TIROS radiation data. *J. Atmos. Sci.*, *21:* 152–6.

Rasool, S. I., 1984: ISLSCP: International Satellite Land-Surface Climatology Project. *Bull., Amer. Meteor. Soc.*, *65:* 143–4.

Rasool, S. I. (ed.), 1987: Potential of remote sensing for the study of global change. *Adv. Space Res.*, *7:* 97pp.

Reed, R. J., 1979: Cyclogenesis in polar air streams. *Mon. Wea. Rev.*, *107:* 38–52.

Reed, R. J., and Blier, W., 1986a: A case study of comma cloud development in the eastern Pacific. *Mon. Wea. Rev.*, *114:* 1681–95.

Reed, R. J., and Blier, W., 1986b: A further study of comma cloud development in the eastern Pacific. *Mon. Wea. Rev.*, *114:* 1696–708.

Reinking, R. F., 1968: Insolation reduction by contrails. *Weather*, *23:* 171–3.

Reiter, E. R., and Whitney, L. F., 1969: Interactions between subtropical and polar-front jet stream. *Mon. Wea. Rev.*, *97:* 432–8.

Reynolds, D. W., and Vonder Haar, T. H., 1977: A bispectral method for cloud parameter determination. *Mon. Wea. Rev.*, *105:* 446–57.

Reynolds, D. W., Vonder Haar, T. H., and Grant, L. O., 1978: Meteorological satellites in support of weather modification. *Bull., Amer. Meteor. Soc.*, *59:* 269–81.

Reynolds, R. W., Folland, C. K., and Parker, D. E., 1989: Biases in satellite-derived sea-surface-temperature data. *Nature*, *341:* 728–31.

Richards, F., and Arkin, P., 1981: On the relationship between satellite-observed cloud cover and precipitation. *Mon. Wea. Rev.*, *109:* 1081–93.

Robinson, D., Kunzi, K., Kukla, G., and Roth, H., 1984: Comparative utility of microwave and shortwave satellite data for all-weather charting of snow cover. *Nature*, *312:* 434–5.

Robinson, D. A., Scharfen, G., Barry, R. G., and Kukla, G., 1987: Analysis of interannual variations of snow melt on Arctic sea ice mapped from meteorological satellite imagery, In: *Large Scale Effects of Seasonal Snow Cover*. Proceedings of the Vancouver Symposium. IAHS Publ., no. 166. pp. 315–27.

Robinson, D. A., Scharfen, G., Serreze, M. C., Kukla, G., and Barry, R. G., 1986: Snow melt and surface albedo in the Arctic Basin. *Geophys. Res. Lett.*, *13:* 945–8.

Robinson, W. D., Chesters, D., and Uccellini, L. W., 1986: Optimized retrievals of precipitable water fields from combinations of VAS satellite and conventional surface observations. *J. Geophys Res.*, *91:* 5305–18.

Robock, A., 1983: Ice and snow feedbacks and the latitudinal and seasonal distribution of climate sensitivity. *J. Atmos. Sci.*, *40:* 986–97.

Robock, A., and Kaiser, D., 1985: Satellite-observed reflectance of snow and clouds. *Mon. Wea. Rev.*, *113:* 2023–9.

Robock, A., and Scialdone, J., 1986: Comparison of Northern Hemisphere snow cover data sets, In: Kukla, G. *et al.* (eds.) Snow Watch '85, Proceedings of the Workshop held 28–30 October 1985 at the Univ. of Maryland, College Park, MD. Glaciological Data Report GD-18, World Data Center for Glaciology (Snow and Ice), CIRES, Univ. of Colorado, Boulder: March 1986. pp. 141–60.

Rockwood, A. A., and Cox, S. K., 1978: Satellite inferred surface albedo over northwestern Africa. *J. Atmos. Sci.*, *35:* 513–22.

Rodgers, E. B., Salomonson, V. V., and Kyle, H. L., 1976: Upper tropospheric dynamics as reflected in Nimbus 4 THIR 6.7 μm data. *J. Geophys. Res.*, *81:* 5749–58.

Rodgers, E., Siddalingaiah, H., Chang, A. T. C., and Wilheit, T., 1979: A statistical technique for determining rainfall over land employing Nimbus 6 ESMR measurements. *J. Appl. Meteor.*, *18:* 978–91.

Rodgers, E. B., Stout, J., Steranka, J., and Chang, S., 1990: Tropical cyclone-upper atmospheric interactions as inferred from satellite total ozone observations. *J. Appl. Meteor.*, *29:* 934–54.

Rogers, C. W. C., 1965: A technique for estimating low-level wind velocity from satellite photographs of cellular convection. *J. Appl. Meteor.*, *4:* 387–93.

References

Rogers, D. M., Magnano, M. J., and Arns, J . H., 1985: Mesoscale Convective Complexes over the United States during 1983. *Mon. Wea. Rev.*, *113:* 888–901.

Rossow, W. B., 1989: Measuring cloud properties from space: a review. *J. Climate*, *2:* 201–13.

Rossow, W. B., Mosher, F., Kinsella, E., Arking, A., Desbois, M., Harrison, E., Minnis, P., Ruprecht, E., Sèze, G., Simmer, C., and Smith, E., 1985: ISCCP cloud algorithm intercomparison. *J. Clim. Appl. Meteor.*, *24:* 877–903.

Rossow, W. B., Garder, L. C., and Lacis, A. A., 1989a: Global, seasonal cloud variations from satellite radiance measurements. Part I: sensitivity of analysis. *J. Climate, 2:* 419–58.

Rossow, W. B., Brest, C. L., and Garder, L. C., 1989b: Global, seasonal surface variations from satellite radiance measurements. *J. Climate*, *2:* 214–47.

Rowles, K., 1978: A guide to satellite picture analysis. *Meteor. Mag.*, *107:* 205–9.

Running, S. W., and Nemani, R. R., 1988: Relating seasonal patterns of the AVHRR vegetation index to simulated photosynthesis and transpiration of forests to different climates. *Remote Sensing Envt.*, *24:* 347–67.

Ruprecht, E., 1985: Statistical approaches to cloud classification. *Adv. Space Res.*, *6:* 151–64.

Sabins, F. F., 1987: *Remote Sensing Principles and Interpretation.* Freeman and Co., NY. 449pp.

Sadler, J. C., 1969: Average Cloudiness in the Tropics from Satellite Observations. *International Indian Ocean Expedition Meteorological Monographs, Number 2.* East-West Center Press, Honolulu, Hawaii.

Sadler, J. C., and Kilonsky, B. J. 1985: Deriving surface winds from satellite observations of low-level cloud motions. *J. Clim. Appl. Meteor.*, *24:* 758–69.

Sadler, J. C., Oda, L., and Kilonsky, B. J., 1976: Pacific Ocean cloudiness from satellite observations. Tech. Rept. Meteorology, University of Hawaii, Honolulu. 137pp.

Saha, K., 1971: Mean cloud distributions over tropical oceans. *Tellus*, *23:* 183–94.

Saito, K., and Baba, A., 1988: A statistical relation between relative humidity and the GMS observed cloud amount. *J. Meteor. Soc. Japan*, *66:* 187–92.

Sakakibara, H., Ishibara, M., and Yanagisawa, Z., 1988: Classification of mesoscale snowfall systems observed in western Hokuriku during a heavy snowfall period in January 1984. *J. Meteor. Soc. Japan*, *66:* 193–9.

Sakellariou, N. K., and Leighton, H. G., 1988: Identification of cloud-free pixels in inhomogeneous surfaces from AVHRR radiances. *J. Geophys. Res.*, *93:* 5287–93.

Salby, M. L., 1989: Climate monitoring from space: asynoptic sampling considerations. *J. Climate, 2:* 1091–105.

Sanders, F., and Gyakum, J. R., 1980: Synoptic-dynamic climatology of the 'bomb'. *Mon. Wea. Rev.*, *108:* 1589–606.

Sardie, J. M., and Warner, T. T., 1983: On the mechanism for the development of polar lows. *J. Atmos. Sci.*, *40:* 869–81.

Sardie, J. M., and Warner, T. T., 1985: A numerical study of the development mechanisms of polar lows. *Tellus*, *37A:* 460–77.

Saunders, R. W., 1986: An automated scheme for the removal of cloud contamination from AVHRR radiances over western Europe. *Int. J. Remote Sensing*, *7:* 867–86.

Saunders, R. W., 1988: Cloud-top temperature/height: a high resolution imagery product from AVHRR data. *Meteor. Mag.*, *117:* 211–21.

Saunders, R. W., and Hunt, G. E., 1980: METEOSAT observations of diurnal variation of radiation budget parameters. *Nature*, *283:* 645–7.

Saunders, R. W., and Hunt, G. E., 1983: Some radiation budget and cloud measurements derived from Meteosat 1 data. *Tellus*, *35B:* 177–88.

Saunders, R. W., Ward, N. R., England, C. F., and Hunt, G. E., 1982: Satellite observations of sea surface temperature around the British Isles. *Bull., Amer. Meteor. Soc., 63:* 267–72.

Savage, R. C., 1978: The radiative properties of hydrometeors at microwave frequencies. *J. Appl. Meteor., 17:* 904–11.

Savage, R. C., and Weinman, J. A., 1975: Preliminary calculations of the upwelling radiance from rainclouds at 37.0 and 19.35 GHz. *Bull., Amer. Meteor. Soc., 56:* 1272–4.

Scharfen, G., Barry, R. G., Robinson, D. A., Kukla, G., and Serreze, M. C., 1987: Large-scale patterns of snow melt on Arctic sea ice mapped from meteorological satellite imagery. *Ann. Glaciol., 9:* 1–6.

Schiffer, R. A., and Rossow, W. B., 1983: The International Satellite Cloud Climatology Project (ISCCP): the first project of the World Climate Research Programme. *Bull., Amer. Meteor. Soc., 64:* 779–84.

Schiffer, R. A., and Rossow, W. B., 1985: ISCCP global radiance data set: a new resource for climate research. *Bull., Amer. Meteor. Soc., 66:* 1498–505.

Schmetz, J., Mihita, M., and van de Berg, L., 1990: METEOSAT observations of longwave cloud-radiative forcing for April 1985. *J. Climate, 3:* 784–91.

Schmetz, J., and Turpeinen, O. M., 1988: Estimation of the upper tropospheric relative humidity field from METEOSAT water vapor image data. *J. Appl. Meteor., 27:* 889–99.

Schmugge, T., 1978: Remote sensing of surface soil moisture. *J. Appl. Meteor., 17:* 1549–57.

Schmugge, T. J., Meneely, J. M., Rango, A., and Neff, R., 1977: Satellite microwave observations of soil moisture variations. *Water Resour. Bull., 13:* 265–81.

Schulz, T. M., and Samson, P. J., 1988: Nonprecipitating low cloud frequencies for central North America: 1982: *J. Appl. Meteor., 27:* 427–40.

Schwerdtfeger, W., and Kachelhoffer, S., 1973: The frequency of cyclonic vortices over the southern ocean in relation to the extension of the pack ice belt. *Antarct. J. US, 8:* 234.

Scialdone, J., and Robock, A., 1987: Comparison of Northern Hemisphere snow cover data sets. *J. Clim. Appl. Meteor., 26:* 53–68.

Scofield, R. A., 1987: The NESDIS operational convective precipitation estimation technique. *Mon. Wea. Rev., 115:* 1773–92.

Scofield, R. A., and Oliver, V. J., 1977: A scheme for estimating convective rainfall from satellite imagery. *NOAA Tech. Memo NESS 86.* US Dept. of Commerce, Washington, DC. 47pp.

Scorer, R. S., 1986: *Cloud Investigation by Satellite.* Ellis Horwood, Chichester, England.

Seaver, W. T., and Lee, J. E., 1987: A statistical examination of sky cover changes in the contiguous United States. *J. Clim. Appl. Meteor., 26,* 88–95.

Segal, M., Schreiber, W. E., Kallos, G., Garratt, J. R., Rodi, A., Weaver, J., and Pielke, R. A., 1989: The impact of crop areas in northeast Colorado on midsummer mesoscale thermal circulations. *Mon. Wea. Rev., 117:* 809–25.

Sellers, P. J., Hall, F. G., Asrar, G., Strebel, D. E., and Murphy, R. E., 1988: The First ISLSCP Field Experiment (FIE). *Bull., Amer. Meteor. Soc., 69:* 22–7.

Sellers, P. J., Rasool, S. I., and Bolle, H-J., 1990: A review of satellite data algorithms for studies of the land surface. *Bull., Amer. Meteor. Soc., 71:* 1429–47.

Sèze, G., and Desbois, M., 1987: Cloud cover analysis from satellite imagery using spatial and temporal characteristics of the data. *J. Clim. Appl. Meteor., 26:* 287–303.

Sèze, G., Drake, F., Desbois, M., and Henderson-Sellers, A., 1986: Total and low cloud amounts over France and southern Britain in the summer of 1983:

comparison of surface-observed and satellite-retrieved values. *Int. J. Remote Sensing*, **7:** 1031–50.

Shapiro, M. A., Fedor, L. S., and Hampel, T., 1987: Research aircraft measurements of a polar low over the Norwegian Sea. *Tellus, 39A:* 272–306.

Shenk, W. E., and Salomonson, V. V., 1985: A simulation study exploring the effects of sensor spatial resolution on estimates of cloud cover from satellites. *J. Appl. Meteor., 11:* 214–20.

Shenk, W. E., Vonder Haar, T. H., and Smith, W. L., 1987: An evaluation of observations from satellites for the study and prediction of mesoscale events and cyclone events. *Bull., Amer. Meteor. Soc., 68:* 21–35.

Shih, C-F., Vonder Haar, T. H., and Campbell, G. G., 1985: Cloudiness over the eastern equatorial Pacific for summer 1983 from geostationary satellite data. *Adv. Space Res., 5:* 197–201.

Shine, K. P., Henderson-Sellers, A., and Slingo, A., 1984: The influence of the spectral response of satellite sensors on estimates of broadband albedo. *Quart. J. Royal Meteor. Soc., 110:* 1170–9.

Short, D. A., and Wallace, J. M., 1980: Satellite-inferred morning-to-evening cloudiness changes. *Mon. Wea. Rev., 108:* 1160–9.

Simpson, J., Adler, R. F., and North, G. R., 1988: A proposed Tropical Rainfall Measuring Mission (TRMM) satellite. *Bull., Amer. Meteor. Soc., 69:* 278–95.

Slingo, A., 1990: Sensitivity of the Earth's radiation budget to changes in low clouds. *Nature, 343:* 49–51.

Smith, G. L., Barkstrom, B. R., and Harrison, E. F., 1987: The Earth Radiation Budget Experiment: early validation and results. *Adv. Space Res., 7:* (3)167–77.

Smith, W. L., Suomi, V. E., Menzel, W. P., Woolf, H. M., Sromovsky, L. A., Revercomb, H. E., Hayden, C. M., Erickson, D. N., and Mosher, F. R., 1981: First sounding results from VAS-D. *Bull., Amer. Meteor. Soc., 65:* 232–6.

Smith, W. L., Suomi, V. E., Zhou, F. X., and Menzel, W. P., 1982: Nowcasting applications of geostationary satellite atmospheric sounding data, In: Browning, K. A. (ed.) *Nowcasting*. Academic Press. pp. 123–35.

Smith, W. L., Woolf, H. M., and Jacob, W. J., 1970: A regression method for obtaining real-time temperature and geopotential height profiles from satellite spectrometer measurements and its application to Nimbus 3 'SIRS' observations. *Mon. Wea. Rev., 98:* 582–603.

Smith, W. L., Woolf, H. M., Hayden, C. M., Wark, D. Q., and McMillin, L. M., 1979: The TIROS-N Operational Vertical Sounder. *Bull., Amer. Meteor. Soc., 58:* 1177–87.

Smith, W. L., Bishop, W. P., Dvorak, V. F., Hayden, C. M., McElroy, J. H., Mosher, F.R., Oliver, V. J., Purdom, J. F., and Wark, D. Q., 1986: The meteorological satellite: overview of 25 years of operation. *Science, 231:* 455–62.

Smull, B. F., and Houze, Jr., R. A., 1985: A midlatitude squall line with a trailing region of stratiform rain: radar and satellite observations. *Mon. Wea. Rev., 113:* 117–33.

Somerville, R. C. J., and Remer, L. A., 1984: Cloud optical thickness feedbacks in the CO_2 climate problem. *J. Geophys. Res., 89:* 9668–72.

Sovetova, V. D., and Grigorov, S. I., 1978: Utilization of satellite information on cloud vortexes to determine pressure characteristics. *Soviet Meteor. Hydrol., 2:* 15–22.

Spencer, R. W., and Christy, J. R., 1990: Precise monitoring of global temperature trends from satellites. *Science, 247:* 1558–62.

Spencer, R. W., and Santek, D. A., 1985: Measuring the global distribution of intense convection over land with passive microwave radiometry. *J. Clim. Appl. Meteor., 24:* 860–4.

Spencer, R. W., Goodman, H. M., and Hood, R. E., 1989: Precipitation retrieval over land and ocean with the SSM/I: identification and characteristics of the scattering signal. *J. Atmos. Oceanic Technol.*, *6:* 254–73.

Spencer, R. W., Olson, W. S., Rongzhang, W., Martin, D. W., Weinman, J. A., and Santek, D. A., 1983a: Heavy thunderstorms observed over land by the Nimbus 7 Scanning Multichannel Microwave Radiometer. *J. Clim. Appl. Meteor.*, *22:* 1041–6.

Spencer, R. W., Hinton, B. B., and Olson, W. S., 1983b: Nimbus-7 37 GHz radiances correlated with radar rain rates over the Gulf of Mexico. *J. Clim. Appl. Meteor.*, *22:* 2095–9.

Spencer, R. W., Martin, D. W., Hinton, B. B., and Weinman, J. A., 1983c: Satellite microwave radiances correlated with radar rain rates over land. *Nature*, *304:* 141–3.

Spillane, K. T., and Yamaguchi, K., 1962: Mechanisms of rain producing systems in southern Australia. *Austral. Meteor. Mag.*, *38:* 1–19.

Staelin, D. H., Kunzi, K. F., Pettyjohn, R. L., Poon, R. K. L., and Wilcox, R. W., 1976: Remote sensing of atmospheric water vapor and liquid water with the Nimbus 5 Microwave Spectrometer. *J. Appl. Meteor.*, *15:* 1204–14.

Starr, D. O'C., 1987: A cirrus-cloud experiment: intensive field observations planned for FIRE. *Bull., Amer. Meteor. Soc.*, *68:* 119–24.

Steffen, K., and Maslanik, J. A., 1988: Comparison of Nimbus 7 Scanning Multichannel Microwave Radiometer radiance and derived sea ice concentrations with Landsat imagery for the North Water area of Baffin Bay. *J. Geophys. Res.*, *93:* 10769–81.

Steffen, K., and Schweiger, A. J., 1990: A multisensor approach to sea ice classification for the validation of DMSP-SSM/I passive microwave derived sea ice products. *Photogramm. Eng. Remote Sensing*, *56:* 75–82.

Stephens, G. L., 1990: On the relationship between water vapor over the oceans and sea surface temperature. *J. Climate*, *3:* 634–45.

Stephens, G. L., and Webster, P. J., 1981: Clouds and climate: sensitivity of simple systems. *J. Atmos. Sci.*, *38:* 235–47.

Stephens, G. L., Campbell, G. G., and Vonder Haar, T. H., 1981: Earth radiation budgets. *J. Geophys. Res.*, *86:* 9739–60.

Steranka, J., Allison, L. J., and Salomonson, V. V., 1973: Application of Nimbus 4 THIR 6.7 μm observations to regional and global moisture and wind field analyses. *J. Appl. Meteor.*, *12:* 386–95.

Steranka, J., Rodgers, E. B., and Gentry, R. C., 1984: The diurnal variation of Atlantic Ocean tropical cyclone cloud distribution inferred from geostationary satellite infrared measurements. *Mon. Wea. Rev.*, *112:* 2338–44.

Stout, J. E., Martin, D. W., and Sikdar, D. N., 1979: Estimating GATE rainfall with geosynchronous satellite images. *Mon. Wea. Rev.*, *107:* 585–98.

Stowe, L. L., Wellemeyer, C. G., Eck, T. F., Yeh, H. Y. M., and the Nimbus-7 Cloud Data Processing Team, 1988: Nimbus-7 global cloud climatology. Part I: Algorithms and validation. *J. Climate*, *1:* 445–70.

Stowe, L. L., Yeh, H. Y. M., Eck, T. F., Wellemeyer, C. G., Kyle, H. L., and the Nimbus-7 Cloud Data Processing team, 1989: Nimbus-7 global cloud climatology. Part II: First year results. *J. Climate*, *2:* 671–709.

Streten, N. A., 1968: A note on multiple image photo-mosaics for the Southern Hemisphere. *Austral. Meteor. Mag.*, *16:* 127–36.

Streten, N. A., 1970: A note on the climatology of the satellite-observed zone of high cloudiness in the central South Pacific. *Austral. Meteor. Mag.*, *18:* 31–8.

Streten, N. A., 1973: Some characteristics of satellite-observed bands of persistent cloudiness over the Southern Hemisphere. *Mon. Wea. Rev.*, *101:* 486–95.

References

Streten, N. A., 1974: Large-scale sea ice features in the western Arctic basin and the Bering Sea as viewed by the NOAA-2 satellite. *Arct. Alp. Res.*, *6:* 333–45.

Streten, N. A., 1975a: Cloud cell size and pattern evolution in Arctic air advection over the North Pacific. *Arch. Met. Geophys. Bioklim.*, *A24:* 213–28.

Streten, N. A., 1975b: Satellite-derived inferences to some characteristics of the South Pacific atmospheric circulation associated with the Niño event of 1972–73. *Mon. Wea. Rev.*, *103:* 989–95.

Streten, N. A., 1977: Some aspects of the satellite observation of frontal cloud over the Southern Hemisphere, In: *Conference Handbook, Two Day Workshop on Fronts*, May 26–27; Melbourne, Australia. (Royal Meteor. Soc., Austral. Branch and Bureau of Meteorology). 3.1.

Streten, N. A., 1978a: Satellite observed cloud cover in relation to sea surface temperature patterns in the Western Australian region. *Austral. Meteor. Mag.*, *26:* 1–17.

Streten, N. A., 1978b: A quasi-periodicity in the motion of the South Pacific cloud band. *Mon. Wea. Rev.*, *106:* 1211–14.

Streten, N. A., and Downey, W. K., 1977: Defence Meteorological Satellite Program (DMSP) imagery: a research tool for the Australian region. *Austral. Meteor. Mag.*, *25:* 25–36.

Streten, N. A., and Kellas, W. R., 1973: Aspects of cloud pattern signatures of depressions in maturity and decay. *J. Appl. Meteor.*, *12:* 23–7.

Streten, N. A., and Kellas, W. R., 1975: Some applications of simple image analysis techniques to Very High Resolution satellite data. *Austral. Meteor. Mag.*, *23:* 7–19.

Streten, N. A., and Troup, A. J., 1973: A synoptic climatology of satellite observed cloud vortices over the Southern Hemisphere. *Quart. J. Royal Meteor. Soc.*, *99:* 56–72.

Strong, A. E., 1986: Monitoring El Niño using satellite based sea surface temperatures. *Ocean-Air Interacs.*, *1:* 11–28.

Strong, A. E., 1989: Greater global warming revealed by satellite-derived sea-surface-temperature trends. *Nature*, *338:* 642–5.

Strong, A. E., and McClain, E. P., 1984: Improved ocean surface temperatures from space — comparisons with drifting buoys. *Bull., Amer. Meteor. Soc.*, *65:* 138–42.

Strong, A. E., McClain, E. P., and McGinnis, D. F., 1971: Detection of thawing snow and ice packs through the combined use of visible and near-infrared measurements from Earth satellites. *Mon. Wea. Rev.*, *99:* 828–30.

Strong, A. E., and Pritchard, J. A., 1980: Regular monthly mean temperatures of Earth's oceans from satellites. *Bull., Amer. Meteor. Soc.*, *61:* 553–9.

Stum, J., Pinty, B., and Ramond, D., 1985: A parameterization of broadband conversion factors for METEOSAT visible radiances. *J. Clim. Appl. Meteor.*, *24:* 1377–82.

Susskind, J., Reuter, D., and Chahine, M. T., 1987: Cloud fields retrieved from analysis of HIRS2/MSU sounding data. *J. Geophys. Res.*, *92:* 4035–50.

Swift, C. T., and Cavalieri, D. J., 1985: Passive microwave remote sensing for sea ice research. *EOS (Trans. Amer. Geophys. Union)*, *66:* 1210–12.

Szejwach, G., 1982: Determination of semi transparent cirrus cloud temperature from infrared radiances: application to METEOSAT. *J. Appl. Meteor.*, *21:* 384–93.

Tang, M., and Reiter, E. R., 1984: Plateau monsoons of the Northern Hemisphere: a comparison between North America and Tibet. *Mon. Wea. Rev.*, *112:* 617–37.

Tapp, R. G., and Barrell, S. L., 1984: The north-west Australian cloud band: climatology, characteristics and factors associated with development. *J. Climatol.*, *4:* 411–24.

Tarpley, J. D., 1979: Estimating incident solar radiation at the surface from geostationary satellite data. *J. Appl. Meteor.*, *18:* 1172–81.

Tarpley, J. D., Schneider, S. R., and Money, R. L., 1984: Global vegetation indices from the NOAA-7 meteorological satellite. *J. Clim. Appl. Meteor.*, *23:* 491–4.

Taylor, V. R., and Stowe, L. L., 1984: Reflectance characteristics of uniform Earth and cloud surfaces derived from NIMBUS-7 ERB. *J. Geophys. Res.*, *89:* 4987–96.

Taylor, V. R., and Winston, J. S., 1968: Monthly and seasonal mean global charts of brightness from ESSA 3 and ESSA 5 digitized pictures, February 1967–February 1968. ESSA Tech. Report NESC 46, National Environmental Satellite Center, Washington, DC. 9pp. plus 17 charts.

Thepenier, R-M., and Cruette, D., 1981: Formation of cloud bands associated with the American subtropical jet stream and their interaction with midlatitude synoptic disturbances reaching Europe. *Mon. Wea. Rev.*, *109:* 2209–20.

Thomas, G., and Henderson-Sellers, A., 1987: Evaluation of satellite derived land cover characteristics for global climate modelling. *Climatic Change*, *11:* 313–47.

Tian, L., and Curry, J. A., 1989: Cloud overlap statistics. *J. Geophys. Res.*, *94:* 9925–35.

Timchalk, A., and Hubert, L. F., 1961: Satellite pictures and meteorological analyses of a developing low in the central United States. *Mon. Wea. Rev.*, *89:* 429–45.

Townshend, J. R. G., Justice, C. O., and Kalb, V., 1987: Characterization and classification of South American land cover types using satellite data. *Int. J. Remote Sensing*, *8:* 1189–207.

Townshend, J. R. G., Choudhury, B. J., Giddings, L., Justice, C.O., Prince, S. D., and Tucker, C. J., 1989: Comparison of data from the Scanning Multifrequency Microwave Radiometer (SMMR) with data from the Advanced Very High Resolution Radiometer (AVHRR) for terrestrial environmental monitoring: an overview. *Int. J. Remote Sensing*, *10:* 1687–90.

Troup, A. J., and Streten, N. A., 1972: Satellite-observed Southern Hemisphere cloud vortices in relation to conventional observations. *J. Appl. Meteor.*, *11:* 909–17.

Tsonis, A. A., 1984: On the separability of various classes from the GOES visible and infrared data. *J. Clim. Appl. Meteor.*, *23:* 1393–410.

Tucker, C. J., Fung, I. Y., Keeling, C. D., and Gammon, R. H., 1986: Relationship between atmospheric CO_2 variations and a satellite-derived vegetation index. *Nature*, *319:* 195–9.

Turner, J., and Warren, D. E., 1989: Cloud track winds in the polar regions from sequences of AVHRR images. *Int. J. Remote Sensing*, *10:* 695–703.

Turpeinen, O. M., and Schmetz, J., 1989: Validation of the upper tropospheric relative humidity determined from METEOSAT data. *J. Atmos. Oceanic Technol.*, *6:* 359–64.

Turpeinen, O. M., Abidi, A., and Belhouane, W., 1987: Determination of rainfall with the ESOC Precipitation Index. *Mon. Wea. Rev.*, *115:* 2699–706.

Twomey, S. A., Piepgrass, M., and Wolfe, T. L., 1984: An assessment of the impact of pollution on global cloud albedo. *Tellus*, *36B:* 356–66.

Uccellini, L. W., Keyser, D., Brill, K. F., and Wash, C. H., 1985: The Presidents' Day Cyclone of 18–19 February 1979: influence of upstream trough amplification and associated tropopause folding on rapid cyclogensis. *Mon. Wea. Rev.*, *113:* 962–88.

Van Loon, H., and Thompson, A. H., 1966: A note on Southern Hemisphere analysis incorporating satellite information. *Notos*, *15:* 91–7.

Velasco, I., and Fritsch, J. M., 1987: Mesoscale convective complexes in the Americas. *J. Geophys. Res.*, *92:* 9591–613.

Velden, C. S., 1989: Observational analyses of North Atlantic tropical cyclones from NOAA polar orbiting satellite microwave data. *J. Appl. Meteor.*, *28:* 59–70.

References

Velden, C. S., and Smith, W. L., 1983: Monitoring tropical cyclone evolution with NOAA satellite microwave observations. *J. Clim. Appl. Meteor.*, *22:* 714–24.

Viezee, W., Endlich, R. M., and Serebreny, S. M., 1967: Satellite-viewed jet stream clouds in relation to the observed wind field. *J. Appl. Meteor.*, *6:* 929–35.

Viezee, W., Shigeishe, H., and Chang, A. T. C., 1979: Relation between West coastal rainfall and Nimbus 6 SCAMS liquid water data over the northeastern Pacific Ocean. *J. Appl. Meteor.*, *18:* 1151–7.

Vonder Haar, T. H., and Suomi, V. E., 1971: Measurements of the Earth's radiation budget from satellites during a five-year period. Part I: Extended time and space means. *J. Atmos. Sci.*, *28:* 305–14.

Vukovich, F. M., Toll, D. L., and Murphy, R. E., 1987: Surface temperature and albedo relationships in Senegal derived from NOAA-7 satellite data. *Remote Sensing Envt.*, *22:* 413–21.

Walsh, J. E., and Ross, B., 1988: Sensitivity of 30-day dynamical forecasts to continental snow cover. *J. Climate*, *1:* 739–54.

Walsh, J. E., Tucek, D. R., and Petersen, M. R., 1982: Seasonal snow cover and short-term climatic fluctuations over the United States. *Mon. Wea. Rev.*, *110:* 1474–85.

Walsh, S. J., 1987: Comparison of NOAA AVHRR data to meteorologic drought indices. *Photogramm. Eng. Remote Sensing*, *53:* 1069–74.

Walter, B. A., 1980: Wintertime observations of roll clouds over the Bering Sea. *Mon. Wea. Rev.*, *108:* 2024–31.

Wang, J. R., Wilheit, T. T., and Chang, L. A., 1989: Retrieval of total precipitable water using radiometric measurements near 92 and 183 GHz. *J. Appl. Meteor.*, *28:* 146–54.

Warren, S. G., 1982: Optical properties of snow. *Revs. Geophys. Space Phys.*, *20:* 67–89.

Warren, S. G., and Clarke, A. D., 1986: Soot from Arctic Haze: radiative effects on the Arctic snowpack, In: Kukla, G. *et al.* (eds.) Snow Watch '85. Proceedings of the Workshop held 28–30 October 1985 at the Univ. of Maryland, College Park, MD. Glaciological Data Report GD-18, World Data Center for Glaciology (Snow and Ice), CIRES, Univ. of Colorado, Boulder: March 1986. pp. 73–7.

Warren, S. G., Hahn, C. J., and London, J., 1985: Simultaneous occurrence of different cloud types. *J. Clim. Appl. Meteor.*, *24:* 658–77.

Watson, C. E., Fishman, J., and Reichle, Jr., H. G., 1990: The significance of biomass burning as a source of carbon monoxide and ozone in the Southern Hemisphere tropics: a satellite analysis. *J. Geophys. Res.*, *95:* 16443–50.

Weaver, R., Morris, C., and Barry, R. G., 1987: Passive microwave data for snow and ice research: planned products from the DMSP SSM/I system. *EOS (Trans. Amer. Geophys. Union)*, *68:* 769, 776–7.

Weinman, J. A., and Davies, R., 1978: Thermal microwave radiances from horizontally finite clouds of hydrometeors. *J. Geophys. Res.*, *83:* 3099–107.

Weinman, J. A., and Guetter, P. J., 1977: Determination of rainfall distributions from microwave radiation measured by the Nimbus 6 ESMR. *J. Appl. Meteor.*, *16:* 437–42.

Welch, R. M., Sengupta, S. K., and Chen, D. W., 1988: Cloud field classification based upon high spatial resolution textural features. 1. Gray level co-occurrence matrix approach. *J. Geophys. Res.*, *93:* 12663–81.

Welch, R. M., Kuo, K. S., Wielicki, B. A., Sengupta, S. K., and Parker, L., 1988: Marine stratocumulus cloud fields off the coast of southern California observed using LANDSAT imagery. Part I: Structural characteristics. *J. Appl. Meteor.*, *27:* 341–62.

Weston, K. J., 1980: An observational study of convective cloud streets. *Tellus, 32:* 433–8.

Wetzel, P. J., Atlas, D., and Woodward, R. H., 1984: Determination of soil moisture from geosynchronous infrared data: a feasibility study. *J. Clim. Appl. Meteor., 23:* 375–91.

Wetzel, P. J., and Woodward, R. H., 1987: Soil moisture estimation using GOES-VISSR infrared data: a case study with a simple statistical method. *J. Clim. Appl. Meteor., 26:* 107–17.

Wexler, R., 1983: Relative frequency and diurnal variation of high cold clouds in the tropical Atlantic and Pacific. *Mon. Wea. Rev., 111:* 1300–4.

Widger, Jr., W. K., 1964: A synthesis of interpretations of extratropical vortex patterns as seen by TIROS. *Mon. Wea. Rev., 92:* 263–82.

Wielicki, B. A., and Welch, R. M., 1986: Cumulus cloud properties derived using Landsat satellite data. *J. Appl. Meteor., 25:* 261–76.

Wigley, T. M. L., 1989: Possible climate change due to SO_2-derived cloud condensation nuclei. *Nature, 339:* 365–7.

Wilheit, T. T., 1986: Some comments on passive microwave measurement of rain. *Bull., Amer. Meteor. Soc., 67:* 1226–32.

Wilheit, T. T., and Chang, A. T. C., 1980: An algorithm for retrieval of ocean surface and atmospheric parameters from the observations of the scanning multichannel microwave radiometer. *Radio Sci., 15:* 525–44.

Wilheit, T. T., Chang, A. T. C., Rao, M. S. V., Rodgers, E. B, and Theon, J. S., 1977: A satellite technique for quantitatively mapping rainfall rates over the oceans. *J. Appl. Meteor., 16:* 551–60.

Winston, J. S., 1967: Planetary-scale characteristics of monthly mean long-wave radiation and albedo and some year-to-year variations. *Mon. Wea. Rev., 95:* 235–56.

Winston, J. S., 1985: Climatic variations in the outgoing longwave radiation of the past decade as observed from NOAA polar orbiting satellites. *Adv. Space Res., 5:* 121–5.

Woodbury, G. E., and McCormick, M. P., 1983: Global distributions of cirrus clouds determined from SAGE data. *Geophys. Res. Lett., 10:* 1180–3.

Woodbury, G. E., and McCormick, M. P., 1986: Zonal and geographical distributions of cirrus clouds determined from SAGE data. *J. Geophys. Res., 91:* 2775–85.

Woodley, W. L., Griffith, C. G., Griffin, J. S., and Stromatt, S. C., 1980: The inference of GATE convective rainfall from SMS-1 imagery. *J. Appl. Meteor., 19:* 388–408.

Woodwell, G. M., Houghton, R. A., Stone, T. A., Nelson, R. F., and Kovalick, W., 1987: Deforestation in the tropics: new measurements in the Amazon Basin using Landsat and NOAA Advanced Very High Resolution Radiometer imagery. *J. Geophys. Res., 92:* 2157–63.

Wright, W. J., 1988: The low latitude influence on winter rainfall in Victoria, south-eastern Australia. I. Climatological aspects. *J. Climatol., 8:* 437–62.

Wu, M. C., 1984: Radiation properties and emissivity parameterization of high level thin clouds. *J. Clim. Appl. Meteor., 23:* 1138–47.

Wylie, D. P., 1979: An application of a geostationary satellite rain estimation technique to an extratropical area. *J. Appl. Meteor., 18:* 1640–8.

Wylie, D. P., and Laitsch, D., 1983: The impacts of different satellite data on rain estimation schemes. *J. Appl. Meteor., 22:* 1270–81.

Wylie, D. P., and Menzel, W. P., 1989: Two years of cloud cover statistics using VAS. *J. Climate, 2:* 380–92.

Wylie, D. P., Hinton, B. B., and Millett, K. M., 1981: A comparison of three satellite-based methods for estimating surface winds over oceans. *J. Appl. Meteor., 20:* 439–49.

References

Yamanouchi, T., Suzuki, K., and Kawaguchi, S., 1987: Detection of clouds in Antarctica from infrared multispectral data of AVHRR. *J. Meteor. Soc. Japan*, *65:* 949–61.

Yang, S-K., Smith, G. L., and Bartman, F. L., 1988: An Earth outgoing longwave radiation climate model. Part II: Radiation with clouds included. *J. Climate*, *1:* 998–1018.

Yarnal, B. Y., and Henderson, K. G., 1989a: A satellite-derived climatology of polar low evolution in the North Pacific. *Int. J. Climatol.*, *9:* 551–66.

Yarnal, B., and Henderson, K. G., 1989b: A climatology of polar low cyclogenetic regions over the North Pacific ocean. *J. Climate*, *2:* 1476–91.

Yasunari, T., 1977: Stationary waves in the Southern Hemisphere mid-latitude zone revealed from average brightness charts. *J. Meteor. Soc. Japan*, *55:* 274–85.

Yates, H., Strong, A., McGinnis, Jr., D., and Tarpley, D., 1986: Terrestrial observations from NOAA operational satellites. *Science*, *231:* 463–70.

Yoo, J-M., and Carton, J. A., 1988: Spatial dependence of the relationship between rainfall and outgoing longwave radiation in the tropical Atlantic. *J. Climate*, *1:* 1047–54.

Zick, C., 1983: Method and results of an analysis of comma cloud development by means of vorticity fields from upper tropospheric satellite wind data. *Meteorol. Rundsch.*, *36:* 69–84.

Zillman, J. W., and Price, P. G., 1972: On the thermal structure of mature Southern Ocean cyclones. *Austral. Meteor. Mag.*, *20:* 34–48.

Zwally, H. J., Parkinson, C. L., and Comiso, J. C., 1983: Variability of Antarctic sea ice and changes in carbon dioxide. *Science*, *220:* 1005–12.

Glossary of terms

30–60 day oscillations Time scale of variations in outgoing longwave radiation for the western tropical Pacific, and which consist of eastward propagating disturbances that may have their origin in the Indian Ocean.

absorption Effect of an atmospheric constituent (e.g., CO_2, water vapor) or Earth surface (e.g., water, soil) in converting electromagnetic energy to other forms of energy. In a non-scattering medium, absorption is the inverse of the reflected radiation.

active (sensing) A sensor consisting of both a receiver (antenna) and a transmitter for electromagnetic energy; commonly in the microwave, or radar, portion of the EMS (cf. *passive sensing*).

advection The quasi-horizontal transport of a quantity (e.g., heat, momentum) by the atmosphere or ocean (cf. *convection*).

aerosol Non-gaseous constituents in the atmosphere (e.g., dust, sea salt, volcanic ash) that are scatterers of electromagnetic energy in particular wavelengths.

albedo The reflectance of a body or surface integrated across all solar wavelengths, and expressed as a fraction or percentage of the incident radiation. In satellite remote sensing, the albedo is typically derived for the Earth's surface in the absence of clouds (surface albedo), and for the Earth and atmosphere including clouds (planetary albedo).

algorithm A statistical or physical model-based technique that inverts satellite-sensed radiances into target information (e.g., sea ice concentration, instantaneous rain rate, atmospheric temperature).

anistrophy Radiation that is scattered preferentially (usually in the forward direction), as distinct from that scattered equally in all directions (*isotropy*).

atmospheric window Those wavelengths for which the atmospheric gases (especially CO_2, O_3, and water vapor) are essentially transparent.

baroclinic State of the atmosphere in which constant pressure and constant density surfaces intersect, and mass overturning results. Baroclinity is associated with strong horizontal temperature gradients and a 'thermal wind' (vertical wind shear).

barotropic State of the atmosphere in which constant pressure and density surfaces do not intersect. In the equivalent barotropic state, horizontal gradients of temperature exist but they lie in the same direction as the pressure/height gradient. Thus, the wind speed only changes with height.

Benard convection Cellular convection due to surface heating and in which cloudy areas are interspersed with clear areas; often having an hexagonal appearance when viewed from above.

bi-directional reflectance Dependence of a satellite-sensed albedo on the sun-sensor-target geometry.

bispectral technique The use of radiances in any two distinct wavelength bands (e.g., visible and thermal IR) to deduce target information. In satellite climatology, it is typically used for automated cloud classification and for inferring precipitation rates from cloud fields

blackbody A body or material that absorbs all incident radiation of a particular wavelength (the reflectance and transmittance is zero), and which reradiates that energy at maximum efficiency (i.e., the *emissivity* is equal to 1.0).

brightness temperature Apparent temperature of a non-blackbody radiator (emissivity <1.0), as measured by a radiometer. It is a term frequently used in passive microwave sensing.

broadband Remote sensing over a comparatively wide range of wavelengths (e.g., 0.4–0.7 μm; 8–13 μm), as distinct from narrow-band sensing.

CISK Conditional Instability of the Second Kind: refers to the augmentation of simple cumulus convection (the 'first kind') by frictional convergence (and rotation) in the boundary layer. It is considered a dominant process in the spin-up of tropical vortices, and some species of 'polar low'.

climate The time-averaged state of the atmosphere at a point or over an area, and in which forcings by solar radiation, the atmospheric constituents, orography, and terrestrial surfaces (land, ocean, ice) are implicit in its variations. In satellite remote sensing, the climatic time scales extend beyond about one week.

climatology The science of climate (mean state; variability), including study of the atmospheric and terrestrial components, as they occur on a range of spatial and temporal scales.

cloud Visual manifestation of the condensation or sublimation of water vapor in the atmosphere. The range of possible cloud types (e.g., stratus, cirrus, cumulus) have associated large differences in radiative properties, such as the emissivity and optical depth. These properties are associated with type-dependent differences in cloud thickness, cloud top temperature, and ice-water content.

cloud forcing The climate sensitivity to clouds; most importantly functions of cloud type and cloud fraction. The cloud forcing is the net effect of the albedo enhancement (shortwave reflectance) and greenhouse enhancement (IR absorption and re-emittance) properties of clouds.

cloud indexing Indirect method of precipitation estimation in which the probabilities of precipitation are given by the satellite determination of particular cloud types at a given point in time.

cloud regime Regions within which relatively homogeneous cloud type and cloud persistence attributes occur on climatological time scales.

comma cloud Satellite-observed configuration of a cyclonic cloud vortex (inverted comma in Southern Hemisphere). While the term may be used in conjunction with frontal wave cyclones, it is increasingly reserved for a species of subsynoptic-scale vortex that is crescent-shaped. cf. *polar low*.

contribution function The altitude of the peak radiance sensed in the water vapor channel; or M(log p).

convection Vertical motion in a fluid (ocean, atmosphere) and which results in the transports of heat and momentum.

convective condensation level (CCL) The pressure altitude in the atmosphere at which condensation (and cumuliform cloud formation) occurs, and in which parcels rise by free convection. It varies widely from time to time and from place to place. On a thermodynamic diagram the CCL is given by the intersection of a saturation mixing ratio through the surface dewpoint temperature and the dry adiabat through the surface free convection temperature with the environmental lapse rate curve.

critical blackness Threshold emissivity value beyond which cirrus clouds have a net warming (greenhouse enhancement) effect on the surface temperature. This value varies according to cloud altitude.

diabatic Non-adiabatic process by which heating or cooling occurs in the atmosphere, and in which heat is exchanged across the boundaries of the system (e.g., by latent heat release and uptake; radiational heating and cooling).

dielectric constant An electrical property of terrestrial surfaces that influences microwave emissions (passive) and returns (active). For example, moisture in soil and salinity changes in sea ice influence the conductivity of electrical energy and, hence, their microwave signatures.

diffuse radiation Reflection from a non-Lambertian surface, or one in which the surface roughness is large in relation to the wavelength of radiation considered. The radiation is reflected in all directions.

El Niño Southern Oscillation (ENSO) The suite of climatic anomalies associated with variations in the tropical Pacific Walker (zonal) circulation. The anomalies are most marked at the extreme times of the oscillation, or the 'warm' (El Niño) and 'cold' (La Niña) events. The anomalies are out-of-phase between the western and eastern tropical Pacific. Strong evidence exists for tropical and extratropical teleconnections to the ENSO.

Electromagnetic Spectrum (EMS) The full range of radiation wavelengths, extending from cosmic rays through radio waves. Only relatively few portions of the EMS are regularly exploited in satellite remote sensing of the Earth and its atmosphere.

emissivity Numerical measure of the efficiency of a body to radiate energy as a function of temperature and wavelength considered, as expressed by Kirchoff's Law. By definition the emissivity of a blackbody is equal to 1.0. It is less than unity for a greybody.

Empirical Orthogonal Function (EOF) analysis A statistical technique similar to Principal Components Analysis, in which fields of highly correlated data are represented by a small number of orthogonal eigenvectors. EOFs are different from principal components in that the lengths of the eigenvectors are not scaled by the square root of the eigenvalue.

evapotranspiration Evaporation of moisture from all terrestrial sources, including water bodies, soil and vegetation; and involving the uptake of latent heat and a lowering of the surface temperature.

feedback Two-way, non-linear interaction between components of the climate system, and in which the net effect may be either positive (amplifying) or negative (damping). The sign of some feedbacks are as yet undetermined and, hence, their impact on the climate system is uncertain.

fractal Structures (e.g., some cloud types; convective precipitation areas) that are essentially scale independent over several orders of magnitude.

Gaia Hypothesis The theory that the terrestrial biota (on land; in the sea) is a functional component of the climate, and is actively involved in its relative stability on very long time scales. It was first promulgated by James Lovelock, an atmospheric chemist.

Gaussian   A distribution of values that is statistically normal (bell-shaped).

geostationary (geosynchronous)   The characteristic orbit of a satellite such that its speed of motion exactly matches the Earth's linear rotation speed at the equator, thus keeping the same sub-satellite point. This occurs at a mean satellite altitude of 35 400 km above the equator.

grey scale   The range of radiances sensed by a satellite in a given wavelength. In the thermal IR the grey scale shows equivalent blackbody temperature.

Highly Reflective Cloud (HRC)   Term used to describe cumulonimbus clouds, particularly in the tropics, and as observed in visible wavelengths. They are usually associated with heavy convective rainfall and negative anomalies of OLR.

hydrometeor   Water or ice particle formed as the result of condensation or sublimation.

imagery   Collection of pixel brightnesses or radiances obtained by a satellite sensor and presented in pictorial form.

inversion   1. Positive change of the environmental lapse rate with height (temperature increase), and which represents the absolute stability case. Inversions may be surface-based, where they result from radiational cooling; or above the surface, due to subsidence, frontal activity, or advection. 2. The retrieval of temperature and moisture profiles from radiation fields using an atmospheric sounder. It represents the inverse problem in radiative transfer; the direct problem being the determination of the radiation from known temperature and absorption information.

irradiance   The total upwelling radiation obtained by a wide field-of-view sensor, integrated over a range of path lengths. May also refer to the receipt of emitted solar radiation at the top of the atmosphere (TOA) or at the Earth's surface.

ITCZ (Intertropic Convergence Zone)   Low-latitude region of confluence of the trade winds from the northern and southern hemispheres, characterized by strong convection and HRC development. The trough represents the ascending branch of the meridional Hadley Cell.

Kirchoff's Law   Fundamental law that states that for a given temperature and wavelength of a blackbody, the ratio of the emitted to absorbed radiation (emissivity) is equal to unity.

Lambertian reflectance   Reflectance from an ideal smooth surface in which the radiation is isotropic.

lead   More-or-less linear opening in sea ice, and in which either water or thin ice are present. Leads may occur within the pack or between the shore-fast and pack ice, as a result of deformational stresses. cf. *polynya*.

lidar   Light Detection And Ranging: active remote sensing in u.v., visible and near IR wavelengths for determination of atmospheric profiles of high vertical resolution; particularly for the boundary layer.

life history technique   Method of estimating precipitation from visible and IR data, and which is based on the evolution of cloud patterns through time using geostationary satellite data.

Meso-scale Convective System Large group of convective cells (thunderstorms) which move as one system. Although meso-scale features, MCSs are large enough to interact with the synoptic atmospheric circulation. In middle latitudes they are a warm season phenomenon.

micrometer (μm) Unit of measurement of wavelengths of electromagnetic radiation and atmospheric particle sizes. 1 μm = 10^{-6} m. cf. *nanometer*.

microwave Portion of the EMS between the thermal IR and radio wavelengths, or about 1 mm–1 m.

Mie scatter Atmospheric scattering in which all wavelengths considered are scattered equally. In visible wavelengths cloud droplets are sufficiently large as to scatter all wavelengths (cf. *Rayleigh scatter*). Accordingly, clouds appear white.

minimum brightness Automated method used to 'cloud screen' and obtain clear-sky surface radiances in visible band satellite data. For a given compositing (multi-day) period the lowest reflectance is saved, on the assumption that, over snow-free surfaces, high reflectances are due to transient cloud effects.

monsoon Seasonal wind reversal that occurs in particular regions of the globe and results from differential heating of land and ocean surfaces. In many such places a natural consequence of the wind change is an associated rainy season (e.g., South Asia).

nadir Point on the ground vertically beneath a satellite sensor. For a scanner system on board a polar orbiter, the nadir is a series of points along a line that describes the subsatellite track.

nanometer (nm) Like the micrometer, a unit of measurement. It is equal to 10^{-9} m, and is commonly used for the shorter wavelengths of electromagnetic radiation.

narrow band Remote sensing over a relatively narrow range of wavelengths (e.g., 0.5–0.6 μm). cf. *broad-band sensing*.

nephanalysis Interpretation (often manual) of cloudiness appearing on satellite imagery and its representation by symbols that pertain to cloud coverage, cloud type, and large-scale cloud pattern characteristics.

non-beam filling In microwave data, the contribution of sub-resolution features to the integrated pixel brightness temperature. An example is the under-estimation of instantaneous precipitation rate in convective situations.

nowcasting The forecasting of local weather conditions for up to a maximum of about 9–12 hours, but most commonly out to about 3 hours. Nowcasting is observation-intensive and involves integration and analysis of conventional, satellite and radar data.

OLR (Outgoing Longwave Radiation) Satellite-measured thermal emission from terrestrial surfaces and from clouds in the window region. OLR is low (high) over cold land and high clouds (hot land surfaces). Anomalies of OLR from longer-term mean values are indicators of climate variations, such as ENSO.

parameterization The approximation or simplification of a complex process in a model. For example, the generation of convective clouds.

passive sensing A sensor that receives reflected or emitted radiation. For example, brightness in the 0.5–0.7 μm band; upwelling radiation at 19 GHz, etc. cf. *active sensing*.

pattern recognition   Subjective classification of clouds or surface features in remotely sensed data, on the basis of their characteristic spatial and spectral signatures.

phenology   Progressive change in leaf or crop development.

pixel   Individual picture element, or smallest piece of information, representing the nominal ground resolution of the remote sensing system.

planetary albedo   The planetary reflectance (of Earth) in solar wavelengths, or a portion of those wavelengths, expressed as a fraction of the incident solar radiation (irradiance). It represents the contributions from terrestrial and cloud surfaces, and the atmosphere.

polarization   Directional dependence of the electrical field vector of electromagnetic radiation, and which increases away from nadir. At a given microwave frequency, the surface emission can be resolved into its horizontal and vertical components. Consideration of these components enhances considerably the target information content.

polar low   Generic term for any subsynoptic-scale cyclonic vortex that develops as the result of instability in a cold air stream. cf. *comma cloud*.

polar orbiter   Orbital configuration in which the satellite platform passes over or close to the poles. A meteorological polar orbiter, with its wide field of view, surveys most points on the Earth's surface twice per day; one an ascending and the other a descending node of the orbit. Due to the Earth's rotation, consecutive orbits step to the westward.

polynya   Any non-linear opening in pack ice or between shore-fast and pack ice. cf. *lead*.

precipitable water   The depth (in cm) of water at the surface of the Earth if the total column water vapor were to be condensed out.

Principal Components Analysis   A statistical methodology, frequently used as a means of reducing large sets of intercorrelated climatological data into smaller sets of statistically-independent modes of variation. cf. Empirical Orthogonal Function analysis.

PVA (Positive Vorticity Advection)   The advection downstream of cyclonic vorticity in the free atmosphere, in association with a short wave or jet maximum. It is significant for surface development by promotion of upper-air divergence.

radar (RAdio Detection And Ranging)   A form of active remote sensing in the microwave region. Ground-based radar gives corroborating information on precipitation rates and intensities; spaceborne (high altitude) radar uses Doppler principles to obtain target information.

radiance   Upwelling electromagnetic radiation in a narrower field-of-view and over a much smaller range of path lengths than the irradiance.

radiant temperature   That temperature sensed by a radiometer and which is strongly determined by the surface emissivity; i.e.,

$$T_{rad} = \varepsilon^{1/4} T_{kin}$$

where T_{kin} = kinetic temperature of the substance.

radiative-convective (model)   One-dimensional climate model with vertical resolution, and in which atmospheric and surface temperatures are computed. Clouds may also be predicted.

radiometer Sensor that measures the upwelling radiation; most commonly in thermal IR wavelengths.

Rayleigh scatter Differential scattering of radiation in a given wavelength band by spherical particles that have radii smaller than approximately 1/10 the wavelength of that radiation. In visible wavelengths, the shorter blue wavelengths are scattered more efficiently than the longer red wavelengths by the atmospheric gases; hence, the sky is blue. cf. *Mie scatter*.

resolution (spatial) The smallest distance between closely-spaced objects that can be detected on an image. It is a function of the system (or sensor) resolution and the platform altitude.

retrieval Determination of target information from spectral radiance data by way of an algorithm. Examples are the surface temperature, cloud amount, sea ice concentration, rain rate, etc.

sampling The frequency of observations in time and space required to characterize a particular target or phenomenon. A bias in either the temporal or spatial dimensions (e.g., as a function of polar orbiter overpass time) reduces the ability of the data to characterize fully the target or phenomenon.

scanner A sensor in which an imaged area is the result of a small field-of-view being swept repeatedly from side-to-side, rather than being viewed instantaneously (e.g., by a vidicon) or by a sensor in a 'dwell' mode.

scattering The diffusion of radiation in a medium (e.g., air, water) by small particles. Scattering may be either dependent (Rayleigh) or independent (Mie) of particle size for the wavelength considered.

scatterometer Non-imaging radar that measures the backscatter of microwave radiation from Earth (land, ocean) surfaces.

sensor Instrument used to detect electromagnetic radiation and which converts the intensities into a signal for recording and display.

sensor drift Change with time in the response of a sensor to a given intensity of electromagnetic radiation.

signature Appearance of a target. It is particularly a function of the wavelength considered and the target physical properties.

solar constant Solar irradiance at the top of the atmosphere on a plane normal to the incident radiation, and at the mean distance of the Earth from the Sun. Its value is close to 1365 W m^{-2}. In satellite remote sensing, the solar constant is monitored to provide onboard calibration of the visible band sensor, and for determining the surface-received solar irradiance. cf. *TOA*.

sounder Vertically pointing sensor that is used to retrieve the vertical temperature and moisture structure of the atmosphere. Sounders typically sample across multiple narrow-bands, each of which is sensitive to absorption and re-emission characteristics of different atmospheric gases and at a different level in the atmosphere. Microwave sounders may be used in cloudy scenes. cf. *lidar*.

spectral dependence Wavelength dependence of a target (e.g., green vegetation in visible and near-IR wavelengths).

specular reflectance The reflectance of radiation (typically in solar wavelengths) from a surface that is smooth with respect to the wavelengths considered (i.e., mirror-like).

Glossary of terms

Stefan-Boltzmann Law   Fundamental law that relates the intensity (flux density) of radiation emitted by an ideal blackbody to the fourth power of its absolute temperature.

streamline   A line drawn tangent to the instantaneous velocity of a fluid (e.g., the wind).

subtropical highs   Semi-permanent warm-cored anticyclones located over ocean surfaces in the subtropics, and which are best developed in summer when the Hadley Cell is most intense.

supervised classification   Classification method performed on a group of pixels and for which ground truth from training sites is available. Unsupervised classifications have no such operator-supplied information.

synoptic   General term to describe an overall view; e.g., of weather conditions and associated atmospheric circulations for a point in time. Synoptic time and space scales are of the order of $5 \times 10^2 - 2 \times 10^3$ km and 10^2 hours.

target   Object or feature at the Earth's surface or in the atmosphere that is of interest in a remote sensing context.

teleconnection   Associations (positive or negative) between atmospheric circulation and related weather/climate conditions for widely separated areas of the globe. The extratropical climate anomalies associated with ENSO constitute a teleconnection.

thermal inertia   The ability of a system or body (e.g., soil, water) to undergo a change in temperature for an input of heat over a given time period.

thermal infrared   Wavelengths of radiation between $8-14 \mu$m, and in which the emission of the Earth-atmosphere system peaks. These are heat wavelengths.

thermal wind (V_T)   The vector resultant for a wind at a lower and upper pressure level (i.e., the mean shear through an atmospheric layer). The thermal wind 'blows' parallel to the thickness contours, and is an indicator of baroclinity. It explains the existence of jet streams.

threshold   Automated satellite-based method of cloud cover determination in individual pixels, and in which the coverage is typically binary (cloud, no cloud).

TOA flux   The irradiance measured at the top of the atmosphere, typically for solar wavelengths before atmospheric attenuation. It is the solar constant value reduced by the solar declination (season and latitude dependent).

trajectory   Time-integrated curve representing the quasi-horizontal path of a fluid particle. For steady state (unchanging) flow, a streamline and trajectory are identical.

transmissivity   Ability of a medium (e.g., air, water) to transmit radiation. In the atmosphere, the transmissivity is reduced in the presence of water vapor and aerosols.

truth   Verification of a target's identity by conventional or additional remotely-sensed information. It is the basis of the supervised classification technique.

upslide precipitation   Advective precipitation process that dominates in middle latitudes, especially in the cool season, and results from large-scale gradual ascent of moist air. It occurs when air moving cyclonically and from lower to higher latitudes ascends more-or-less along sloping isentropic (equal potential temperature) surfaces.

verification   Similar to truth. Also, the validation of a model or algorithm using observed data.

vidicon   Electronic camera system (non-film) flown on earlier TIROS and Landsat platforms. It uses a principle similar to that of a television camera tube. cf. *scanner*.

Walker Circulation   Zonal (latitude constrained; west-to-east varying) tropospheric circulation, especially that in the tropical Pacific. Variations in the Walker Circulation are linked with those in the upper ocean during ENSO events.

Wien's Law   Fundamental law that relates the wavelength of maximum radiation emission of a blackbody to its absolute temperature. The wavelength gets shorter as the temperature increases.

zenith angle (solar)   Angle subtended from the local vertical by the Sun.

Index